Computational Molecular Science and Engineering Forum

Core Programming Topic at the 2011 AIChE Annual Meeting

Minneapolis, Minnesota, USA
16-21 October 2011

ISBN: 978-1-61839-740-9

Printed from e-media with permission by:

Curran Associates, Inc.
57 Morehouse Lane
Red Hook, NY 12571

Some format issues inherent in the e-media version may also appear in this print version.

Printed by Curran Associates, Inc. (2012)

For permission requests, please contact AIChE
at the address below.

AIChE
3 Park Avenue
New York, NY 10016-5991

Phone: (203) 702-7660
Fax: (203) 775-5177

www.aiche.org

Additional copies of this publication are available from:

Curran Associates, Inc.
57 Morehouse Lane
Red Hook, NY 12571 USA
Phone: 845-758-0400
Fax: 845-758-2634
Email: curran@proceedings.com
Web: www.proceedings.com

TABLE OF CONTENTS

(18a) Database Driven Discovery of Materials and Catalysts

Thomas Bligaard, Center for Atomic-scale Materials Design, Technical University of Denmark, Kgs. Lyngby, Denmark

We will present our approach to data-rich discovery of materials.

See more of this Session: Computational Catalysis I - Databases and Informatics
See more of this Group/Topical: Catalysis and Reaction Engineering Division

(18b) Fuels and Chemicals From Biomass: Challenges and Opportunities From a Systems Perspective

Prodromos Daoutidis, Chemical Engineering and Materials Science, University of Minnesota, Minneapolis, MN

Biorefineries are envisaged to produce green transportation fuels and chemicals by upgrading biomass in a sequence of processing steps. Although considerable emphasis has been given so far to the "upstream" conversion of biomass to intermediate platforms (sugars or syngas), progress in "downstream" conversion to chemicals and intermediates is still lagging. Due to the oxygen present in biomass and the diversity of raw materials derived from biomass, the necessary downstream reaction and separation processes are generally different from existing ones based on fossil fuels. Furthermore, there is limited data available on physical properties of such molecules, and on their full array of chemical transformations, and their kinetics and thermodynamics. These challenges lead to several emerging opportunities for systems and computational research, spanning several scales: elucidation of the underlying chemistry at the mechanistic level, modeling and design at the reactor and process level, and an enterprise-level analysis of process economics and environmental considerations.

This talk will discuss recent results on: i) the elucidation of the chemical transformations involved in biomass conversion using RING, a recently developed reaction network generation and analysis tool, and ii) the design and optimization of novel reaction-separation processes for biomass-based chemical synthesis. It will also highlight several avenues for research in computational chemistry and chemical informatics, microkinetic and multiscale modeling, and process design, optimization and control, which can collectively have a major impact on the realization of the ambitious concept of an integrated biorefinery.

See more of this Session: Computational Catalysis I - Databases and Informatics
See more of this Group/Topical: Catalysis and Reaction Engineering Division

(18c) Materials Design Workbench: An Open-Access Database for Materials and Molecular Design

Glen Allen Ferguson[1], Jeff Greeley[2], Peter Zapol[1], Larry A. Curtiss[3], Svetlozar Evtimov Nestorov[4], Ian T. Foster[5], David Landis[6], Jens Strabo Hummelshj[7], Karsten W. Jacobsen[8], Thomas Bligaard[9] and Jens Nrskov[10], (1)Material Science Department, Argonne National Laboratory, Argonne, IL, (2)Center for Nanoscale Materials, Argonne, IL, (3)Materials Science Division, Argonne National Laboratory, Argonne, IL, (4)Computer Science, University of Chicago, Chicago, IL, (5)Mathematics and Computer Science and Department of Computer Science, Argonne National Laboratory and University of Chicago, Argonne, IL, (6)Department of Physics CAMD, Technical University of Denmark, Kongens Lyngby, Denmark, (7)SLAC National Accelerator Laboratory, SUNCAT, Menlo Park, CA, (8)Center for Atomic-scale Materials Physics, Dept. of Physics, Technical University of Denmark, Lyngby, Denmark, (9) Center for Atomic-scale Materials Design, Technical University of Denmark, Kgs. Lyngby, Denmark, (10)Chemical Engineering, Stanford University and SUNCAT, Stanford, CA

The use of modern computing resources and efficient computational chemistry software packages has resulted in the production an extraordinary amount of computational data related to materials and catalytic problems. Very few such data, however, are available for public browsing; they tend to be stored locally and are rarely publically disseminated. A significant reason for this is absence of an easy to use, open-access public database for computational materials and molecular data. To address this problem, we are implementing an open-access database, the Materials Design Workbench (MDW), that will store data from computational chemistry software. The data are in a form that can be conveniently searched, visualized, analyzed, and retrieved by users.

The implementation of the Materials Design Workbench is a collaboration between scientists at Argonne National Laboratory, the University of Chicago, the Center for Sustainable Energy through Catalysis at the Stanford Linear Accelerator Center, and the Technical University of Denmark. The implementation consists of several modules. The central piece is a MySQL database that stores information from computational chemistry software and associated experimental data. To facilitate entering computational data into the database, a stand-alone python interface named the Computational Materials Repository (CMR) has also been developed. One of the most important roles of the CMR is to convert the output from calculations into XML format files. These XML files can then be modified to add additional data and uploaded to the database. Two additional applications are under development for visualization and analysis of data that have been entered into the database. The first is a client-side application that runs in any HTML 5 enabled web browser. The second is a server-side web interface.

In this talk, we will discuss two important parts of the MDW. The first is how the MySQL database stores and organizes the computational data. The second, which will include a preview, is the server-side application. This application, which is in development at the University of Chicago and Argonne National Laboratory, currently consists of a web interface for viewing and retrieving structures from the database. A user can query the database using any of the stored criteria and can display the results as a single molecule or as a collection of molecules. These collections can be analyzed, saved for later analysis, or downloaded. Users can also attach citations or other any other desirable data to their calculations. It is also possible to restrict access to data sets to particular users and groups or to make the data globally readable. These features, combined with the powerful visualization and analysis software that is also being developed, make MDW an attractive solution for the storage, retrieval and analysis of materials and molecular data.

See more of this Session: Computational Catalysis I - Databases and Informatics
See more of this Group/Topical: Catalysis and Reaction Engineering Division

(18d) Model-Based Design of Catalysts

W. Nicholas Delgass[1], Mahdi Abu-Omar[2], James M. Caruthers[1], Fabio Ribeiro[3], William F. Schneider[4], Kendall T. Thomson[1], Grigori Medvedev[1], Jun Wang[3], John Clay[4], Mary Jones[2], Wen-Sheng Lee[1], Jorge Pazmino[3], Kaiwalya Sabnis[1], Mayank Shakhar[1], Stephen D. Stamatis[1], Jeffery M. Switzer[1], D. Keith Steelman[2], Nicholas Travia[2], W. Damion Williams[1] and Silei Xiong[1], (1)School of Chemical Engineering, Purdue University, West Lafayette, IN, (2)Chemistry Department, Purdue University, West Lafayette, IN, (3)Chemical Engineering, Purdue University, West Lafayette, IN, (4) Chemical and Biomolecular Engineering, University of Notre Dame, Notre Dame, IN

This team of researchers is working on a model-based approach to catalyst design called *Discovery Informatics*. In this concept, the first objective is to identify descriptors that that can be linked to the kinetic performance of a family of catalysts through a quantitative *forward model*. This demanding task requires a number of computational tools. Since a microkinetic description is essential for the forward model, we have focused on tools for model discrimination and accurate extraction of rate constants from kinetic data. For single site olefin polymerization, which is our homogeneous catalysis test bed, we have developed a parallelized population balance modeling tool which allows molecular weight distributions from gel permeation chromatography to be included in a multi-response data set that is used to verify mechanism and provide rate constants for each of the necessary elementary steps. For the water gas shift (WGS) reaction on supported metals, our heterogeneous catalysis test bed, we have included Bayesian statistical analysis that provides specific accounting for error and Markov Chain Monte Carlo analysis to give probability distributions for parameters and their correlations with each other. While descriptors to relate to rate constants can sometimes be identified through linear correlations from statistically designed experiments, we have turned in this developmental stage to density functional theory to help identify descriptors, to better define reaction pathways, and, particularly in the case of heterogeneous systems, to compute energies that are difficult to measure.

Progress in building forward models will be discussed in terms of hexene polymerization over non-Cp coordination complexes of Zr and Hf and WGS over supported Au and Pt. In the homogeneous polymerization system we have discovered that the pendant ligand, NMe_2 versus THF, has a large effect on rate constant for chain transfer, while the change from Zr to Hf, lowers the propagation rate constant by a factor of 20. We will interpret this data in light of DFT results and our previous finding that the ion pair separation energy is an important descriptor[i]. For the WGS catalysts, we have identified the corner and perimeter atoms in contact with the support as the active sites for supported Au, all surface atoms as active for supported Pt, and, in spite of these differences, that the support plays an important role for activation of water for both Au and Pt catalysts. Models integrating these findings will be presented.

The value of a successful forward model is that it can be used in a guided stochastic search of the descriptor space to identify optimum catalyst candidates. Tools and information needed and obstacles which must be overcome to close this *inverse* cycle will also be discussed.

[i] T. A. Manz, K. Phomphrai, G. Medvedev, B. B. Krishnamurthy, S. Sharma, J. Haq, K. A. Novstrup, K. T. Thomson, W.N. Delgass, J. M. Caruthers, and M. M. Abu-Omar, *J. Am. Chem. Soc.*, *129(13)*, 3776-3777 (2007).

See more of this Session: Computational Catalysis I - Databases and Informatics
See more of this Group/Topical: Catalysis and Reaction Engineering Division

(37a) Mechanistic Study of Methane Hydrate Nucleation Under Realistic Conditions

Brandon C. Knott, Michael F. Doherty and Baron Peters, Chemical Engineering, UC Santa Barbara, Santa Barbara, CA

The formation of methane hydrate clathrates is of great interest in many scientific and industrial contexts, including as a hindrance in oil and natural gas pipelines and as a potential energy resource. We employ semigrand canonical Monte Carlo (SGCMC) simulations to study the nucleation of methane hydrate clathrates utilizing a coarse-grained model that represents each molecule as a single particle.[1] SGCMC simulations facilitate the study of nucleation at low (and constant) driving force. This is desirable because these are the conditions under which most scientific and industrial systems of interest operate. We seek to shed light on the hydrate formation mechanism under the conditions at which they form in nature and important industrial contexts. We study nucleation pathways along two distinct coordinates: 1) the clustering of methane molecules regardless of water structure and 2) the number of clathrates (water) cages formed.

[1]L.C. Jacobson and V. Molinero, J. Phys. Chem. B **114**, 7302 (2010).

See more of this Session: Multiscale Modeling I
See more of this Group/Topical: Catalysis and Reaction Engineering Division

(37b) Development of a Coarse -Grained Model, and Its Application to Free Energy Studies In Proteins

Spyridon Vicatos, Anna Rychkova, Anatoly Dryga and Arieh Warshel, Chemistry, University of Southern California, Los Angeles, CA

Simplified Coarse Grained protein models have been in use since 1975 and are being used extensively in folding studies. In this work, we provide a brief description of our Coarse Grained (CG) model and present its application on the free energy calculations of large biological systems. This model has been recently used for the study of the free energy associated with transmembrane (TM) helices insertions (Rychkova et al, PNAS 107,17598-17603). Our CG model, after recent modifications for the calculation of electrostatic interactions, was used for the study of TM/translocon system even though the structural characteristics are still not clear. The CG model has been used for the study of voltage activated ion channels. Meaningful converging microscopic simulations for ion channels are not yet available and macroscopic treatments involve major uncertainties in key features. The CG model has been extended, in order to perform simulations of membrane/protein systems under external potential (Dryga et al, BBA, under preparation). Finally, the CG model shows promising results, when used for the calculation of protein folding energies ΔG_{fold} and changes of folding energies upon mutations $\Delta\Delta G$.

See more of this Session: Multiscale Modeling I
See more of this Group/Topical: Catalysis and Reaction Engineering Division

(37c) Coarse-Graining Proteins Using Relative Entropy Theory

M. Scott Shell, Scott Carmichael and Aviel Chaimovich, Department of Chemical Engineering, University of California Santa Barbara, Santa Barbara, CA

The development of accurate coarse-grained molecular models is central to many efforts in soft-condensed matter, materials, and biophysics that seek to understand large length and time scale properties. We recently proposed a broad new statistical-mechanical framework for such problems based the relative entropy, a phase-space functional measuring the "information" lost upon coarse graining and hence the (inverse) fitness of a given coarse-grained model [1]. We suggest that minimization of the relative entropy provides a universal variational principle for coarse-graining.

Here, we show how this approach provides a quantitative and systematic approach to the development of coarse-grained models of proteins. We first describe the development of stable, efficient, and robust numerical coarse-graining algorithms based upon relative entropy minimization [2]. In particular, these approaches are able to locate optimal models in very high dimensional parameter spaces, here for one-, two-, and three-bead (per amino acid) models of peptides. We show that the relative entropy framework offers exquisite control over coarse-graining errors, and is able to generate coarse peptide models whose structural correlations and folding behavior are in excellent agreement with more expensive all-atom simulations. We also demonstrate how these models are well-suited to large-scale simulations of peptide self-assembly. Second, we show that the relative entropy approach provides an a priori prediction of coarse-graining errors (i.e., errors in the properties predicted by coarse-grained models), and gives a systematic strategy for designing coarse-grained potentials [3]. In particular, we use this theory to develop simple analytical energy-landscape models of protein folding. Here, an ensemble of predicted structures for a large protein is projected onto a coarse-grained funnel landscape. The relative entropy quantifies which structure, when considered at the bottom of the funnel, best describes the energy-structure relationships, in turn predicting which structure is nearest to the true native structure. We present results on several model protein systems and from a recent community-wide blind structure prediction competition.

[1] M. S. Shell, "The relative entropy is fundamental to multiscale and inverse thermodynamic problems," J. Chem. Phys. 129, 144108 (2008).

[2] A. Chaimovich and M.S. Shell, "Anomalous waterlike behavior in spherically-symmetric water models optimized with the relative entropy," Phys. Chem. Chem. Phys 11, 1901 (2009).

[3] A. Chaimovich and M. S. Shell, "Coarse-graining errors and optimization using a relative entropy framework," J. Chem. Phys. 134, 094112 (2011).

A Multi-scale Modeling Study of MDH-Catalyzed Methanol Oxidation: The Effect of the Ion in the Enzyme Active Site

Purnima Kharidehal and Daniela S. Mainardi*

*Institute for Micromanufacturing, Chemical Engineering Program, Louisiana Tech University, Ruston, LA 71272

ABSTRACT

A multi-scale modeling approach is presented here to investigate methanol oxidation by Ca^{2+}- and Mg^{2+}-containing Methanol Dehydrogenase (MDH) enzymes. A combination of tools involving Density Functional Theory (DFT), Transition State Theory (TST), and Kinetic Monte Carlo (kMC) are combined in order to model enzyme kinetics. Information regarding energy barriers and pre-exponential factors needed to determine reaction rates involved in each step of the Modified-Hydride Transfer methanol oxidation mechanism by MDH is obtained from DFT and TST respectively. This microscopic reaction rates are then provided as inputs in the kMC program, and the methanol oxidation process is modeled on a 2-D reactive surface representing the enzyme active site. A grid representing possible methanol adsorption, diffusion and reactions sites were mapped onto the projected active site image in a 2-D surface for a random substrate approach onto the reactive enzyme surface. Comparisons regarding enzyme kinetics were performed for methanol oxidation by Ca^{2+}- and Mg^{2+}-containing MDH.

INTRODUCTION

Methanol Dehydrogenase (MDH) enzyme is a bacterial quinoprotein that oxidizes methanol and other primary alcohols to their corresponding aldehydes [1]. The active site of MDH contains a Pyrroloquinoline Quinone (PQQ) cofactor, a divalent calcium cation, a catalytic base ASP-303 and twelve other amino acids, and water molecules[1]. It has been found that the Ca^{2+} ion in the active site of MDH is bonded to the nitrogen atom molecule (N6), the C5 quinone oxygen (O5), the oxygen of the carboxylate (O10) of the PQQ, the oxygen atoms (O12 and O13) from the carboxylate of Glu-177, and the oxygen atoms O11 and O14 of Asn-261 and Asp-303, respectively (Figure 1) [2-8]. Results from X-ray studies, distances from the Ca^{2+} ion to the N6, O5, and O10 atoms of PQQ are in the 2.32–2.47, 2.25–2.77, and 2.30–2.44 Å range, respectively[2-8] .Ca^{2+} ion in the active site holds the PQQ in place and also acts as a Lewis acid during the oxidation mechanism[5]. From previous studies, it was observed that the energy barrier for the oxidation of methanol was influenced by the atomic size of the metal ion in the active site of MDH[9].

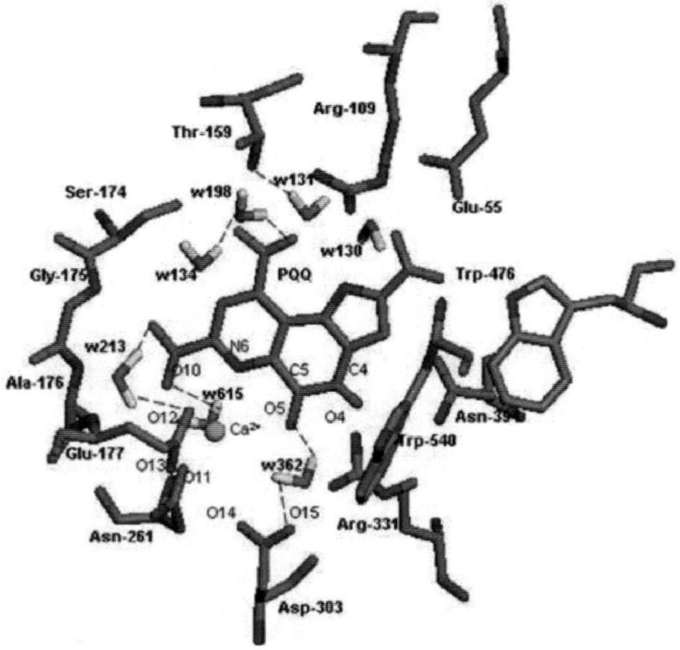

Figure 1: Active site of Methanol Dehydrogenase enzyme (MDH) [7].

Oxidation Mechanism for MDH

In the literature, two mechanisms for the oxidation of methanol by the naturally occurring Ca^{2+}-containing MDH have been commonly proposed; the three step addition-elimination (A-E) mechanism and the four step Hydride transfer (H-T) mechanism[6, 10]. The rate determining step for methanol oxidation by naturally occurring Ca^{2+}-MDH is the formation of formaldehyde[6, 10]. No agreement has been reached in the literature regarding which mechanism (A-E or H-T) actually operates in nature; however the Hydride Transfer (H-T) mechanism seems to be the preferred one. Hence, in this study the Hydride transfer mechanism is considered[11].

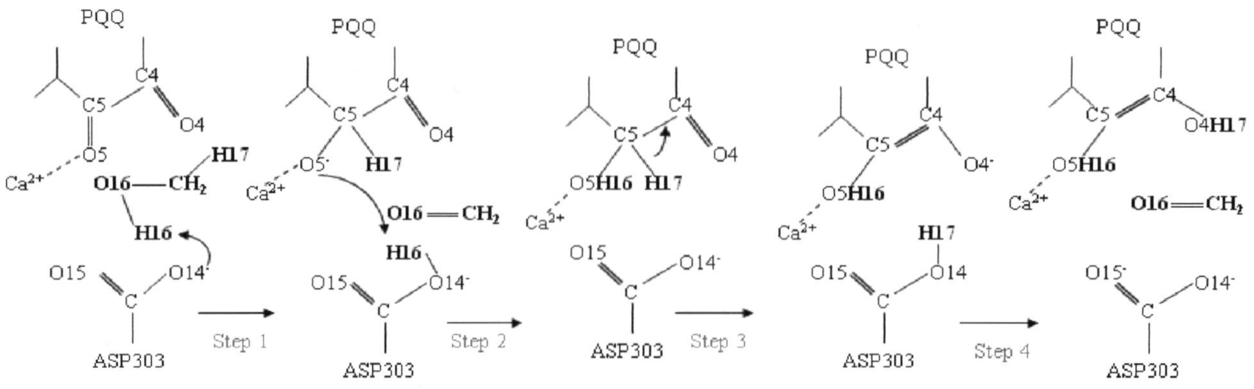

Figure 2: Four step Hydride-Transfer (H-T) methanol electro-oxidation mechanisms by Ca^{2+}-containing Methanol Dehydrogenase Enzyme[10].

Several experimental and theoretical studies have been reported that other bivalent metal ions can replace the natural ones in the active site of MDH enzymes[12]. Spectroscopic and semi-empirical calculations have shown that the binding constant of Sr^{2+} and Ba^{2+} to PQQ are weaker than that of Ca^{2+}; which can be attributed to the size of the calcium ion as that best fits in the binding pocket of PQQ in the active site. It has been found that the binding of the larger metal ions to the PQQ cause a distortion, making the coordination smaller than that of the calcium ion[12]. To observe the structural variations of the active site and kinetics of MDH enzyme on the methanol oxidation, the MDH active site Ca^{2+} ion was replaced with Ba^{2+} and Sr^{2+} and reported that there were no major differences between these enzymes in the interactions between PQQ and the metal ions; however, when tested upon the oxidation of methanol, the activation energy decreased in both modified MDH enzymes[12]. Sr^{2+}-MDH showed a decrease in activation energy (7.6 kcal/mol) with respect to that of Ca^{2+}-MDH (8.5 kcal/mol); but the lowest activation energy was found in the case of Ba^{2+}-MDH (3.5 kcal/mol), which was unexpected since Barium ion (weaker Lewis acid) should decrease the activity of the enzyme and hence the activation should be larger[12]. Thus, it shows that as the ionic radii increased (Ba^{2+}(1.35Å)>Sr^{2+}(1.12Å)>Ca^{2+}(0.99Å)) the activation energy decreased ($E_{act}Ba^{2+}$-MDH < $E_{act}Sr^{2+}$-MDH < $E_{act}Ca^{2+}$-MDH).

Kinetic Monte Carlo

Methanol oxidation by MDH enzymes is investigated using a multi-scale approach, as implemented in a Dynamic (kinetic) Monte Carlo (kMC) program which permits the investigations of length scales up to meters and time-scales up to the range of seconds. The kMC program, Carlos

4.1, developed by John Lukkien and A. P. J Jansen is described here[13, 14]. The approach followed here starts by first describing the system of the catalytic reactions by a stochastic model where an MDH catalytic surface (active site of the enzyme) is created for methanol oxidation to take place, and the lattice-gas models are considered to define the catalytic surface. The surface is defined by a discrete, regular two dimensional array of lattice points called "sites" (Figure 3) [13-15]. These sites with characteristic values and the lattice together are termed as a "*configuration*" [13-15]. Thus, a change from one configuration to another can be seen as a reaction. The species can diffuse, react, associate and dissociate once they are added onto the lattice following a set of probabilities and rules. As the simulation proceeds, the system undergoes dynamic evolution until a quasi –equilibrium steady state is reached.

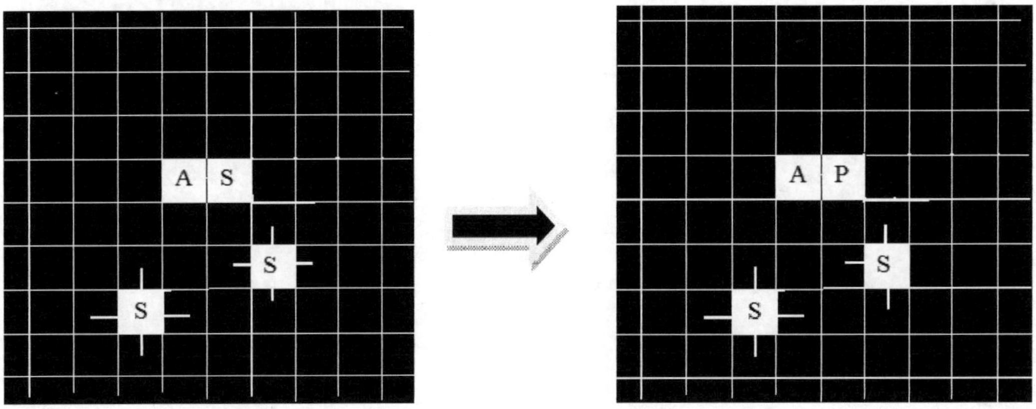

Figure 3: The sites occupied by the substrate molecules (methanol, S) on the Lattice (black grid) after following a random walk. When the substrate molecules reach the most reactive region of the enzyme active site (A), they get converted into product molecules (formaldehyde, P) [16].

Monte Carlo simulations can be implemented for steady-state as well as dynamic systems[13-15]. In a steady state system, where the system does not vary with time, a simple pattern is followed (Figure 3), where a lattice site is chosen randomly first and then the probabilities are calculated to determine whether the reaction at that site is possible. Once the reaction probabilities are calculated, the next site is chosen and the same procedure is repeated. After each grid site has been visited at least once on average, the simulation completes one Monte Carlo time step. In the case of dynamic systems (vary with time), there is an uncertainty in calculating the reaction probabilities with respect to the reaction steps. In order to simulate the dynamical interpretation, Monte Carlo methods by Fichthorn and Weinberg are used[17]. Microscopic reaction rates are needed as input, and these can be obtained either from experiments or *Ab Initio* calculations, and as a result, kMC methods yield time evolution of individual reaction rates (difficult to determine experimentally); which compete with each other as a function of external variables such as pressure and temperature. The rates are specified as probabilities and the time evolution of surface configuration is given by a Master Equation[14] (1), which is derived from first principles as Monte Carlo is a probabilistic approach.

$$\frac{dP(c,t)}{dt} = \sum_{c' \neq c} [P(c',t) r_{c',c} - P(c,t) r_{cc'}] \qquad \text{Eq. (1)}$$

where $P(c,t)$ is the probability of finding the system in configuration c, at time t, and $r_{cc'}$ is the microscopic rate of the reaction that transfer c into c'. There are methods which are used internally by the Carlos program to solve this equation: First-Reaction Method (FRM), Variable Step Size Method (VSSM), Random Selection Method (RSM) [14]. The First-Reaction Method (FRM) [14] is used here, which is appropriate for cases where reaction constants vary with time. According to this method, when the system is in a given configuration c, the set of all possible reactions is determined, and a time of occurrence $t_{c'c}$ is generated for each reaction i compatible with configuration c, according to equation (2).

$$\exp\left[-\int_{t}^{t_{c'c}} r^{i}_{c'c}(t')\cdot dt\right] = \delta \qquad \text{Eq. (2)}$$

In equation (2), $r^{i}_{c'c}$ is the time dependent rate of reaction i, and δ is a random number selected uniformly in the interval (0,1). Then, the reaction with the smallest time is selected, the configuration is changed accordingly, and the time is incremented to t. Finally, the set of possible reactions is generated according to the new configuration c', and the procedure is repeated. The microscopic rate r_i are related to macroscopic parameters as stated by equation (3),

$$r_i = v_i \exp\left[-\frac{E_{ai}}{RT}\right] \qquad \text{Eq. (3)}$$

where v_i is the prefactor, and E_{ai} is the activation energy of a reaction i.

Transition State Theory

Activation energies are required in order to calculate the rate constants for the processes to be modeled. For the MDH-assisted methanol oxidation study, the activation energies are obtained using Density Functional Theory (DFT) calculations on MDH models. The calculation of rate constants from the activation energies is conducted using Transition State Theory (TST) for the prediction of pre-exponential factors and Equation (3) [18]. According to the TST, the reactants need to overcome an energy barrier in order to form products and it postulates equilibrium (Boltzmann) energy distribution, which relates the reaction rate of a reactions with its Gibbs free energy by the following expression:

$$k = \frac{k_B T}{h} \exp\left(-\frac{\Delta G}{RT}\right) \qquad \text{Eq.(4)}$$

where k is the rate constant (s^{-1}), k_B is the Boltzmann's constant (3.29×10^{-24} cal/K), h is the Planck's constant (6.626×10^{-34} Js), T is the absolute temperature (298.15 K at room temperature), R is the universal gas constant (8.314 J/Kmol) and ΔG is the Gibbs free energy of activation[18].

METHODOLOGY

Density Functional Theory

The $DMOL^3$ module[19] of the Materials Studio® software by Accelrys, Inc. is used in this work[20]. It is a quantum mechanical package that models the electronic structure and energetics of molecules, solids and surfaces using Generalized Gradient Approximation (GGA) functionals within the Density Functional Theory formalism[21, 22]. The Becke-Lee-Yang-Parr (BLYP) exchange correlation functional[23, 24] and the double numerical with polarization (DNP) basis set, the best set available in $DMOL^3$, were used for performing the optimization calculations. The DNP basis set considers a polarization *d* function on heavy atoms and a polarization *p* function on hydrogen atoms. It can be compared to split-valence double zeta 6-31G** basis set in size, but is more accurate than the Gaussian basis sets of the same size. The errors are expected to be in the second decimal for calculated bond lengths (Å) and in the order of 2–5 (kcal/mol) in energies when BLYP/DNP theory level is used; which has been proven to be acceptable for the study of Ca^{2+}- and Ba^{2+}-MDH [25].

Small models of enzyme active sites are used to study the enzyme H-T reaction mechanism (Figure 2)[26-33]. Accordingly for treating the neglected enzyme environment, a homogeneous polarizable medium with some assumed dielectric constant is used, and density functional theory calculations are then performed to obtain reaction energy barriers[26-28, 30-33]. A homogeneous polarizable medium was considered with a continuum solvation model known as COSMO (conductor-like screening model) as implemented in the $DMOL^3$ module, to estimate solvation effects of the rest of the enzyme that are not included in the model[34-36]. This model works in such a way that the solute molecule forms a cavity within the dielectric continuum of permittivity that represents the solvent. To model the protein surroundings, the standard value for the dielectric constant ε was chosen to be 4 and so, that the final energy includes the $DMOL^3$/COSMO electrostatic energy, the geometry optimizations were performed in the presence of this solvation model[26-28, 30-33]. To confirm that the stationary points on the potential energy surface of the systems are in fact local minima (all real frequencies) or transition states (only one imaginary frequency), the harmonic vibrational frequency calculations were performed.

Kinetic Monte Carlo Modeling

To explore the kinetics and construction of the lattice representing the enzyme active site, it is important to discuss the assumptions and boundary conditions of the systems, since the reaction mechanics are relative to these aspects. The following assumptions are made to describe the model and to reduce complexity based on the assumptions reported [16]:

1. On the surface of the lattice, the substrate molecules follow a random walk.
2. The individual substrate molecules do not interact with each other.
3. There are no interactions between the substrate molecules and the obstacles.
4. The substrate molecules move towards the active site by the shortest available distance following the random walk motion.
5. The excluded volume condition is maintained i.e. at any instant of time one lattice site cannot be occupied by more than one molecule of the same or different species.

The well-known Michaelis-Menten equilibrium enzyme kinetics [37] equation is considered:

$$E + S \rightarrow ES \rightarrow E + P \qquad \text{Eq.(5)}$$

where E denotes the enzyme, S denotes the substrate (methanol), ES is the intermediate complex (enzyme-substrate), and P is the product (formaldehyde). Since the equation (5) is assumed to be diffusion limited, the formation of an intermediate ES is not taken into account and the substrate is almost instantaneously converted to product at the active site (A).Therefore, equation (5) reduces to a simpler reaction between the active site (A) and the substrate (S) molecule.

$$A + S \xrightarrow{k_{(t)}} A + P \qquad \text{Eq. (6)}$$

To carry out the simulations, a 64 × 64 lattice with cyclic boundary conditions was used. An important aspect to consider while modeling enzymatic reactions is the distribution of the substrate and obstacles. Previous works reported on enzyme kinetics modeling have used random sampling to distribute the species on the lattice and their reports showed that the enzyme kinetics were influenced by aspects like obstacle density. Thus, according to the approach described by Dandala *et al.*[16], a biased approach is used in sampling the lattice and to investigate its effect on the kinetics of methanol oxidation by MDH. The center of masses for all the amino acids in MDH are calculated and plotted with the active site amino acids as the center; which is in a way similar to the coarse graining method where the molecules are represented as beads. The representation of the MDH active site as beads is shown in Figure 4.

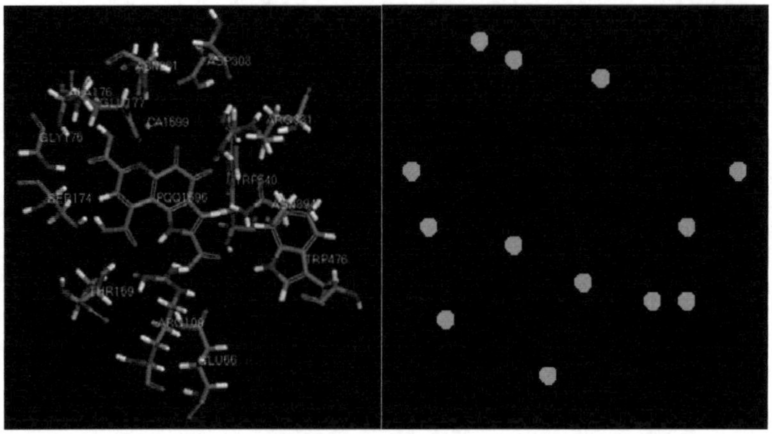

Figure 4. MDH enzyme Active Site (left) showing the various amino acids and the active site model showing the amino acids plotted as dots on the lattice according to their center of masses (right) [16].

The rate constants of H-T process are given as input at the start of each simulation and these are calculated from the activation energies obtained through the DFT calculations. The obstacle and substrate concentrations are specified at the start of each simulation and the simulation time is counted in Monte Carlo steps. For each simulation, the obstacle concentration is fixed while the substrate concentration decreases as the simulation proceeds due to the conversion of substrate molecules to product.

The simulation starts off with the substrate molecules (red) diffusing (making a random walk) on the lattice with the obstacles distributed (black) on it (Figure 5). Active site amino acids (green) are distributed at the center of the lattice and when the substrate molecules come in contact with any one of the active site amino acids, a product (blue) is formed which also diffuses on the lattice. The substrate molecule (red) entering the active site regime to react can be seen in Figure 5 [16]. Through this model, it is not only possible to visualize the enzymatic reactions that take place inside the enzyme but also to get useful information, such as the concentration profiles of substrate and product with respect to Monte Carlo time, obstacle density, reaction rate constants and the positions of obstacles on the lattice [16].

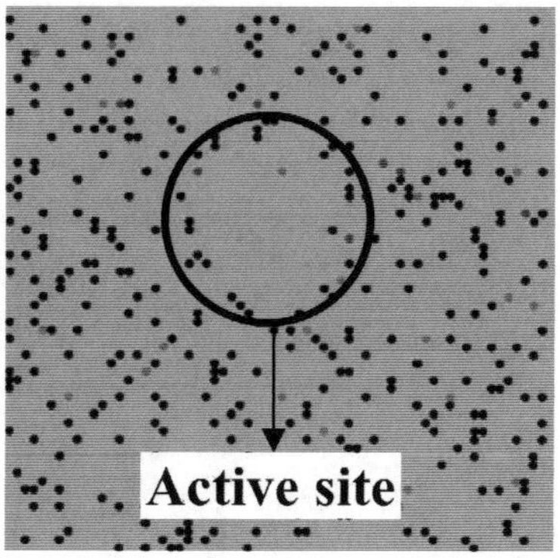

Figure 5. The green dots represent the active site cluster plotted on the two dimensional lattice with the Substrate (Methanol molecules) as red dots and Product (Formaldehyde Molecules) as blue dots making a random walk. The black dots represent the Obstacles (other amino acids in MDH) [16].

RESULTS AND DISCUSSION

DFT Calculations

In order to investigate the proposed H-T methanol oxidation mechanism, from previous studies Mg^{2+}-MDH active site models were selected[38]. The active site model consists of the PQQ molecule, Mg^{2+} ion, aspartic acid, glutamic acid, asparagine and three water molecules (Figure 6). According to the Himo *et al.* approach[39], unprotonated (charged) amino acids can be modeled by their corresponding protonated (neutral) amino acids. This was implemented and tested by Himo *et al.* [39] and well justified by the fact that proteins and enzymes have low dielectric constants ($\varepsilon \sim 4$), indicating that charge separation, if present, should be very small. These authors have shown that, by using these charge-neutral models, accurate enzymatic reaction energies and barriers were yielded. Thus, in our model we used the "Charge-neutral models" for amino acids[39]. The active site model corresponding to Mg^{2+} ion instead of the naturally occurring Ca^{2+} in the MDH active site was tested upon the H-T methanol oxidation mechanism. The reactant, intermediates and product for the four steps of the mechanism were optimized at the BLYP/DNP theory level and included solvation effects with a dielectric constant of 4 to represent the protein environment. The transition states were obtained for each step of the mechanism and the energies, representative bond distances, and energy barriers were compared to the ones obtained for the naturally occurring MDH.

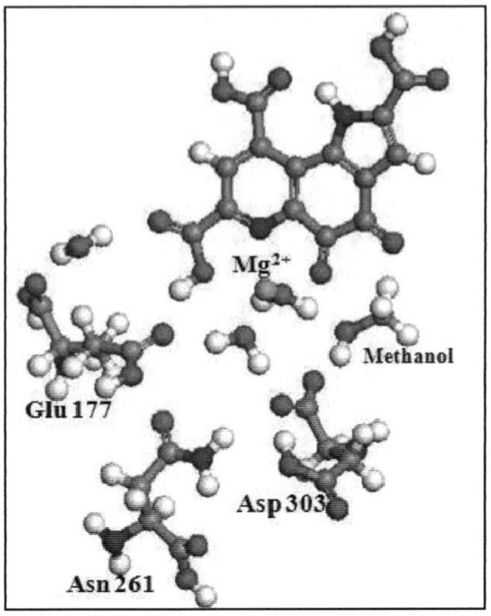

Figure 6: Active site model for Mg^{2+}-containing MDH.

To obtain the minimum energy configuration the reactant, intermediates and product were energy minimized and the ground state conformations were found. The energies obtained for the different orientations were compared and the minimum energy structure was considered for obtaining the transition states (Figure 7). The geometries of the reactant, intermediates and product of Mg^{2+}- containing MDH are in good agreement with the experimentally observed configurations of Ca^{2+}-MDH[38].

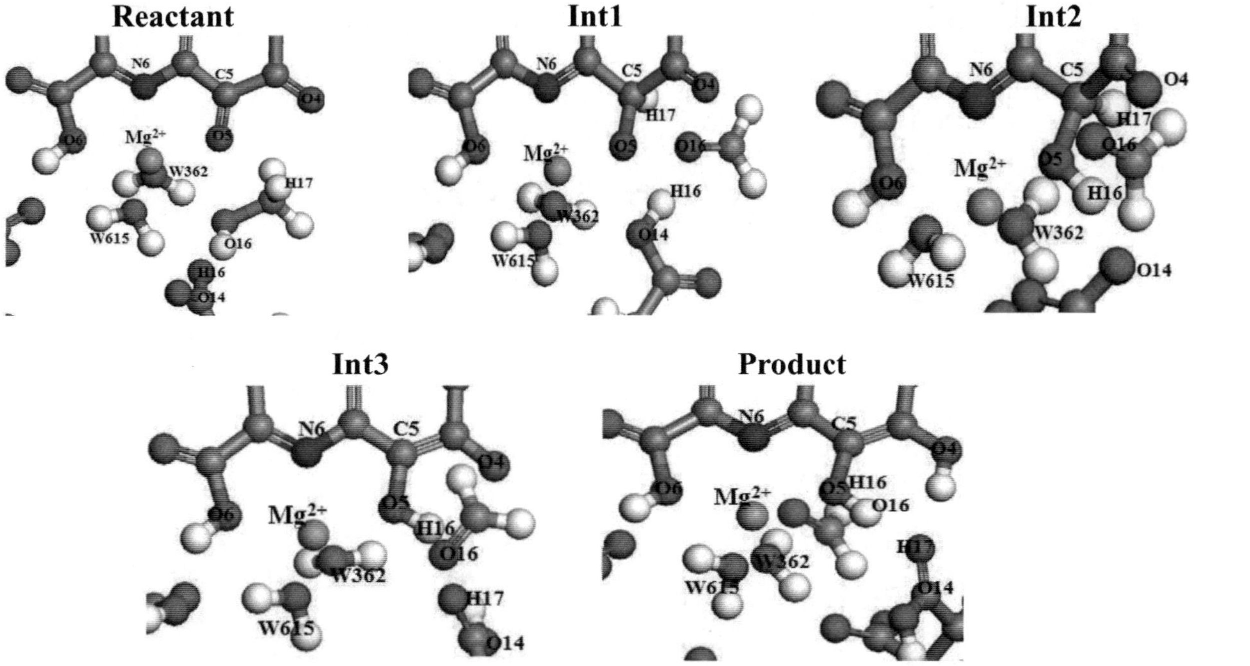

Figure 7: Minimum energy structures corresponding to Mg^{2+}-MDH models corresponding to the H-T methanol oxidation mechanism of Figure 2 obtained at BLYP/DNP theory level.

Rate Determining Step of the Hydride Transfer Mechanism by Mg^{2+}-MDH:

In this step, the hydride (H17) transfers from methanol to the C5 of PQQ and the H16 proton abstraction from methanol to Asp-303 occurs simultaneously forming formaldehyde (Figure 7). Thus, the intermediate (INT1) is evolved through the first transition state (TS1). At BLYP/DNP theory level on Mg^{2+}-MDH, the C5-H17 bond length was reduced from 3.63Å (Reactant) to 2.769Å (TS1) and finally to 1.121Å (INT1) and the bond length O14-H16 evolved from 1.768Å to 1.041Å (TS1) to 1.013Å (INT1) showing the abstraction of the proton H16 by Asp-303. For Mg^{2+}-MDH, the free energy barrier for TS1 was calculated to be 21.2 Kcal/mol (19.2 Kcal/mol for Ca^{2+}-MDH) [38] with respect to the reactant (Figure 8). Mg^{2+} did not polarize the C5-O5 bond as Ba^{2+}[38], thus resulting in less negative charge on the oxyanion O5 of PQQ. The hydrogen bonding (W362-O5) increased from 3.085Å (Reactant) to 3.347Å (TS1) to 3.169Å (INT1); hence the barrier is higher for this step compared to that of Ca^{2+}-MDH.

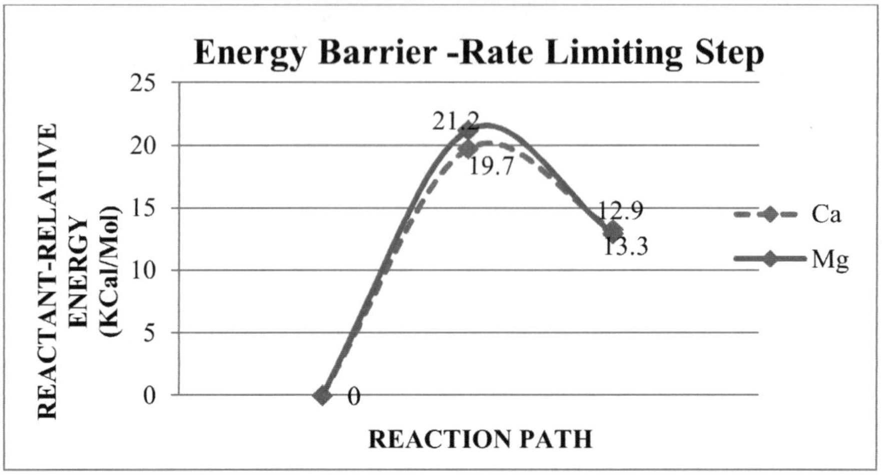

Figure 8: Energy Barriers for the Rate Determining Step in the H-T mechanism by MDH active site models with the two ions.

Kinetic Monte Carlo Simulations

Using Equation (4) and the free energy barrier values from the DFT calculations, the rates for the mechanism steps are obtained. The rate constant obtained for the case of Mg^{2+}-MDH is lower than the rate constant in the case of Ca^{2+}-MDH for the rate determining step as the energy barrier in the case of Mg-MDH is higher than that of Ca^{2+}-MDH. These rate constants are given as input and the substrate (methanol) and obstacle (amino acids that are not part of the active site) concentrations are also specified at the start of each simulation. The influence of the number of amino acids (obstacle density) on the diffusion of methanol towards the active site is investigated (Figure 9 for Ca^{2+}-MDH and Figure 10 for Mg^{2+}-MDH). This will provide useful information like the variations in the product formation rate or substrate consumption as the number of obstacle is changed.

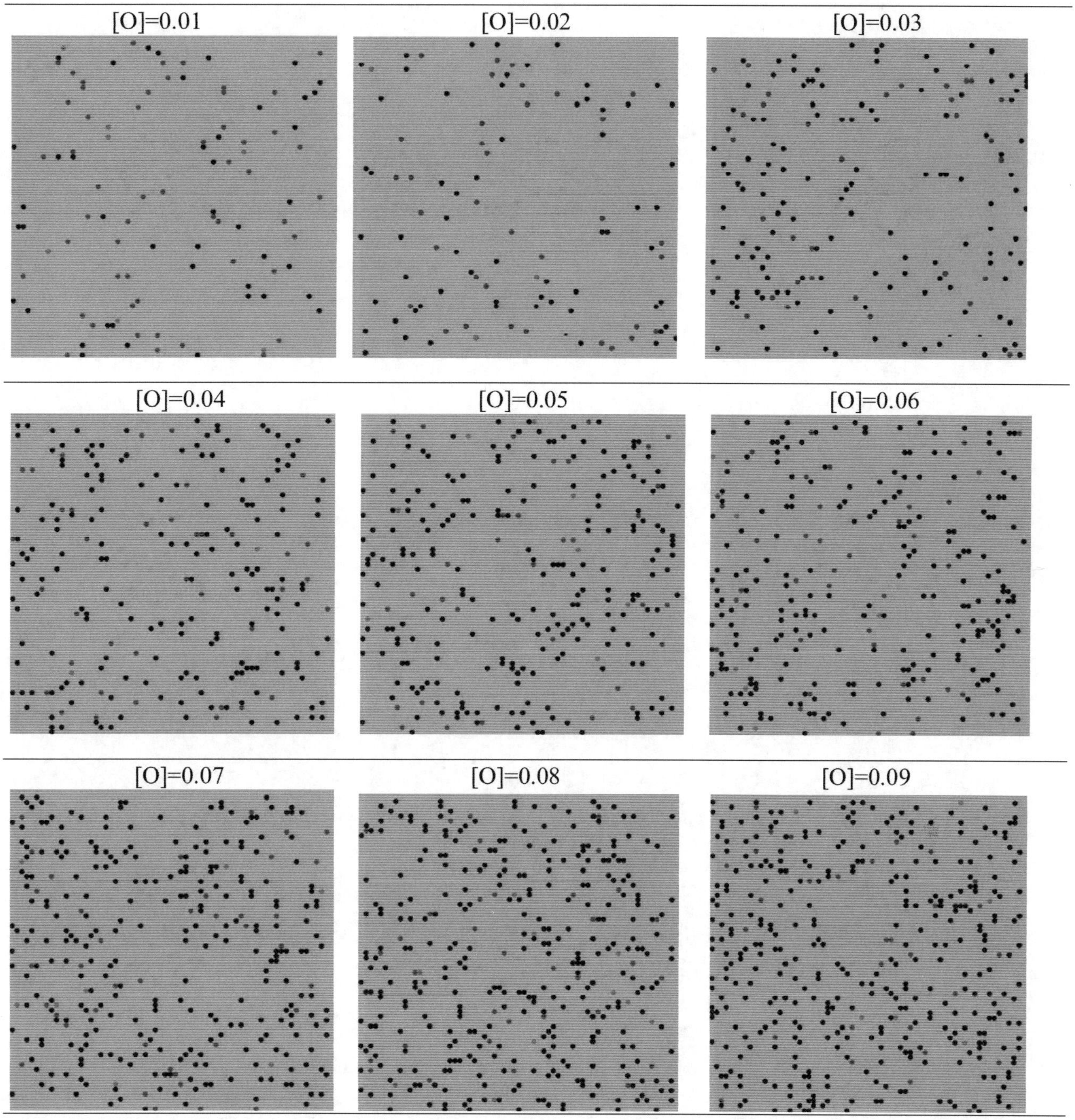

Figure 9. Lattice configurations showing the Substrate (Methanol molecules) as red dots and Product (Formaldehyde Molecules) as blue dots with increasing Obstacle concentration (other amino acids in MDH), [O], at t=1500MC steps for Ca^{2+}-containing MDH and the green dots represent the active site cluster.

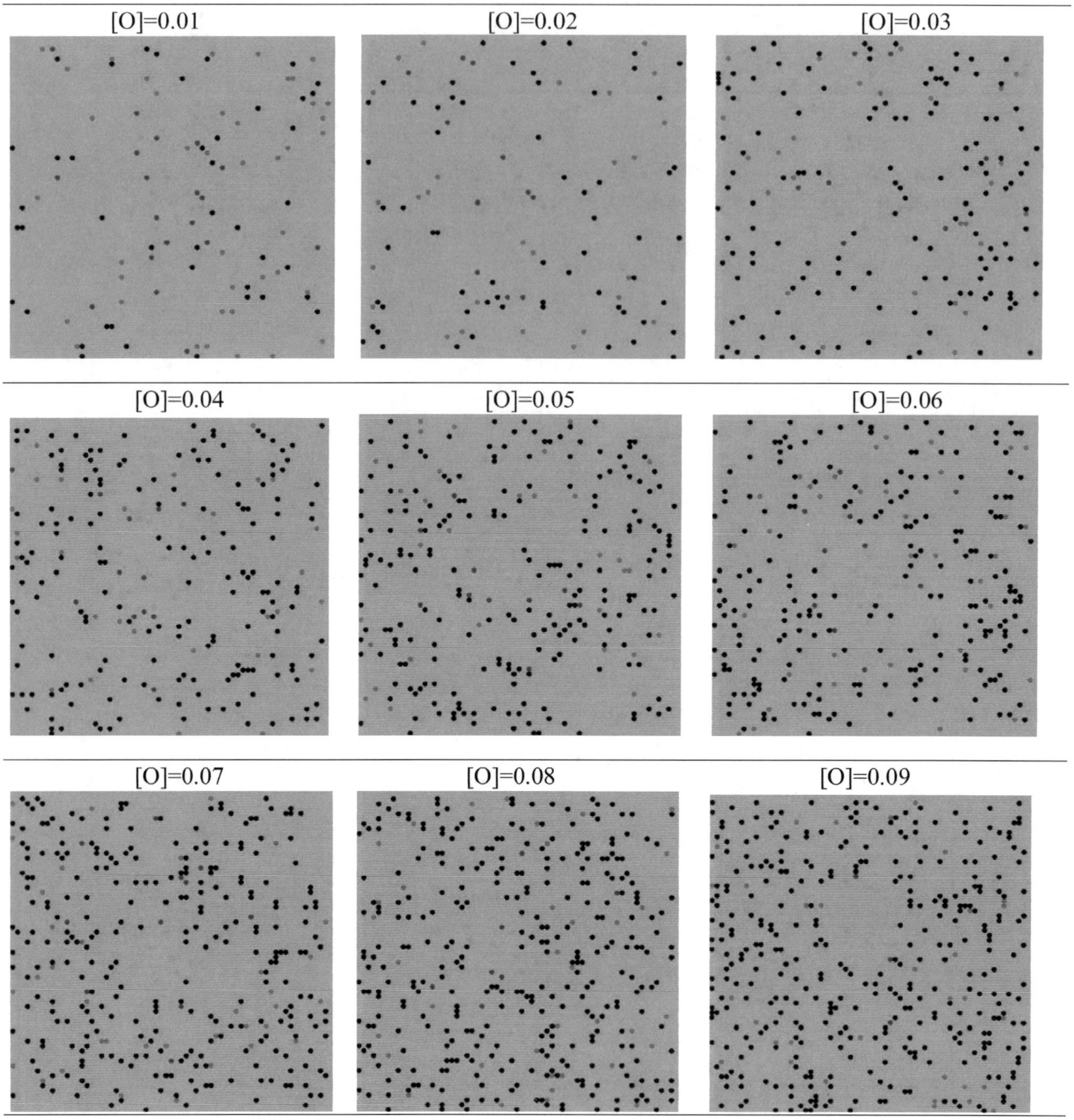

Figure 10. Lattice configurations showing the Substrate (Methanol molecules) as red dots and Product (Formaldehyde Molecules) as blue dots with increasing Obstacle concentration (other amino acids in MDH) at t=1500 MC steps for Mg^{2+}-containing MDH and the green dots represent the active site cluster.

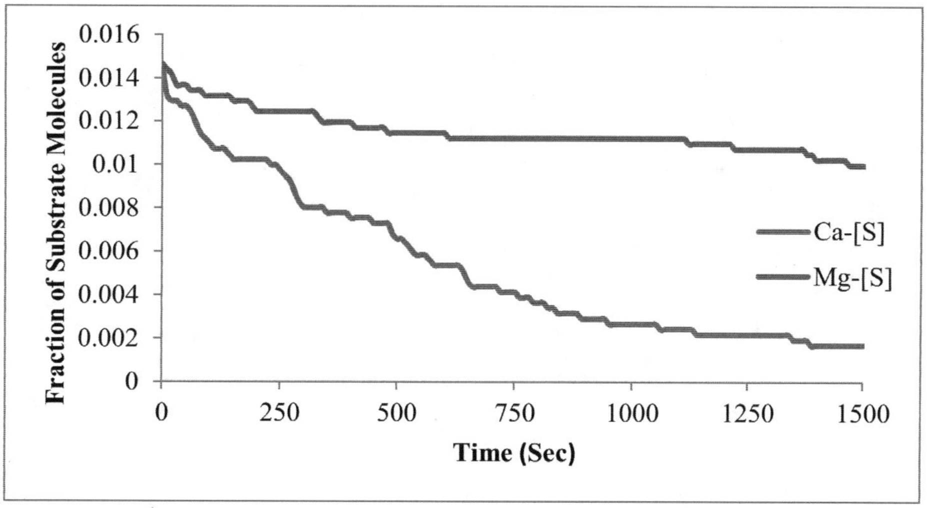

Figure 11. Fraction of Substrate (methanol) molecules formed for MDH active site models with the two ions with respect to time.

From Figure 11, the concentration of substrate molecules decreased with respect to time, implying $\frac{-d[S]}{dt}$=*rate.* For both the MDH active site models, the substrate consumption decreased. As observed from the simulations (Figure 9 and Figure 10) for increasing obstacle concentrations ([O]) i.e. the number of amino acids in MDH, the formation of formaldehyde is reduced in the case of Mg^{2+}-MDH and increased in the case of Ca^{2+}-MDH proving that Ca^{2+}-containing MDH is more reactive than Mg^{2+}-containing MDH. Thus, this implies that the amino acids around MDH also play an important role for the product formation (Figure 12). As shown in Figure12, after t=1500 MC steps for [O]=0.01 and [O]=0.09, the average of the fraction of product molecules is higher for the case of Ca-MDH.

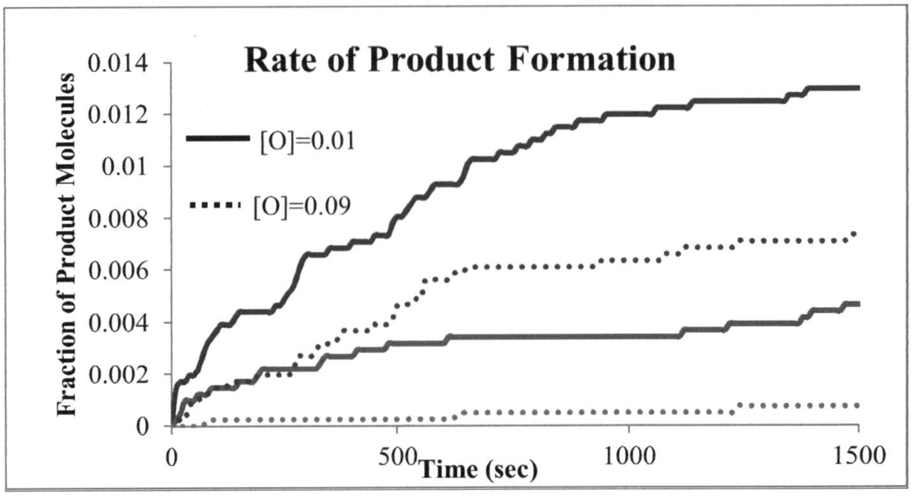

Figure 12. Rate of Product (Formaldehyde Molecules) formation for the two ions MDH active site models with respect to time for different Obstacle (amino acids) concentrations. Blue represents Ca-MDH and red represents Mg-MDH.

When the rate of formation of product molecules was plotted for increasing obstacle concentrations for both the ions MDH active site models, the Ca^{2+}-containing MDH has more product formation for [O]=0.01 and [O]=0.09 than the Mg^{2+}-containing MDH.

CONCLUSION

Using the generalized gradient approximation within the density functional theory formalism at the BLYP/DNP theory level methanol oxidation by a PQQ-containing MDH enzyme was investigated using Mg^{2+} in place of Ca^{2+} in the active site. The proposed hydride transfer mechanism was tested for Mg^{2+}-MDH active site model and it was found that the geometries of the intermediates, reactant and product agreed well with the experimental results. The free energy barrier for the rate limiting step (step1) for the Mg^{2+}-MDH was found to be 1.5kcal/mol more than the Ca^{2+}-MDH. The obtained energy barriers for the methanol oxidation reaction mechanism in Mg^{2+}-MDH were used as input for calculating the rate constants for the rate limiting step and these were used as input for the kMC program. The diffusion of methanol towards the active site is investigated for an increasing number of amino acids (obstacle density) and thus observed that Ca-containing MDH is more reactive than the modified MDH.

ACKNOWLEDGEMENTS

Financial support provided by NSF through the "Louisiana EPSCoR Research Infrastructure Improvement" grant (EPS 1003897) is acknowledged.

REFERENCES

[1] C. Anthony, *Methanol Dehydrogenase, a PQQ-Containing Quinoprotein Dehydrogenase*. New York.: Kluwer Academic/ Plenum Publishers, 2000.

[2] Z. X. Xia, Y. N. He, W. W. Dai, S. White, G. Boyd, and F. S. Mathews, "Detailed Active Site Configuration of a New Crystal Form of Methanol Dehydrogenase from Methylophilus W3a1 at 1.9 Å Resolution," *Biochem J.*, vol. 38, pp. 1214-1220, 1999.

[3] Z. X. Xia, Y. N. He, W. W. Dai, S. White, G. Boyd, and F. S. Mathews, "Determination of Gene sequence and Three-dimensional Structure at 2.4 Å resolution of Methanol Dehydrogenase from *Methylophilus W3A1*," *J. Mol. Biol.*, vol. 259, pp. 480-501, 1996.

[4] Z. X. Xia, W. W. Dai, J. P. Xiong, Z. P. Hao, V. L. Davidson, S. White, and F. S. Mathews, "The Three-Dimensional Structures of Methanol Dehydrogenase from Two Methylotrophic Bacteria at 2.6 Å Resolution," *J. Biol. chem.*, vol. 267, pp. 22289-22297, 1992.

[5] S. White, G. Boyd, F. S. Mathews, Z. X. Xia, W. W. Dai, Y. S. Zhang, and V. L. Davidson, "The Active Site Structure of Calcium Containing Methanol Dehydrogenase," *J. Biochem*, vol. 32, pp. 12955-12958, 1993.

[6] P.R. Afolabi, K. Amaratunga, O. Majekodunmi, S. L. Dales, R. Gill, D. Thompson, J. B. Cooper, S. P. Wood, P. M. Goodwin, and C. Anthony, "Site-Directed Mutagenesis and XRay-Crystallography of the PQQ-Containing Quinoprotein Methanol Dehydrogenase and Its Electron Acceptor, Cytochrome CL," *Biochem J.*, vol. 40, pp. 9799-9809, 2001.

[7] P. A. Williams, L. Coates, F. Mohammed, R. Gill, P. T. Erskine, A. Coker, S. P. Wood, C. Anthony, and J. B. Cooper*, "The atomic resolution structure of methanol dehydrogenase from Methylobacterium extorquens," *Acta. Crystall. Section D*, vol. D61, pp. 75-79, 2005.

[8] M. Ghosh, C. Anthony, K. Harlas, M. G. Goodwin, and C. C. F. Blake, "The Refined Structure of the Quinoprotein Methanol Dehydrogenase from Methylobacterium Extorquens at 1.94 Å," *Struc. (London)*, vol. 3, pp. 1771-1787, 1995.

[9] N. B. Idupulapati and D. S. Mainardi, "Coordination and binding of ions in Ca2+- and Ba2+-containing methanol dehydrogenase and interactions with methanol," *Journal of Molecular Structure: THEOCHEM*, vol. 901, pp. 72, 2009.

[10] C. Anthony and P. Williams, "The Structure and Mechanism of Methanol Dehydrogenase," *Biochimica. Biophysica. Acta*, vol. 1647, pp. 18-23, 2003.

[11] Y.-J. Zheng, Z.-x. Xia, Z.-w. Chen, F. S. Mathews, and T. C. Bruice, "Catalytic mechanism of quinoprotein methanol dehydrogenase: A theoretical and x-ray crystallographic investigation," *Proceedings of the National Academy of Sciences of the United States of America*, vol. 98, pp. 432, 2001.

[12] M. G. Goodwin and C. Anthony, "Characterization of a Novel Methanol Dehydrogenase Containing a Ba2+ Ion at the Active Site," *Biochem J.*, vol. 318, pp. 673-679, 1996.

[13] J. J. Lukkien, J. P. L. Segers, P. A. J. Hilbers, R. J. Gelten, and A. P. J. Jansen, "Efficient Monte Carlo methods for the simulation of catalytic surface reactions," *Physical Review E - Statistical Physics, Plasmas, Fluids, and Related Interdisciplinary Topics*, vol. 58, pp. 2598-2610, 1998.

[14] R. M. Nieminen and A. P. J. Jansen, "Monte Carlo simulations of surface reactions," *Applied Catalysis A: General*, vol. 160, pp. 99-123, 1997.

[15] A. P. J. Jansen, "Monte Carlo simulations of chemical reactions on a surface with time-dependent reaction-rate constants," *Computer Physics Communications*, vol. 86, pp. 1-12, 1995.

[16] A. P. J. J. a. D. S. M. N.K.R.Dandala, "A Multi-scale Modeling Approach for Studying MDH-Catalyzed Methanol Oxidation," in *Multiscale Modeling: From Atoms to Devices*, T. C. Pedro Derosa, Ed. Florida: CRC Press Inc, 2010, pp. 91-112.

[17] K. A. Fichthorn and W. H. Weinberg, "Theoretical foundations of dynamical Monte Carlo simulations," *The Journal of Chemical Physics*, vol. 95, pp. 1090-1096, 1991.

[18] J. B. Foresman, "*Exploring chemistry with electronic structure methods*," *Gaussian, Inc, Pittsburgh, PA,1996*, 1996.

[19] Accelrys Inc., "$DMOL^3$ User Guide," (San Diego, 2003).

[20] Accelyrs, "Materials Studio," 2006.

[21] R. M. Dreialer and E. K. U. Gross, *Density Functional Theory: An Approach to Quantum Many Body Problem*. Berlin: Springer, 1990.

[22] W. Koch and M. C. Holthausen, *A Chemist's Guide to Density Functional Theory*, Second ed: Wiley-CVH, 2001.

[23] C. Lee, W. Yang, and R. Parr, "Accurate and simple analytic representation of the electron-gas correlation energy," *phys. Rev. B*, vol. 37, pp. 786, 1998.

[24] A. D. Becke, "Density-functional thermochemistry III. The role of exact exchange," *J. Chem. Phys*, vol. 88, pp. 2547, 1988.

[25] N. B. Idupulapati and D. S. Mainardi, "A DMOL3 Study of the Methanol Addition-Elimination Oxidation Mechanism by Methanol Dehydrogenase Enzyme," *Mol. Sim.*, vol. 34, pp. 1057-1064, 2008.

[26] R. Zhen Liao, J. Yu, F. M. Raushel, and F. Himo, "Theoretical Investigation of the Reaction Mechanism of the Dinuclear Zinc Enzyme Dihydroorotase," *Chem. Eur. J.*, vol. 14, pp. 4287-4292, 2008.

[27] R. Sevastik and F. Himo, "Quantum chemical modeling of enzymatic reactions: The case of 4-oxalocrotonate tautomerase," *Bioorg. Chem.*, vol. 35, pp. 444-457, 2007.

[28] P. Velichkova and F. Himo, "Theoretical Study of the Methyl Transfer in Guanidinoacetate Methyltransferase," *J. Phys.Chem. B*, vol. 110, pp. 16-19, 2006.

[29] M. Leopoldini, N. Russo, and M. Toscano, "The Preferred Reaction Path for the Oxidation of Methanol by PQQContaining Methanol Dehydrogenase: Addition–Elimination versus Hydride-Transfer Mechanism," *Chem. Euro. j*, vol. 13, pp. 2109-2117, 2007.

[30] L. Noodleman, T. Lovell, W. Han, J. Li, and F. Himo, "Quantum Chemical Studies of Intermediates and Reaction Pathways in Selected Enzymes and Catalytic Synthetic Systems," *Chem. Rev.*, vol. 104, pp. 459-508, 2004.

[31] K. H. Hopmann and F. Himo, "Theoretical Study of the Full Reaction Mechanism of Human Soluble Epoxide Hydrolase," *Chem. Eur. J.*, vol. 12, pp. 6898-6909, 2006.

[32] K. H. Hopmann and F. Himo, "Quantum Chemical Modeling of the Dehalogenation Reaction of Haloalcohol Dehalogenase," *J. Chem. Theory Comput.*, vol. 4, pp. 1129-1137, 2008.

[33] P. Georgieva and F. Himo, "Density functional theory study of the reaction mechanism of the DNA repairing enzyme alkylguanine alkyltransferase," *Chemical Physics Letters*, vol. 463, pp. 214, 2008.

[34] B. M. R. Cammi., J. Tomasi, *J. Phys.Chem*, vol. 103, pp. 9100, 1999.

[35] B. M. R. Cammi., J. Tomasi, *J. Phys.Chem A*, vol. 104, pp. 5631, 2000.

[36] G. S. M. Cossi, N. Rega,V. Barone, *J. Chem. Phys*, vol. 43, pp. 117, 2002.
[37] A. Cornish-Bowden, *Fundamentals of enzyme kinetics*. London: Portland Press, Ltd., 2004.
[38] N. B. Idupulapati and D. S. Mainardi, "Quantum Chemical Modeling of Methanol Oxidation Mechanisms by Methanol Dehydrogenase Enzyme: Effect of Substitution of Calcium by Barium in the Active Site," *The Journal of Physical Chemistry A*, vol. 114, pp. 1887, 2010.
[39] F. Himo, "Catalytic Mechanism of Benzylsuccinate Synthase, a Theoretical Study," *The Journal of Physical Chemistry B*, vol. 106, pp. 7688, 2002.

(37e) Dynamics and Control of Aggregate Thin Film Surface Morphology for Improved Light Trapping: Implementation On a Large-Scale Kinetic Monte-Carlo Simulation

Jianqiao Huang, Xinyu Zhang, G. Orkoulas and Panagiotis D. Christofides, Department of Chemical and Biomolecular Engineering, University of California, Los Angeles, Los Angeles, CA

Photovoltaic (solar) cells are an important source of sustainable energy. Thin-film silicon solar cells are currently among the most important and widely used solar cells and their share of the overall solar cell market is steadily increasing (e.g.,[1]). Research on optical modeling of thin-film silicon solar cells indicates that the scattering properties of the thin film interfaces directly influence the light trapping process and the efficiencies of thin-film silicon solar cells (e.g., [2],[3]). However, no efforts have been seen in improving the conversion efficiency of thin-film solar cells via the regulation of the thin film surface morphology during the manufacturing process by appropriately tailoring the surface slope and roughness to desired specifications. Thus, it is desirable to develop systematic approaches to manufacture thin film solar cells with optimal conversion efficiencies via computational modeling and real-time model-based control of the manufacturing process. Recently, we have initiated an effort towards modeling and control of thin film surface morphology to optimize the light reflectance and transmittance properties of thin films. In this direction, we have studied the dynamics and lattice size dependence of surface mean slope [4] and have developed predictive control algorithms to regulate both surface roughness and slope of KMC simulators with small domain size in one [5] and two [6] spatial dimensions. However, control of thin film surface morphology within a domain comparable to the visible light wavelength has remained unsolved.

This work focuses on the application of a model predictive controller to a large-scale kinetic Monte-Carlo model. The dynamics of the evolution of the thin film surface height profile are modeled by an Edwards-Wilkinson-type equation (a second-order stochastic partial differential equation) in one spatial dimension. The thin film surface morphology is characterized in terms of surface roughness and surface slope. Analytical solutions of the expected surface roughness and surface slope are obtained by solving the Edwards-Wilkinson equation and are used in the controller design. The model parameters of the Edwards-Wilkinson equation are estimated from large scale kinetic Monte-Carlo (KMC) simulations. This parameter dependence on the deposition rate is used in the formulation of the predictive controller to predict the influence of the control action on the surface roughness and slope at the end of the growth process. The cost function of the controller involves penalties on both surface roughness and mean slope from set-point values as well as constraints on the magnitude and rate of change of the control action. The controller is applied to a large-scale one-dimensional KMC model with a domain size comparable to the wavelength of visible light. Simulation results show that the proposed controller successfully regulates surface roughness and slope to set-point values at the end of the deposition that yield desired levels of thin film reflectance and transmittance.

[1] M. A. Green. Thin-film solar cells: Review of materials, technologies and commercial status. Journal of Materials Science: Materials in Electronics, 18:15--19, 2007. [2] M. Zeman. Optical and electrical modeling of thin film silicon solar cells. Journal of Material Research, 4:889-898, 2008.
[3] A. Poruba and A. Fejfar. Optical absorption and light scattering in micro-crystalline silicon thin ¯ lms and solar cells. Journal of Applied Physics, 88:148-160, 2000.
[4] J. Huang, G. Hu, G. Orkoulas, and P. D. Christofides. Dynamics and lattice-size dependence of surface mean slope in thin film deposition process. Industrial & Engineering Chemistry Research, 50: 1219--1230, 2011.
[5] X. Zhang, G. Hu, G. Orkoulas, and P. D. Christofides. Predictive control of surface mean slope and roughness in a thin film deposition process. Chemical Engineering Science,65:4720-4731, 2010.
[6] X. Zhang, G. Hu, G. Orkoulas, and P. D. Christofides. Multivariable Model Predictive Control of Thin Film Surface Roughness and Slope for Light Trapping Optimization, Industrial & Engineering Chemistry Research, 49: 10510-10516, 2010.

See more of this Session: Multiscale Modeling I
See more of this Group/Topical: Catalysis and Reaction Engineering Division

(37f) A Predictive Control Approach for Thin Film Growth of Photovoltaic Systems

Ali Rahnamoun and **Antonios Armaou**, Chemical Engineering, The Pennsylvania State University, University Park, PA

A Predictive Control approach for thin film growth of photovoltaic systems

Ali Rahnamoun and Antonios Armaou

Department of Chemical Engineering, The Pennsylvania State University,University Park, PA 16802

When one considers the limited fuel resources of energy, the importance of photovoltaic systems in the future energy provision becomes obvious. The limited conversion efficiensy of these systems however severely limits their use currently. During the last two decades, there have been lots of studies to enhance the efficiency of photovoltaic systems. The efficiency of such systems is highly dependent on controling their microstructure within strict limits. Two important factors affecting this microstructure are the surface roughness and porosity of the produced thin film solar cells. Due to the lack of accurate sensors that can timely provide microstructure information, model based approaches to control the quality of these photovoltaic systems have been pursued. The models that predict film microstructure however, pose significant challenges both from an analysis and control point of view [1]. Such difficulties are attributed in part to the unavailability of closed form models to describe the process evolution at molecular-level detail and their computationally intensive nature that prevents their real time implementation.

To circumvent the mentioned limitations, one of the previous approaches includes identified stochastic partial differential equation models to design the controller [2]. This approach however assumes specific structure to the nonlinear stochastic terms. Another approach relied on the off-line and subsequently the on-line identification of bilinear models for the process, which are then used for the controller synthesis [3].

The objective of the present work is to develop a model predictive control algorithm to simultaneously regulate the surface roughness and porosity of thin film solar cell to obtain higher efficiency. To simulate the evolution of thin film growth, Kinetic Monte Carlo simulation is utilized. Two processes in this simulation include deposition with considering shadowing effect and migration of the deposited particles. For achieving the closed loop model of the simulated system, fuzzy system identification is utilized [4]. The proposed method explicitly accounts for the presence of stochastic terms in the process and the issue of noise in the measurements. The kalman filter is applied to the model to reach to modeling results with more accuracy. The developed model is then employed with a model predictive framework to drive the process at the desired process objective. The simulation results demonstrate the effectiveness of the nonlinear model predictive controller designed to regulate the surface roughness and porosity of the thin film solar cell to achieve higher efficiency.

REFERENCES:

[1] P. D.Christofides, A.Armaou , "Control of Multiscale and Distributed process systems-preface," *Computers &Chemical Engineering*., **29**,687-688, 2005.

[2] Y Lou, P. D.Christofides, "Nonlinear Feedback Control of Surface Roughness Using a Stochastic PDE: Design and Application to a Sputtering Process," *Industrial & Engineering Chemistry Research*., **45**, 7177-7189, 2006.

[3] A.Varshney, A.Armaou, "Multiscale Optimization using hybrid PDE /KMC process systems with applications to thin film growth", *Chemical Engineering Science*., **60**, 6780-6794, 2005.

[4] A.Rahnamoun, A.Armaou, "A Nonlinear Predictive Control of Processes with Multiscale Objectives using a Fuzzy-System Identification Approach",American Control Conference 2011.

See more of this Session: Multiscale Modeling I
See more of this Group/Topical: Catalysis and Reaction Engineering Division

(37g) Moment Closure for Chemical Reactions

Yiannis Kaznessis and **Fatemeh Ghasemi**, Department of Chemical Engineering and Materials Science, University of Minnesota, Minneapolis, MN

We present a systematic procedure for constructing moment closure functions of arbitrary order for the stochastic models of chemical reaction systems. In the stochastic formulation of chemical reactions, the dynamics of the first M-order moments of the species populations generally do not form a closed system of differential equations, in the sense that the time-derivatives of first M-order moments generally depend on moments of order higher than M. However, for analysis purposes, these dynamics are often made to be closed by approximating the needed derivatives of the first M-order moments by nonlinear functions of the same moments. These functions are called the moment closure functions. We will discuss a new approach to construct moment closure functions. Our method is based on calculating the correlation function of individual moments. We use several examples, with linear and nonlinear sets of reactions, to demonstrate the applicability and the limitations of the scheme.

See more of this Session: Multiscale Modeling I
See more of this Group/Topical: Catalysis and Reaction Engineering Division

(37h) Multiscale Analysis and Simulation of Bubbly Flow

Ning Yang[1], Zongying Wu[2], Jianhua Chen[1] and Jinghai Li[1], (1)State Key Laboratory of Multiphase Complex Systems, Institute of Process Engineering, Chinese Academy of Sciences, Beijing, China, (2) Department of Chemical Engineering, Xi'an Jiaotong University, Xi'an, China

Bubble columns are considered as favorable reactors in process engineering. Scale-up and optimization of these reactors require an extensive understanding of gas-liquid two-phase flow in order to gain further insight into mixing and transfer properties. However, the gas-liquid flow and/or gas-liquid-solid flow in bubble columns is quite complicated not only because of its versatile macro-scale flow regimes, but due to those intricate structures occurring at meso and micro scales, posing great challenges for CFD simulation of multi-phase flow in these reactors. Simulation based on Eulerian-Eulerian method, though usually employed in commercial CFD packages, is still not powerful enough to capture real physics due to our limited understanding on the multi-scale flow structure and hence inadequate development of closure laws including drag, lift and turbulence models. The aim of this study is to use the dual-bubble-size (DBS) model to develop a closure law for the ratio of effective drag coefficient to bubble diameter, and then apply this model in simulation of gas-liquid two-phase flow, mass transfer and chemical reaction for FT synthesis. In addition to mass and momentum conservation equations, the stability condition reflecting the compromise between two different dominant mechanisms serves as another constraint for the model, and therefore it is expected to offer a more accurate drag closure for CFD simulation. Prior to this integration, we investigate the respective effects of bubble diameter and correction factor and find that the effect of bubble diameter is limited, whereas the correction factor due to the bubble swarm effect is eminent and appropriate correction factor has to be selected for different correlations of standard drag efficient to be in accord with experiments. By contrast, the new model can well predict the radial gas holdup distribution, the total gas holdup as well as the two-phase flow field without the need of adjusting model parameters, and then it is used for a scale-up process of F-T systhesis, showing its great potential and advantage in understanding the complex nature of multiscale structure of gas-liquid flow in bubble columns.

See more of this Session: Multiscale Modeling I
See more of this Group/Topical: Catalysis and Reaction Engineering Division

(52a) A Fast Hybrid Molecular Dynamics Method for Estimating Gas Diffusion In Polymeric Glasses

Yingying Jiang, Department of Chemical Engineering, University of Texas at Austin, Austin, TX

A new hybrid Monte Carlo/Molecular Dynamics method is developed for estimating the slow diffusional processes of light gases in glassy polymers. Diffusion coefficients, as small as 10^{-9} cm^2/s are estimated for penetrants in four different polymers at 298 K. In all cases, agreement between literature experimental data and values obtained from the fast hybrid molecular dynamics method ranges from good to excellent.

See more of this Session: Recent Advances In Molecular Simulation Methods I
See more of this Group/Topical: Computational Molecular Science and Engineering Forum

(52b) Solvent Effect On the Ultrafast Spectra In Water Solution: Explicit and Implicit Solvation

Yun-Kui Li[1], Han-Yu Wu[2], Quan Zhu[1], Ke-Xiang Fu[1] and Xiang-Yuan Li[1], (1)College of Chemical Engineering, Sichuan University, Chengdu, China, (2)College of Chemistry, Sichuan University, Chengdu, China

In the framework of continuous medium theory, a new form of the electrostatic free energy of nonequilibrium solvation has been derived by the method of constrained equilibrium state[1]. This form has been further deduced to solve the electron transfer[2] and hydrated electron[3] problems, and in this work to treat the solvent effect on the ultrafast UV/Vis absorption spectra in water solution with respect to both explicit and implicit solvation strategies[4]. For the explicit solvation, the newly inferred energy expression was implemented by means of modifying the nonequilibrium module of the averaged solvent electrostatic potential/molecular dynamics (M-ASEP/MD) program, which permits an atomic-level description of the solvent and the mutual polarization between the solute and solvent molecules. For the implicit solvation, an approximation of point dipole and sphere cavity was employed with considering the solvent as a uniform and continuous polarizable dielectric medium. In addition, a new method of determination of the cavity radius was proposed, taking the short-range dispersion/repulsion and long-range electrostatic interactions between the solute and solvent molecules into account with the aid of the MD simulations. Sequentially, the two solvation strategies that are based on our nonequilibrium solvation theory were performed to study the solvatochromic shift of formaldehyde, a representative chromophore molecule, in water solvent and verified to show self-consistence with each other.

References

1 X. Y. Li, F. C. He, K. X. Fu, W. J. Liu, *J. Theor. Comput. Chem.* **2010**, *9*, *Supp. 1*, 23–37.

2 X. Y. Li, Q. D. Wang, J. B. Wang, J. Y. Ma, K. X. Fu, F. C. He, *Phys. Chem. Chem. Phys.* **2010**, *12*, 1341–1350.

3 X. J. Wang, Q. Zhu, Y. K. Li, X. M. Cheng, X. Y. Li, K. X. Fu, F. C. He, *J. Phys. Chem. B* **2010**, *114*, 2189–2197.

4 Y. K. Li, Q. Zhu, X. Y. Li, K. X. Fu, X. J. Wang, X. M. Cheng, *J. Phys. Chem. A* **2011**, *115*, 232.

See more of this Session: Recent Advances In Molecular Simulation Methods I
See more of this Group/Topical: Computational Molecular Science and Engineering Forum

(52c) Precise Simulation of Freezing Transitions

Michael Nayhouse[1], Ankur M. Amlani[1] and G. Orkoulas[2], (1)Chemical and Biomolecular Engineering, UCLA, Los Angeles, CA, (2)Department of Chemical and Biomolecular Engineering, University of California, Los Angeles, Los Angeles, CA

Despite recent advances, precise simulation of freezing transitions continues to be a challenging task. The main difficulties in simulations of fluid-solid coexistence are the very dense, nearly-incompressible nature and the different symmetry of the two coexisting phases. Precise simulation of freezing is a topic of considerable theoretical and practical importance. In protein solutions and colloidal suspensions for instance, it is typically observed that fluid-fluid demixing is metastable against crystallization.

In this work, a simulation method, based on a modification of the constrained cell model of Hoover and Ree, is proposed and tested on hard-sphere and Lennard-Jones systems. In the fully occupied constrained cell model of Hoover and Ree, each particle is confined within its own Wigner-Seitz cell. Constant pressure simulations of the constrained cell model indicate, as Hoover and Ree first pointed out, that it exhibits a point of mechanical instability at about 60% expansion from close packing. Below that point, the solid is mechanically unstable since without the confinement imposed by the cell walls it will rapidly disintegrate to a disordered fluid-like phase.

In their original work associated with freezing of hard-sphere systems, Hoover and Ree considered a modified cell model by introducing an external field of variable strength. High values of the external field variable favor configurations with one particle per cell and thus stabilize the solid phase. Normal system behavior is restored in the limit of vanishing external field. Hoover and Ree argued that as the strength of the field varies, one might be able to go continuously from the fluid to the solid phase and vise versa. They did not, however, exclude the possibility of critical/Curie type of terminal point that separates continuous from discontinuous behavior.

In the present work, the modified cell model is simulated under constant-pressure conditions using tempering and histogram reweighting techniques. Simulation results on hard-sphere and Lennard-Jones systems indicate that the transition from the fluid to the solid is continuous below the mechanical stability point. Beyond the point of mechanical stability, however, the passage from the fluid to the solid occurs via a first-order phase transition at a finite value of the external field. As the pressure increases, the magnitude of the field that sustains coexistence decreases. The previous observations suggest that the fluid-solid transition of the system under consideration can be studied by analyzing the field-induced transition of the modified cell model in the limit of vanishing external field. The coexistence pressure and densities for hard-spheres and Lennard-Jones systems are determined via finite-size scaling techniques for first-order phase transitions. The results so obtained are in close accord with available data from other techniques.

(52d) Small Molecule Solvation Free Energy: Enhanced Conformational Sampling Using Expanded Ensemble Molecular Dynamics Simulation

Andrew S. Paluch, Department of Chemical and Biomolecular Engineering, University of Notre Dame, Notre Dame, IN, David L. Mobley, Department of Chemistry, University of New Orleans, New Orleans, LA and Edward J. Maginn, Chemical & Biomolecular Engineering, University of Notre Dame, Notre Dame, IN

Knowledge of the underlying free energy of biologically active species in solution is crucial to understanding their thermodynamic phase behavior, which is essential for rational drug design. We will present an efficient expanded ensemble method to calculate the solvation free energy of small molecules with complex topologies by molecular dynamics simulation. By computing the solvation free energy of ibuprofen in water and small alcohols at 300 K and 1 bar and comparing to reference simulation results, we will validate the methodology. In the present study, the carboxylic acid functional group of ibuprofen is subject to free energy barriers of rotation of 14-20 k_BT, leading to sampling difficulties with conventional molecular dynamics simulations. While several advances have been made to overcome such weaknesses, we will demonstrate how this shortcoming is easily overcome by using an expanded ensemble methodology to facilitate the sampling of phase space.

See more of this Session: Recent Advances In Molecular Simulation Methods I
See more of this Group/Topical: Computational Molecular Science and Engineering Forum

(52e) Efficient, Precise and Accurate Methods of Calculating Solid-Phase Free Energies by Molecular Simulation

Tai Boon Tan, Andrew J. Schultz and David A. Kofke, Chemical and Biological Engineering, University at Buffalo, The State University of New York, Buffalo, NY

Free energy is an important element in evaluating the stability of a solid crystal. A stable form of crystalline solid occupies an atomic/ molecular packing that gives the lowest free energy. The existing reliable free-energy calculation methods usually involve lengthy thermodynamic integration procedures. To improve the calculation efficiency, we present a novel approach that enables us to calculate the absolute free energy of a crystalline solid by applying targeted free-energy perturbation staged using overlap sampling. The underlying idea of this method is based upon the knowledge of phase-space relation between the two perturbing systems and introduction of appropriate scaling parameter (to the atomic displacement and/ or molecular orientation) to improve the overlap of configuration space of the two systems and eventually the free-energy results.

We have examined this idea to compute the change in free energy with temperature of the soft-sphere crystalline solid. In this method, the free energy difference between nearby temperatures is calculated via overlap-sampling free-energy perturbation with the Bennett's optimization. Coupled to this is a harmonically targeted perturbation that displaces the atoms in a manner consistent with the temperature change, such that for a harmonic system the free-energy difference would be recovered with no error. A series of such perturbations can be assembled to bridge larger gaps in temperature. An absolute free energy is then computed by implementing the series to near-zero temperature, where the harmonic model becomes very accurate. This method is shown to provide very precise and accurate results. An extension of this method is the absolute free-energy calculation of the realistic linear-molecular nitrogen model, specifically the orientationally-ordered α- and orientationally-disordered β-phase structures. Through this method, the absolute free energies for both the orientationally-ordered and disordered structures are successfully calculated and the coexistence curve for the two phases is traced from zero to 0.12 GPa.

See more of this Session: Recent Advances In Molecular Simulation Methods I
See more of this Group/Topical: Computational Molecular Science and Engineering Forum

(52f) Dissipative Particle Dynamics Simulations Involving Chemical Reactions

John K. Brennan[1], Martin Lisal[2], Joshua D. Moore[1] and Sergei Izvekov[1], (1)Weapons and Materials Research Directorate, U.S. Army Research Laboratory, Aberdeen Proving Ground, MD, (2)E. Hala Laboratory of Thermodynamics, Academy of Sciences of the Czech Republic, Prague, Czech Republic

Many applications of solid composite materials involve chemical energy release. For example, energetic materials, composites of high-energy, high-density crystallites and polymer binder, are used as explosives, propellants and detonators. Likewise, microstructured metallic systems are of interest in many different areas, including soldering in electronic-parts assembly and thermites for biocidal applications [1]. Energy exchange mechanisms for these materials occur over many length and time scales, where the energy release rate is intimately controlled by the nano- and microstructure of the composite, such as the crystallite size distribution, porosity, defects, and passivation layer.

For such systems, modeling and simulation is a useful tool, since it is not hampered by the experimental challenges caused by the steep thermal and chemical gradients. Rather, limitations are due to the approximations made in the models, computational resources, and the available methodologies. For example, continuum scale modeling can depict macro-scale events; however, it lacks the modeling fidelity to properly include material microstructure that influences the overall behavior. While atomistic-level simulations can provide more detailed descriptions, these simulations cannot access the length and times scales necessary to adequately model a microstructurally-heterogeneous material. These well-recognized limitations at either scale have motivated us to develop particle-based simulation models and methods to bridge these spatial and temporal modeling regimes while ensuring multiscale consistency. The approach must be capable of capturing the known thermo-mechanical responses for these systems, including phase transitions, structural rearrangements, and chemical reactions.

In this talk, we will present the status of this ongoing project. To date, progress has been encouraging, where we find the constant-energy and constant-enthalpy Dissipative Particle Dynamics (DPD) methods [2-4] coupled with accurate coarse-grain models [5, 6] to be viable tools for simulating condensed-phase reactive events. The focus of this talk is on our recent work in developing a method within the DPD framework that can capture chemical reactivity. The approach, inspired by the work of Maillet and co-workers [7], treats the coarse-grain particle as a *microreactor* as opposed to a methodology that requires a reactive-type interaction potential. An additional progress variable is assigned to each particle that monitors the "extent-of-reaction" associated with reactions occurring within each particle. Particles are allowed to shrink and swell to account for volume changes associated with reactivity. Current applications are for simple, model reacting systems, however, extensions to complex reacting systems are underway, where both top-down and bottom-up parameterization is being utilized.

This effort is supported by the Institute for Multi Scale Reactive Modeling of Insensitive Munitions

(MSRM), which is a multi-team effort led by the U.S. Army Research Laboratory and the U.S. Army Armament Research, Development and Engineering Center, involving various other national laboratories and academic groups totaling over 20 scientists. ML acknowledges funding through a Cooperative Agreement with the U.S. Army Research Laboratory.

References

1. National Academy of Sciences Report, *Advanced Energetic Materials*, National Research Council, National Academy Press (2004).

2. J. Bonet Avalos and A.D. Mackie, *Europhysics Letters*, **40**, 141 (1997).

3. P. Español, *Europhysics Letters*, **40**, 631 (1997).

4. M. Lísal, J.K. Brennan and J. Bonet Avalos, *J. Chem. Phys.*, submitted (2011).

5. S. Izvekov, P.W. Chung, and B.M. Rice, *J. Chem. Phys.*, **133**, 064109 (2010).

6. S. Izvekov, P.W. Chung, and B.M. Rice, *J. Chem. Phys.*, submitted (2011).

7. J.B. Maillet, L. Soulard and G. Stoltz, *Europhys. Lett.*, **78**, 68001 (2007).

See more of this Session: Recent Advances In Molecular Simulation Methods I
See more of this Group/Topical: Computational Molecular Science and Engineering Forum

(52g) Rigid Body Constraints In Molecular Dynamics Simulations Optimized for Graphics Processing Units

Trung D. Nguyen[1], Carolyn L. Phillips[2], Joshua A. Anderson[1] and Sharon C. Glotzer[1], (1)Department of Chemical Engineering, University of Michigan, Ann Arbor, MI, (2)Applied Physics, University of Michigan, Ann Arbor, MI

We have designed and implemented massively parallel rigid body constraint algorithms in HOOMD-Blue[1], an open-source general purpose molecular dynamics simulation package optimized for graphics processing units (GPUs). In the HOOMD-Blue code package, rigid constraints can be used seamlessly with non-rigid parts of the system (e.g., point particles, bonded particles and walls), and with an array of integration methods (e.g. NVE, NVT, NPT, and Brownian Dynamics). We have also incorporated the FIRE energy minimizer[2], reformulated to be applicable to mixed systems of rigid bodies and non-rigid particles. For our typical simulations, the GPU implementation running on a single NVIDIA® GTX 480 card is twice as fast as LAMMPS[3] running on 32 CPUs.

References

[1] HOOMD-Blue, http://codeblue.umich.edu/hoomd-blue/.
[2] E. Bitzek, P. Koskinen, F. Gaehler, M. Moseler, P. Gumbsch, Structural relaxation made simple, Physical Review Letters 2006, 97 170201.
[3] LAMMPS, http://lammps.sandia.gov/.

See more of this Session: Recent Advances In Molecular Simulation Methods I
See more of this Group/Topical: Computational Molecular Science and Engineering Forum

(72f) Large Scale Distributed Computing Calculations of Binding Free Energies Using Folding@Home and GPU Simulations of Proteins Using OpenMM

Michael R. Shirts, Chemical Engineering, University of Virginia, Charlottesville, VA

We first present our efforts to use distributed computing to calculate high precision free energies of binding using distributed computing. We discuss our simulations of small molecule binding to FKBP, where we obtain binding free energies less than 1 kcal/mol RMS from experiment for molecules of typical drug size, with sampling validated by the fact that relative free energies with the same system differ from absolute free energies of binding by less than 0.3 kcal/mol. We also present our results with the SAMPL3 blind test set, using distributed computing generalized ensemble simulations to identify binders to trypsin out of a set of 540 small molecules, and to compute the free energies of binding for known binders. In both cases, 10,000's of volunteer computers on the Folding@Home distributed computing system were used. We also present our efforts to study local protein unfolding using implicit solvent simulations on GPUs using the open source OpenMM API, which allows simulations up to 70x faster than standard packages on CPU's.

See more of this Session: Beyond Standard Hardware: GPUs, Cloud Computing and Crowdsourcing
See more of this Group/Topical: Computational Molecular Science and Engineering Forum

(72b) Sintering Rate and Mechanism of TiO_2 Nanoparticles by Molecular Dynamics Using GPUs

Beat Buesser, Arto J. Grhn and Sotiris E. Pratsinis, Department of Mechanical and Process Engineering, ETH Zurich, Zurich, Switzerland

Sintering Rate and Mechanism of TiO_2Nanoparticles by Molecular Dynamics

B. Buesser, A.J. Grhn and S.E. Pratsinis

Particle Technology Laboratory, Institute of Process Engineering,

Department of Mechanical and Process Engineering, ETHZrich, 8092 Zrich, Switzerland

Titania (TiO_2)nanoparticles have many attractive applications in photovoltaic[1] and photocatalytic[2] processes, to name a few.The performance of TiO_2 nanoparticles depends considerably on theirsize and composition which are determined by the sintering rate during their synthesis.The sintering rate is crucial for designing gas-phase processes with controlledproduct particle size, structure, composition and eventual performance in anumber of applications[3].

Seto *et al.*[4] have proposed a sinteringrate similar to Kobata *et al.*[5] for TiO_2 andvalidated it by accounting for the detailed fluid mechanics of their hot wallreactor forming rather large TiO_2 nanoparticles (d_p= 10 - 100 nm). Little is known, however, for the sintering rate of small TiO_2nanoparticles (d_p < 10 nm) as it is difficult to reliablymeasure it.

On the otherhand molecular dynamics (MD) simulations have been used to study the sinteringto full coalescence of metallic and metalloid nanoparticles[6,7] though much less has beendone for ceramic ones as their force fields and potentials are difficult todetermine. Koparde and Cummings[8] have investigated thesintering of two TiO_2 nanoparticles, with CPU's reaching up to 1 nsin residence time, by tracking the shrinkage of the center-to-center distanceand the growth of the sintering neck using the force field of Matsui and Akaogi[9], while comparingMD with phenomenological sintering models and investigating the melting of TiO_2nanoparticles[10].

Here[11], sintering of rutile TiO_2nanoparticles is investigated by MD on graphical processing units (GPU)[12] of a consumer graphiccard that was originally developed for the visualization of highly realisticcomputer games. The acceleration achieved by the GPU, compared to a single CPU(Figure 1), allowed simulating TiO_2 nanoparticle sintering fromadhesion and neck growth to finally full coalescence over several hundrednanoseconds for the first time and determining the sintering rate of very smallTiO_2 nanoparticles (d_p < 5 nm) by monitoringthe evolution of their surface area (Figure 2). For the smallest particlediameters, the MD-obtained sintering rates were smaller than that predicted bytheory developed for larger particles[4,5]. Ions on the particlesurface exhibited higher net displacement than bulk ones revealing that surfacediffusion is the dominant sintering mechanism of TiO_2 nanoparticles.An expression for the sintering rate of rutile TiO_2 nanoparticleshas been extracted from MD, bridging the gap of knowledge from a few moleculesto several nanometers, the key size range for nanoparticle properties andperformance. This MD-derived sintering rate could facilitate the use ofphenomenological models[13] to design processes forlarge scale manufacture and processing of small nanoparticles[14]. The low costs for consumerGPU hardware and low energy consumption combined with extremely highcomputational power make GPU's suitable for scientific computing.

Financial support from the Swiss National ScienceFoundation (SNF) grant # 200021-119946/1 is gratefully acknowledged.

References

1. Gratzel M. Photoelectrochemicalcells. *Nature.* 2001; 414, (6861), 338-344.

2. Formenti M, Juillet F, MeriaudeauP, Teichner SJ, Vergnon P. Preparation in a hydrogen-oxygen flame of ultrafinemetal oxide particles. Oxidative properties toward hydrocarbons in the presenceof ultraviolet radiation. *J. Colloid Interface Sci.* 1972; 39, (1),79-89.

3. Strobel R, Pratsinis SE. Flameaerosol synthesis of smart nanostructured materials. *J. Mater. Chem.* 2007;17, 4743-4756.

4. Seto T, Shimada M, Okuyama K.Evaluation of sintering of nanometer-sized titania using aerosol method. *AerosolSci. Technol.* 1995; 23, (2), 183-200.

5. Kobata A, Kusakabe K, Morooka S.Growth and transformation of TiO_2 crystallites in aerosol reactor. *AIChEJ.* 1991; 37, (3), 347-359.

6. Arcidiacono S, Bieri NR, PoulikakosD, Grigoropoulos CP. On the coalescence of gold nanoparticles. *Int. J.Multiphas. Flow.* 2004; 30, (7-8), 979-994.

7. Zachariah MR, Carrier MJ. Moleculardynamics computation of gas-phase nanoparticle sintering: a comparison withphenomenological models. *J. Aerosol Sci.* 1999; 30, (9), 1139-1151.

8. Koparde VN, Cummings PT. Moleculardynamics simulation of titanium dioxide nanoparticle sintering. *J. Phys.Chem. B.* 2005; 109, (51), 24280-24287.

9. Matsui M, Akaogi M. Moleculardynamics simulation of the strucutral and physical properties of the fourpolymorphs of TiO_2. *Mol. Simulat.* 1991; 6, 239-244.

10. Koparde VN, Cummings PT. Sintering oftitanium dioxide nanoparticles: a comparison between molecular dynamics andphenomenological modeling. *J. Nanopart. Res.* 2008; 10, (7), 1169-1182.

11. Buesser B, Grhn A, Pratsinis SE.Sintering Rate and Mechanism of TiO_2 Nanoparticles by MolecularDynamics. *J. Phys. Chem. C.* 2011; accepted,

12. Garland M, Le Grand S, Nickolls J,Anderson J, Hardwick J, Morton S, Phillips E, Yao Z, Volkov V. ParallelComputing Experiences with CUDA. *Micro IEEE.* 2008; 28, (4), 13-27.

13. Koch W, Friedlander SK. The effect ofparticle coalescence on the surface area of a coagulating aerosol. *J.Colloid Interface Sci.* 1990; 140, (2), 419-427.

14. Pratsinis SE. Aerosol-basedtechnologies in nanoscale manufacturing: from functional materials to devicesthrough core chemical engineering. *AIChE J.* 2010; 56, (12), 3028-3035.

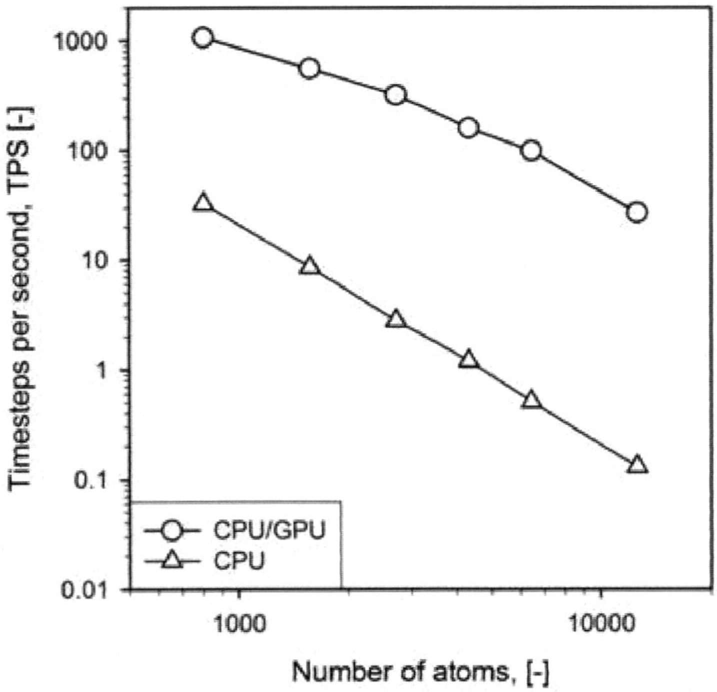

Figure 1 The Number of timesteps for molecular dynamics of TiO_2nanoparticles undergoing sintering on GPU (circles) and CPU (triangles) asfunction of number of atoms.

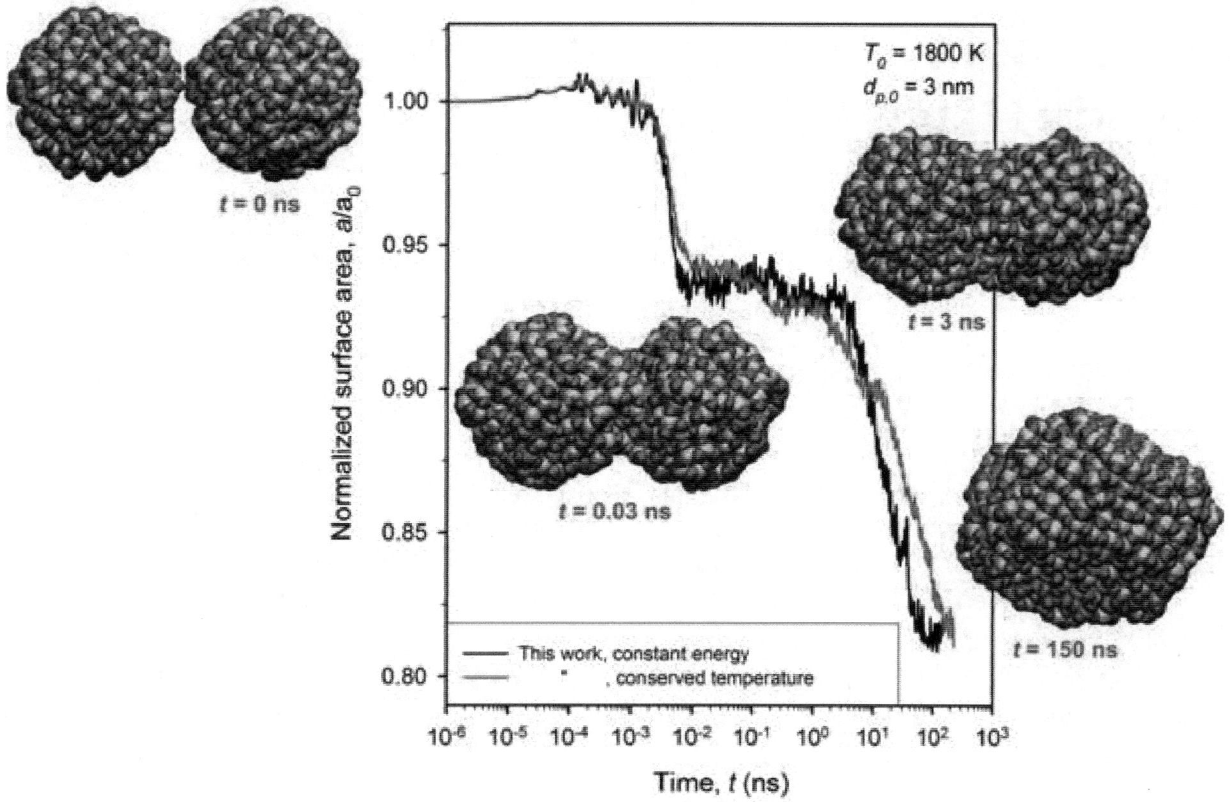

Figure2 Evolution ofsurface area of two TiO_2 nanoparticles (d_0 = 3 nm)at 1800 K.

See more of this Session: Beyond Standard Hardware: GPUs, Cloud Computing and Crowdsourcing
See more of this Group/Topical: Computational Molecular Science and Engineering Forum

(72c) Simulation of Nano-Confined Fluids and Colloids Using GPUs

Christopher R. Iacovella, Chemical and Biomolecular Engineering, Vanderbilt University, Nashville, TN and Peter T. Cummings, Center of Nanophase Materials Sciences, Oak Ridge National Laboratory, Oak Ridge, TN

The nano-confinement of fluids and/or colloids is a particularly interesting problem with applications ranging from lubrication in computer hard drives to gas adsorption in industrial processes. Understanding the fundamental properties, dynamics, structures, and phase transitions of nano-confined systems is of great importance to the design and engineering of these varied systems and is an area where simulation has been widely applied [1]. In this work we perform MD simulations of nano-confined fluids using the HOOMD-Blue simulation package, optimized to run on graphical processing units (GPUs) [2]. We find single GPU performance equivalent to 64+ commodity processors, allowing for the rapid exploration of a wide range of parameters and the simulation of a large number of statistical measures for each statepoint. Here, we report the behavior and GPU performance of a variety of different fluid models, ranging from simple isotropic Lennard-Jones spheres to anisotropic fluids to detailed polymer models.

* Funding and computational support provided by the U.S. Civilian Research and Development Foundation, Award Number UKC1-9201-LV-09 and the Keeneland GPU cluster as part of the the National Institute for Computational Sciences, project-ID UT-TNEDU014.

[1] Cummings PT, Docherty H, Iacovella CR, Singh JK, Phase transitions in nanoconfined fluids: The evidence from simulation and theory, *AICHE Journal*, 56 (4), 2010
[2] Anderson JA, Lorenz, CD, Travesset A, General purpose molecular dynamics simulations fully implemented on graphics processing units, *Journal of Computational Physics*, 227, (2008)

See more of this Session: Beyond Standard Hardware: GPUs, Cloud Computing and Crowdsourcing
See more of this Group/Topical: Computational Molecular Science and Engineering Forum

(72d) GPU Accelerated Monte Carlo Simulations In the Gibbs and Canonical Ensembles

Jason R. Mick[1], Jeffrey J. Potoff[2], Eyad Hailat[3], Vincent Russo[3] and Loren Schwiebert[3], (1)Chemical Engineering and Materials Science, Wayne State University, Detroit, MI, (2)Chemcial Engineering and Materials Science, Wayne State University, Detroit, MI, (3)Computer Science, Wayne State University, Detroit, MI

The use of molecular simulation to study complex physical phenomena at the atomic level has grown exponentially over the last decade with increasing CPU power and the development of parallel molecular dynamics codes that scale efficiently over thousands of processors. As a result, the simulation of 100,000 atom systems has become routine, state of the art calculations are now possible on systems of over 1 million atoms, examples of which include the tobacco mosaic virus[1] and the ribosome[2]. Despite our current ability to simulate such systems, there is a demand to simulate increasingly larger systems for longer timescales.

Graphics processors (GPUs) have begun to play a crucial role in extending molecular simulation to even larger length and time scales [4-6]. In comparison to CPUs, GPUs possess a highly parallel structure allowing for more effective processing of large data sets. For this reason, intrinsic parallel traits prevalent in particle/molecular simulations may be exploited for more efficient computation on the GPU [7,8]. Each multiprocessor in the CUDA programming model has the ability of executing a significant number of threads in parallel. These threads, organized into individual blocks, work cooperatively to efficiently share and synchronize operations.

In this work a GPU-accelerated Monte Carlo simulation engine, capable of performing simulations in the canonical, isobaric-isothermal and Gibbs ensembles[6], is presented. The code uses a modular architecture and is written in C++. Sections of the code designed to run on the GPU, such as the calculation of pairwise interactions, utilize the Compute Unified Device Architecture (CUDA) from NVIDIA Corp. In comparison to optimized serial, CPU-bound code, our GPU-accelerated Monte Carlo simulation engine provides an order of magnitude reduction in wall-clock time for a system containing 65,536 particles interacting via Lennard-Jones potentials. Similar performance increases are observed for simulations in the Gibbs ensemble.

1. Freddolino, P.L., et al., *Molecular dynamics simulations of the complete satellite tobacco mosaic virus.* Structure, 2006. **14**(3): p. 437-49.

2. Sanbonmatsu, K.Y., S. Joseph, and C.S. Tung, *Simulating movement of tRNA into the ribosome during decoding.* Proc Natl Acad Sci U S A, 2005. **102**(44): p. 15854-9.

3. Stone, J.E., et al. *Accelerating molecular modeling applications with graphics processors.* Journal of Computational Chemistry, 2007. **28**(16): p. 2618-40.

4. *AMBER 11 NVIDIA GPU ACCELERATION SUPPORT.* http://ambermd.org/gpus/ (accessed April 29, 2011)

5. Brown, W.M., *Porting LAMMPS to GPUs.* SOS 14 Conference.

6. Panagiotopoulos, A.Z., *Direct determination of phase coexistence properties of fluids by Monte Carlo simulation in a new ensemble.* Molecular Physics. 1987, **61**(4):p. 813-26.

7. Anderson, J.A., et al., *General purpose molecular dynamics simulations fully implemented on graphics processing units.* Journal of Computational Physics, 2008. **227**(10): p. 5342-59.

8. Liu, W., et al., *Molecular dynamics simulations on commodity GPUs with CUDA.* Proceedings of the 14th International Conference on High Performance Computing, 2007. p. 185-96.

See more of this Session: Beyond Standard Hardware: GPUs, Cloud Computing and Crowdsourcing
See more of this Group/Topical: Computational Molecular Science and Engineering Forum

(72e) HOOMD-Blue, General-Purpose Many-Body Dynamics On the GPU

Joshua A. Anderson[1], **Carolyn L. Phillips**[2], Trung D. Nguyen[1] and Sharon C. Glotzer[3], (1)Chemical Engineering, University of Michigan, Ann Arbor, MI, (2)Applied Physics, University of Michigan, Ann Arbor, MI, (3)Department of Chemical Engineering, University of Michigan, Ann Arbor, MI

HOOMD-blue[1] is an open source code for performing molecular dynamics and related many-body dynamics simulations on graphics processing units (GPUs). Current GPUs are widely available, inexpensive, and capable of teraflops of floating-point computation and hundreds of gigabytes per second of memory bandwidth. In typical benchmarks, HOOMD-blue on a current generation GPU is about 80-100 times faster than on a single CPU core. HOOMD-blue has a growing international user and developer base.

HOOMD-blue functionality includes standard molecular dynamics code features and options, including NVT, NPT, NVE and Brownian dynamic integrators and diverse pair, bond, and angle potentials. Release 0.10.0 includes rigid body constraints [2], which are commonly used in a wide range of molecular modeling applications from the atomistic scale, modeling the bonds in molecules such as water, carbon dioxide, and benzene, to the colloidal scale, modeling macroscopic rods, plates and patchy nanoparticles. HOOMD-blue also includes long-range electrostatics computed with the PPPM (particle-particle/particle-mesh) method [3]. Simulation methods such as dissipative particle dynamics[4], energy minimization[2], and constraint forces have also been added.

In this talk, we present an overview of HOOMD-blue and its capabilities, discuss the GPU algorithms that drive it, and demonstrate the performance that it delivers.

1] Find HOOMD-blue online at: http://codeblue.umich.edu/hoomd-blue/

2] Nguyen TD, Phillips CL, Anderson JA, and Glotzer SC, "Rigid body constraints realized in massively-parallel molecular dynamics on graphics processing units," submitted.

3] Contributed by Barr S, Mertmann P, and Panagiotopoulos AZ.

4] Phillips CL, Anderson JA, and Glotzer SC, "Pseudo-Random Number Generation for Brownian Dynamics and Dissipative Particle Dynamics Simulations on GPU Devices", submitted.

See more of this Session: Beyond Standard Hardware: GPUs, Cloud Computing and Crowdsourcing
See more of this Group/Topical: Computational Molecular Science and Engineering Forum

GPGPU *vs.* Free Volume: Using Graphics Hardware to Define, Characterize, and Visualize Negative Space in Polymers

Frank T. Willmore (presenting author), David S. Simmons, Jack F. Douglas
Polymers Division, National Institute of Standards and Technology, Gaithersburg, MD

Introduction

General Purpose Graphical Processing Units (GPGPUs) provide a tool for the efficient decomposition of spatial domains in a variety of applications. The highly parallel computational nature of this tool allows for the efficient computation of energies that can be used for the characterization of unoccupied or 'free' space in materials as it relates to the permeability and transport properties of materials used in the preservation of proteins and other soft materials of interest to National Institute of Standards and Technology (NIST) Polymers Division.

From Pixels to Voxels

Originally designed for the rendering of 2D images of pixels, GPGPUs excel at performing the non-branching single-instruction multiple-data (SIMD) type of function evaluation for uniformly spaced sets of points on a grid. This technology is proving useful for 3D rendering, and *via* new programming models, has been generalized for the evaluation of non-visual functionals most notably in quantum chemistry.[1][2][3]

Advances in computing hardware now make it possible to form efficient discrete representations of three-dimensional systems at considerable detail. The amount of memory required to represent a system at a fixed level of detail grows as the linear dimension cubed. Contrast this to the exponential growth in time of the number of cores on a given die and of the amount of RAM on a compute node. A cube of voxels of dimension *1024 x 1024 x 1024*, or approximately a billion points is now representable in memory. A simple one-byte scalar grayscale model at this resolution requires 1GB of RAM. A double-precision vector field requires 24GB of RAM.

Free Volume Intensity as a Material Property

As a model application for GPU-based voxelization, we present a metric for measuring free volume in the context of glass-forming materials. This energetically defined free volume 'intensity' provides a direct thermodynamic measure of the propensity for a space to be occupied, a mechanism to capture analytically the spatial and temporal variation of free volume. The free volume intensity (FVI) is defined as,

---

[1] Ufimtsev, I.S. & Martinez, T.J. Quantum Chemistry on Graphical Processing Units. 2. Direct Self-Consistent-Field Implementation. *J. Chem. Theory Comput.* **5**, 1004-1015 (2009).

[2] Ufimtsev, I.S. & Martinez, T.J. Quantum Chemistry on Graphical Processing Units. 3. Analytical Energy Gradients, Geometry Optimization, and First Principles Molecular Dynamics. *J. Chem. Theory Comput.* **5**, 2619-2628 (2009).

[3] Ufimtsev, I.S. & Martínez, T.J. Quantum Chemistry on Graphical Processing Units. 1. Strategies for Two-Electron Integral Evaluation. *J. Chem. Theory Comput.* **4**, 222-231 (2008).

$$FVI = e^{-\beta\psi_{ins}},$$

where ψ_{ins} represents the insertion energy for a test particle. FVI is related to the Widom insertion factor[4] B by the integral:

$$B = \frac{1}{V}\int FVI\, dV$$

The Widom insertion factor, like the Bondi 'empty' volume,[5] is an average over the whole material, and can be interpreted as a bulk property, namely as the solubility of an additional molecule and is rigorously related to the chemical potential:

$$\mu_i = -k_B T \ln\left(\frac{\mathbf{B}_i}{\rho_i \lambda^3}\right)$$

where ρ_i is the number density of species i and λ represents the thermal wavelength.

The free volume intensity parameter is intrinsically a function of the chosen force field. We illustrate the domain decomposition for two models. The first system is a well-characterized[6] model glass former consisting of bead-spring chains and which employs a simple force field model of a truncated and shifted Lennard-Jones 6-12 potential. Beads on a chain are joined by a harmonic potential. The test particle interacts *via* a Berthelot combining rule. All metrics of the system are dimensionless/Lennard-Jones units. The second is an atomistic model of polydimethylsiloxane (PDMS) which employs the non-bond interaction of the condensed-phase optimized molecular potentials for atomistic simulation studies[7] (COMPASS) force field. In both instances, the attractive component of the forcefield is omitted, to generate a purely repulsive potential. In this way, the free volume intensity is bounded between 0 and 1.

Like the insertion factor, the free volume intensity can be defined for the insertion of any species. In this work we define the insertion of a zero diameter particle with unit value of the energetic parameter epsilon. Because it is defined for all of space, the free volume intensity is a measure which compliments metrics which are defined for the bulk system (e.g. Bondi volume[5]), or in terms of specific free volume elements (rattle volumes[8], energetic cavity sizing[9] (CESA), Voronoi volumes).[10] Examining an isosurface of the FVI allows for shape analysis of free volume elements, or FVE's. The discrete

---

[4] Widom, B. Some Topics in the Theory of Fluids. *J. Chem. Phys.* **39**, 2808 (1963).

[5] Bondi, A. Free Volumes and Free Rotation in Simple Liquids and Liquid Saturated Hydrocarbons. *The Journal of Physical Chemistry* **58**, 929-939 (1954).

[6] Jain, T.S. & de Pablo, J.J. Influence of confinement on the vibrational density of states and the Boson peak in a polymer glass. *J. Chem. Phys.* **120**, 9371 (2004).

[7] Sun, H. COMPASS: An ab Initio Force-Field Optimized for Condensed-Phase ApplicationsOverview with Details on Alkane and Benzene Compounds. *The Journal of Physical Chemistry B* **102**, 7338-7364 (1998).

[8] Sastry, S., Truskett, T.M., Debenedetti, P.G., Torquato, S. & Stillinger, F.H. Free volume in the hard sphere liquid. *Molecular Physics* **95**, 289-297 (1998).

[9] in `t Veld, P.J., Stone, M.T., Truskett, T.M. & Sanchez, I.C. Liquid Structure via Cavity Size Distributions. *The Journal of Physical Chemistry B* **104**, 12028-12034 (2000).

[10] Rycroft, C.H., Grest, G.S., Landry, J.W. & Bazant, M.Z. Analysis of granular flow in a pebble-bed nuclear reactor. *Phys. Rev. E* **74**, 021306 (2006)

implementation of the method readily generates voxels which can be used for shape analysis as well as rendering.

Interactions of each of the atoms in the simulation box were calculated with all other atoms in the simulation box as well as with images of atoms in the 26 nearest periodic replicas of the simulation box. The potential was truncated but not shifted. For the Lennard-Jones model, the smallest box used was 11.502σ which corresponds to an error in the repulsive component of the potential of $O(10^{-13})$ dimensionless Lennard-Jones units. The implicit error relative to unity for single precision floating point operation under IEEE standard 754-2008 is $O(10^{-7})$ and as such, the systematic error introduced by truncation is neglected.

For the LJ system, there were 50 chains of 32 beads each, for a total of 1600 atoms. Running on a single node of the Texas Advanced Computing Center (TACC) Lonestar sytem and utilizing a single GPGPU (NVidia M2070) in single precision, the wall time for calculation of potential energy for all data points on a $256 \times 256 \times 256$ grid was 71 seconds.

Figures

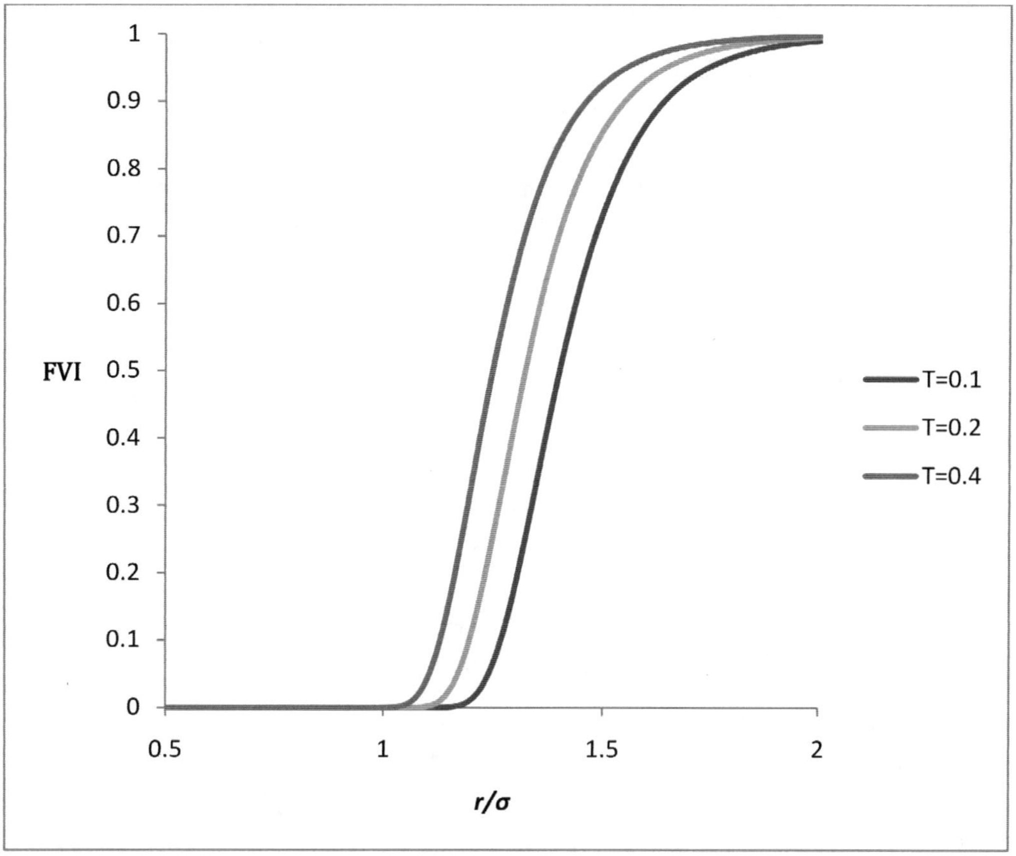

Figure 1. Free volume intensity isotherms for interaction of test particle with a single center of diameter σ at a separation r for model glass former. As temperature increases, the effective repulsive diameter of the component atoms decreases, consistent with greater population of higher energy states/overlapping of the test particle.

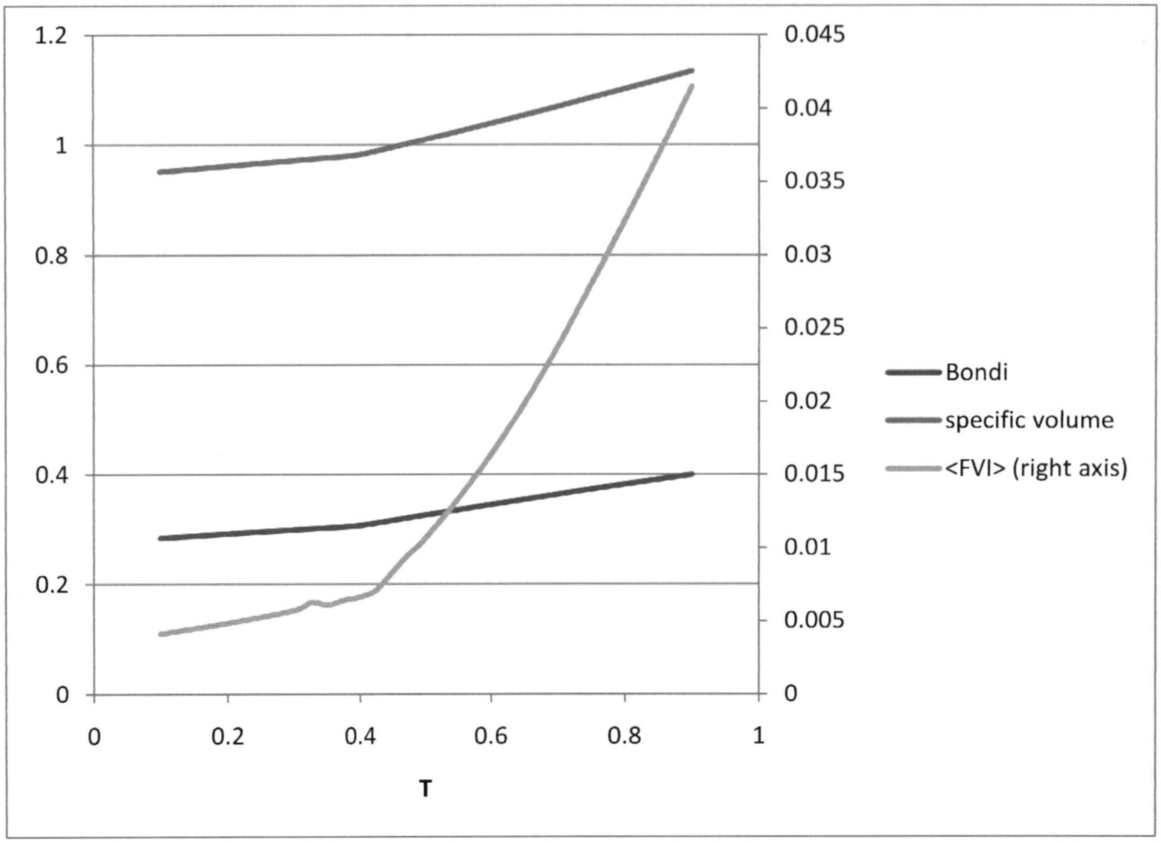

Figure 2. Free volume intensity reflects changes in density as well as an increased accessibility of higher energy states, it displays a stronger sensitivity to changes in temperature than Bondi volumes or specific volume.

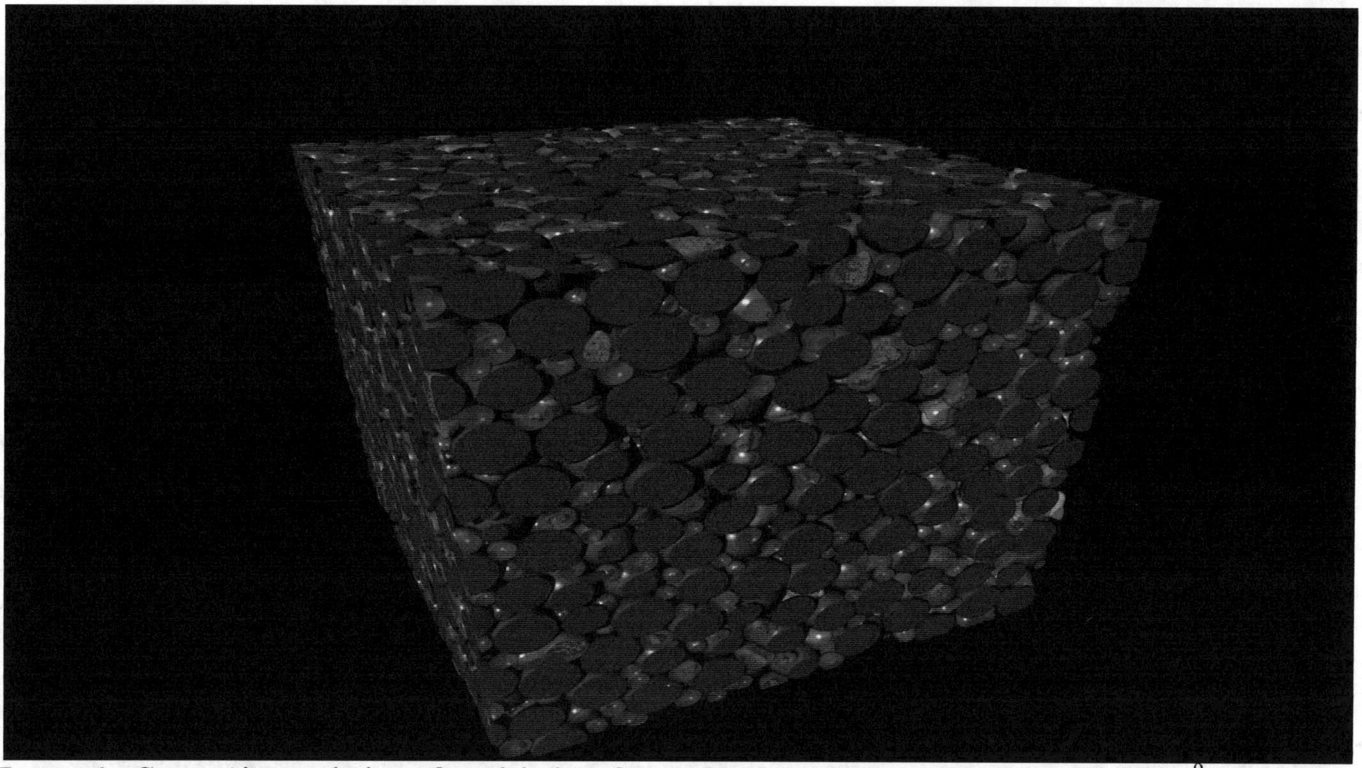

Image 1. Composite rendering of model glass former beads (red), energetically sized CESA[9] cavities (gray) and FVI isosurface (yellow) at a dimensionless temperature $T = 0.2$.

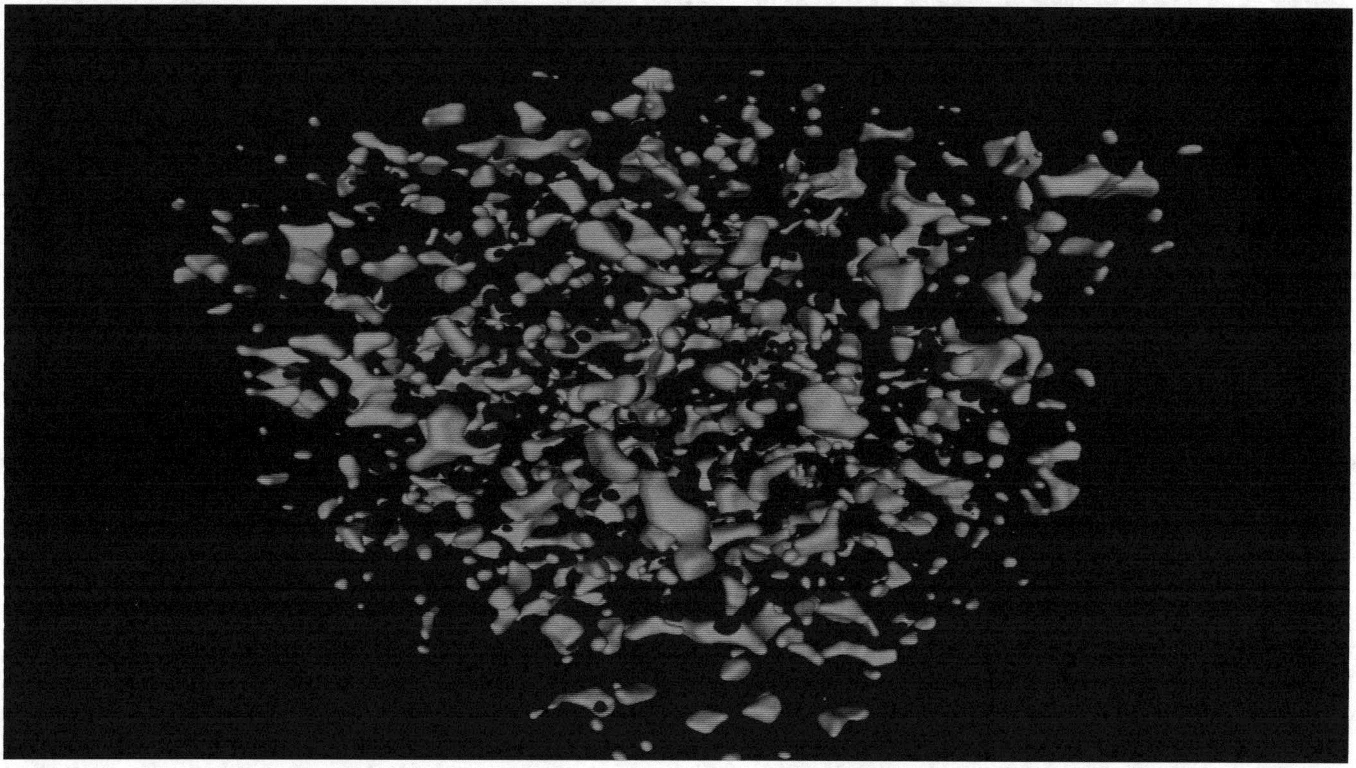

Image 2. Isosurface of FVI = 0.1 for model glass former with dimensionless temperature $T = 0.5$.

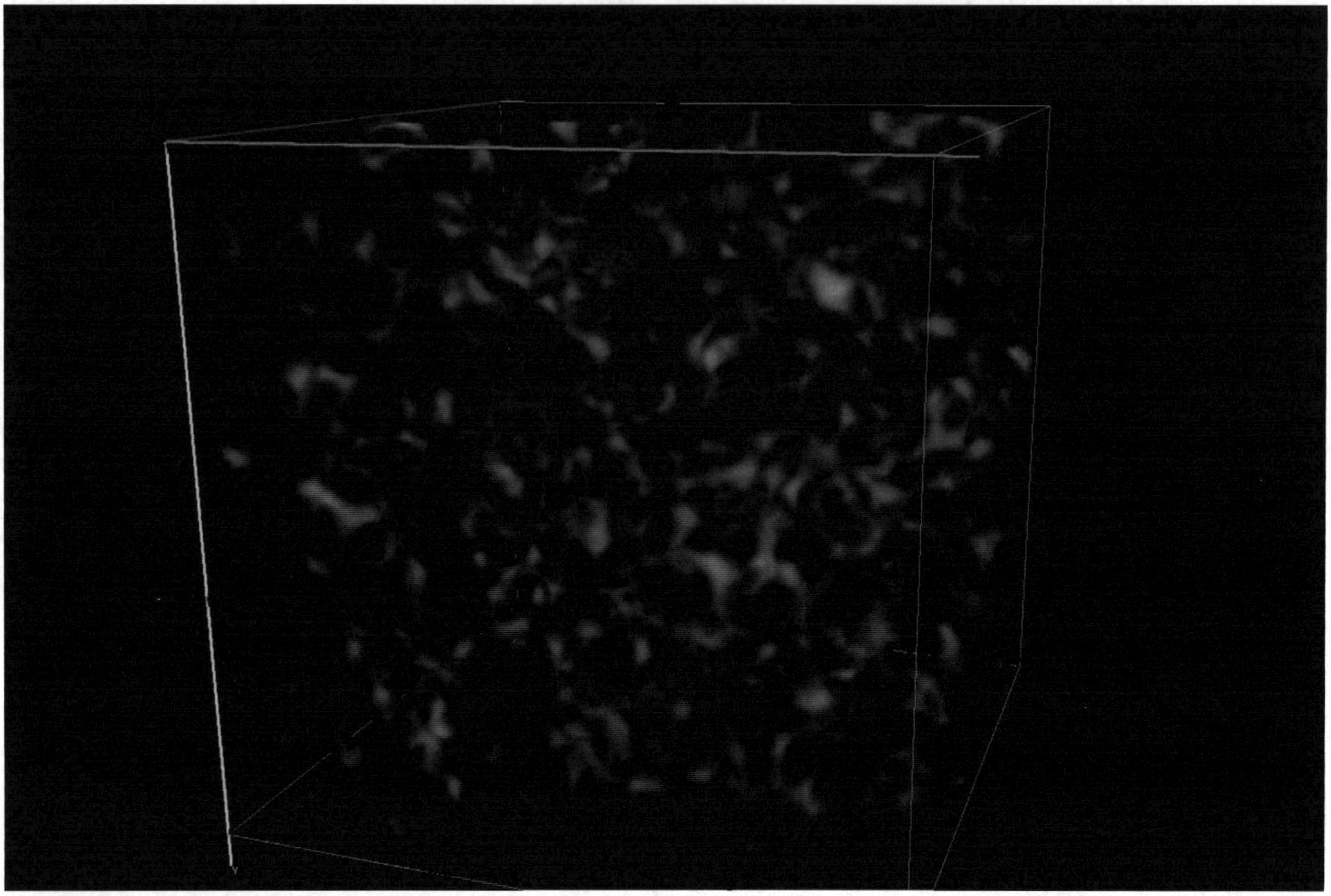

Image 3. Grayscale rendering of FVI for model glass former at dimensionless temperature $T = 0.5$.

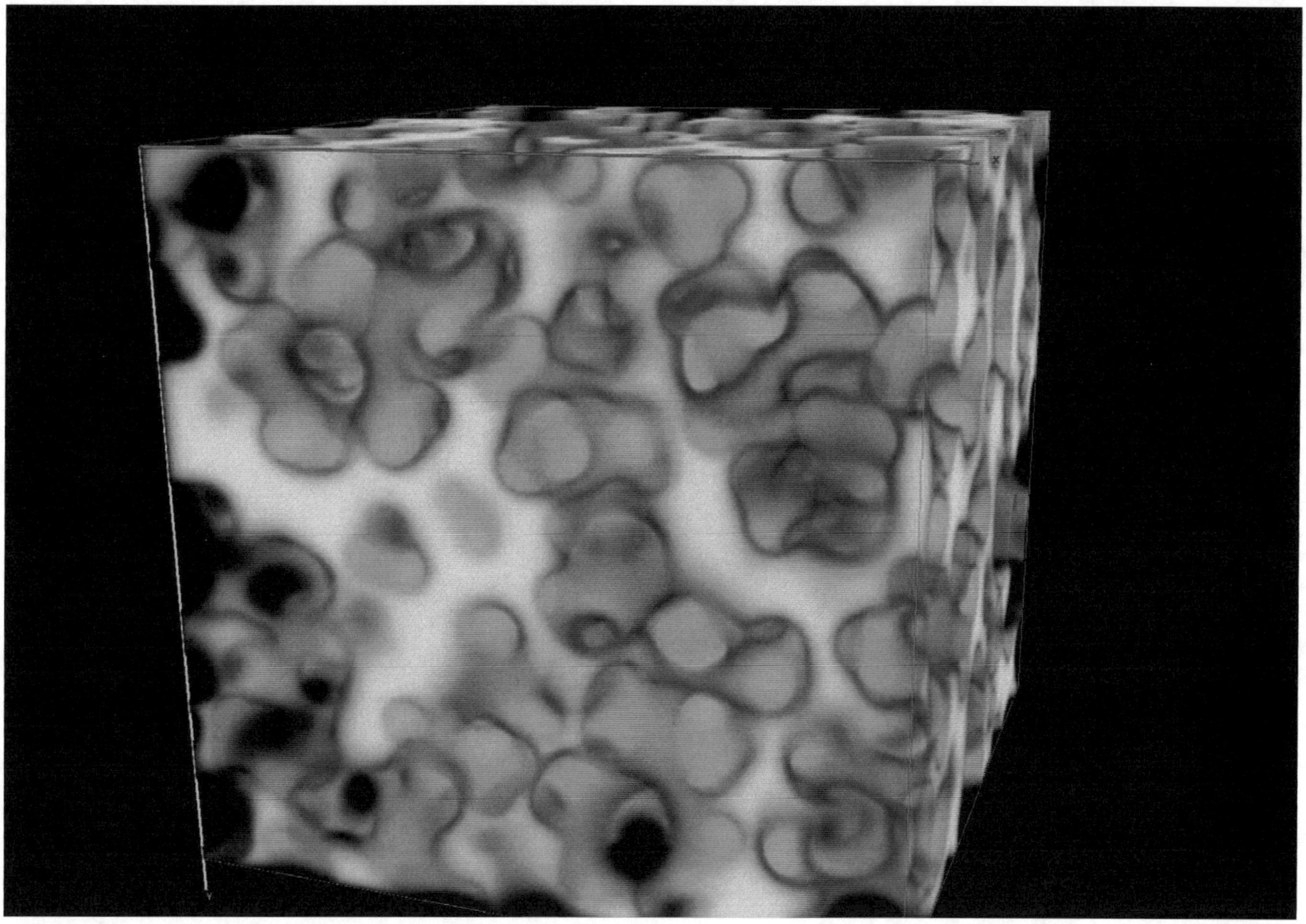

Image 4. Grayscale rendering of FVI for atomistic simulation of configuration of PDMS at $T = 298K$ reveals significant connectivity of free space.

Images 1-2 were rendered using Povray.[11] Images 3-4 were rendered using V3D.[12]

DISCLAIMERS:

Certain commercial equipment, instruments, or materials are identified in this paper in order to specify the experimental procedure adequately. Such identification is not intended to imply recommendation or endorsement by the National Institute of Standards and Technology, nor is it intended to imply that the materials or equipment identified are necessarily the best available for the purpose.

---

[11] POV-Ray: Documentation: 1.5.2 Citing POV-Ray in Academic Publications. at <http://www.povray.org/documentation/view/3.6.0/203/>

[12] Peng, H., Ruan, Z., Long, F., Simpson, J.H. & Myers, E.W. V3D enables real-time 3D visualization and quantitative analysis of large-scale biological image data sets. *Nat Biotech* **28**, 348-353 (2010).

ACKNOWLEDGEMENTS:

The authors acknowledge the Texas Advanced Computing Center (TACC) at The University of Texas at Austin for providing HPC and visualization resources that have contributed to the research results reported within this paper.

This work was made possible in part by the National Research Council's Postdoctoral Fellowship Program.

(72g) Statistical Learning Theory for Protein Dynamics

Kevin R. Haas and Jhih-Wei Chu, Chemical Engineering, University of California, Berkeley, Berkeley, CA

Proteins and allosteric enzymes undergo large-scale conformational transitions and exhibit wide variations in their functionality and reactivity. Our research aims to uncover the illusive underlying mechanisms and dynamics that govern the physical properties of these biomolecules. We developed a novel statistical learning theory to extract from the traditionally convoluted data of single-molecule FRET and pulling experiments, free energy surfaces and diffusion profiles along the physical coordinate employed for measurements. Our method generalizes the Hidden Markov Model (HMM) to continuous state-space Markovian dynamics in order to extract the probabilistic information of the actual dynamic trajectory along the focused coordinate from indirect measurements of photon emission trajectories and the temporal histories of pulling forces.

With a highly resolved and accurate picture of system dynamics from indirect measurements, we can optimize the statistical likelihood for the underlying free energy surface and diffusion profile. We begin with an abstract path-integral representation of the HMM and show how though a self-consistent application of functional optimization can lead to the derivation of Expectation-Maximization (EM) equations. The EM protocol creates progressively better approximations of the observed dynamic trajectory and converged dynamic parameters from single-molecule experimental data.

We apply this method to the dual helical lid opening transitions of tyrosine phosphatase PtpB observed with FRET to understand the proteins resilience to oxidative degradation. We observe the equilibrium probability density and reaction rate of lid dynamics and the coupling between the two domains. We propose and investigate the feasibility of a 2-D experimental technique to probe this coupling to understand mechanism and the accuracy the reaction coordinate.

The thrust of our research is two-fold. First, all-atom molecular dynamics simulations and free energy calculations will be applied to acquire atomic details of protein dynamics. Advanced computational methods such as the calculation of minimum free energy paths would allow us to analyze microscopic mechanisms with full atomic details. The free-energy profiles calculated from the bottoms up will then be contrasted with those determined from the statistical learning of indirect experimental measurements. The matching of bottoms-up and tops-down analyses will provide unprecedented knowledge and understanding of protein machines for modulating their actions and functions.

See more of this Session: Beyond Standard Hardware: GPUs, Cloud Computing and Crowdsourcing
See more of this Group/Topical: Computational Molecular Science and Engineering Forum

(72h) GPU-Based Acceleration In Linear Algebra, Bioinformatics, and Risk Analysis

Panagiotis Vouzis and Nick Sahinidis, Chemical Engineering, Carnegie Mellon University, Pittsburgh, PA

The recent introduction of general-purpose GPUs has created an opportunity and a need to harness their processing power for a wide variety of scientific-computing applications. In this talk, we summarize our recent results on GPU-based computing in applications coming from linear algebra, bioinformatics, and risk analysis. Solving linear systems of equations [1], carrying out BLAST queries [2], and running Monte Carlo simulations [3] exhibit inherent parallelism, but have significant differences, which pose different challenges in their parallel GPU-based implementation. We review and contrast these differences and discuss how each implementation succeeded in gaining significant speedups in comparison to conventional hardware.

[1] P. D. Vouzis, N. V. Sahinidis, "GPU-BLAST: using graphics processors to accelerate protein sequence alignment," Bioinformatics, Vol. 27, no. 2, pages 182-188, 2011.

[2] J. Elble, N. V. Sahinidis, P. D. Vouzis, "GPU computing with Kaczmarz's and other iterative algorithms for linear systems," Parallel Computing, Vol. 36, no. 5-6, 2010.

[3] Y. Zhang, P. D. Vouzis, N. V. Sahinidis, "GPU simulations for risk assessment in CO2 geologic sequestration," Computers & Chemical Engineering, 2011 (accepted).

See more of this Session: Beyond Standard Hardware: GPUs, Cloud Computing and Crowdsourcing
See more of this Group/Topical: Computational Molecular Science and Engineering Forum

MD Simulations with Petaflops Sustainable Performance for Trans-Scale Study on Transport and Reaction Processes

Wei Ge, Chaofeng Hou, Ji Xu, Guofei Shen, Feiguo Chen, Xiaowei Wang, Wenlai Huang, Ying Ren, Yongshen Han and Jinghai Li
Institute of Process Engineering, Chinese Academy of Sciences, Beijing, China

One of the fundamental challenges to chemical engineering is the vast scale difference between molecular structures that define the properties or functions of the chemical products and the reactors or equipments that actually produce these materials, it typically ranges from 10^{-10}m and 10^{-15}s to 10^{1}m and 10^{3}s, and can be even wider. In particular, bottlenecks exist at the so-called meso-scales [1] where neither detailed dynamical study nor lumped statistical study seems to be sufficient or matured. Nano- and micro-fluidics, turbulent eddies and particle clusters in multiphase reaction systems are among the most well-known meso-scale structures.

The supercomputing capabilities available currently have provided an unique opportunity to breaking through some of these bottlenecks. In this presentation, we will demonstrate that, through consistent designing of the physical and mathematical models and the computer software and hardware [2], rigorous molecular dynamics (MD) can now reach truly Petaflops sustainable performance and micron scales in three dimensions or even millimeter scales in at least one dimension, which are reasonably "macro-scales" for statistical studies. For example, using up to 1728 GPUs of the *Mole-8.5* system at IPE (http://www.top500.org/list/2010/06/100), a complete influenza virion, H1N1, constructed by 300 million atoms or radicals including the aqueous solution, is simulated at a speed of 0.77 ns per day [3]. Virtual experiments, such as the docking of the drugs can be conducted. In another example, using all CPUs and GPUs of the Tianhe-1A system (http://www.top500.org/system/10587), the world's fastest computer currently, the silane pyrolysis and silicon deposition processes for the production of high-purity crystalline silicon were simulated with more than 100 billion atoms. The sustainable performance for the simulation of the bulk of silicon crystal has reached 1.87 Petaflops in single precision using all 7168 GPUs of Tianhe-1A [4].

Based on the strategy demonstrated in these examples, general purpose software and hardware platforms for high performance computing (HPC) of MD simulation can be established, which will proven to be a very powerful tool for exploring meso-scale structures, especially the fascinating nano- and micro-scale structures in chemical engineering.

References:

[1] Li J, Ge W, Kwauk M, 2009. Meso-scale phenomena from compromise -- a common challenge, not only for chemical engineering. *arXiv:0912.5407*.

[2] The EMMS group, 2011. Meso-scale oriented simulation towards virtual process engineering (VPE) ¯ the EMMS Paradigm. *Chemical Engineering Science*. In press.

[3] Xu J, Wang X, He X et al., 2011. Application of Mole-8.5 HPC system ¯ Probing the whole influenza virion at the atomic level. *Chinese Science Bulletin*. In press.

[4] Ge W, Hou C, Xu J et al., 2011. Trans-scale Simulation of Silicon Deposition Process on Tianhe-1A. Submitted to *Supercomputing 2011*. Nov. 12-18, Seattle, USA.

(81a) Adsorption and Activation of n-Alkanes On a Pdo(101) Thin Film

Jason F. Weaver[1], Aravind Asthagiri[2], Jose Hinojosa Jr.[3], Can Hakanoglu[1], Abbin Antony[1] and Jeffery M. Hawkins[1], (1)Chemical Engineering, University of Florida, Gainesville, FL, (2)Department of Chemical & Biomolecular Engineering, The Ohio State University, Columbus, OH, (3)Department of Chemical Engineering, University of Florida, Gainesville, FL

The formation of palladium oxide (PdO) is thought to be responsible for the exceptional activity of supported Pd catalysts toward the complete oxidation of alkanes under oxygen-rich conditions. In this talk, I will discuss our recent investigations of the adsorption and activation of *n*-alkanes on a PdO(101) thin film that is prepared by oxidizing Pd(111) in ultrahigh vacuum (UHV) using an oxygen atom beam. We find that alkanes adsorb relatively strongly on the PdO(101) surface by forming σ-complexes along the rows of coordinatively unsaturated (cus) Pd atoms, and that this adsorbed state acts as the precursor for initial C-H bond cleavage. For example, propane and *n*-butane undergo facile C-H bond cleavage on PdO(101) below about 215 K in UHV. I will also discuss density functional theory calculations which show that dative bonding between the alkane and cus-Pd atoms weakens the coordinated C-H bonds, thereby lowering dissociation barriers significantly. Finally, I will discuss recent results on C-H bond selectivity in the initial activation of propane on PdO(101) as well as the activity of binary Pd-M oxides toward alkane activation.

See more of this Session: Computational Catalysis II
See more of this Group/Topical: Catalysis and Reaction Engineering Division

(81b) Multiscale Modeling of the Water-Gas Shift Reaction At the Three Phase Boundary of Pt/TiO2 and Pt/CeO2 Catalysts

Salai C. Ammal[1], Sara Aranifard[2] and **Andreas Heyden**[2], (1)Chemical Engineering, University of South Carolina, Columbia, SC, (2)Department of Chemical Engineering, University of South Carolina, Columbia, SC

For heterogeneously catalyzed reactions with more than one key surface intermediate, it is likely that multiphase catalysts have a significant advantage over conventional monophase catalysts since each phase can potentially be adjusted independently to activate a key reaction step. At the same time, our understanding of bifunctional multiphase systems is relatively poor. It is the objective of this paper to significantly enhance our molecular understanding of heterogeneous catalysis at the three-phase boundary (TPB) of a gas-phase, a reducible oxide surface, and a noble metal cluster. In particular, we intend to illustrate the specific role of the TPB in determining the activity and selectivity of TiO_2 and CeO_2 supported Pt catalysts for the water-gas shift (WGS, $CO + H_2O$ -> $CO_2 + H_2$) reaction.

We performed an *ab initio* atomistic thermodynamics analysis of the Pt/TiO_2 (110) interface in various gas-phase environments and conclude: (1) In an oxidizing atmosphere, oxygen vacancies in TiO_2 should not play any role and the surface Pt atoms will likely be oxidized. (2) Under WGS reaction conditions, formation of oxygen vacancies at the Pt/TiO_2 interface is thermodynamically favorable and the Pt clusters are surface oxygen free. Furthermore, surface Pt atoms are covered by CO and only interfacial Pt atoms are available for reaction.

Next, we investigated the mechanism of the WGS at the Pt/TiO_2 (110) interface. The redox and carboxyl pathways were studied and the results can be summarized as follows: Pt sites away from the TPB primarily act as a CO reservoir. The oxide support plays an essential role in the O-H bond breaking and H_2 formation process. CO poisoning is reduced on Pt atoms at the TPB. Water activation and dissociation at the TPB becomes exothermic and is facile. Finally, there seems to be no significant advantage of the TPB for COOH dissociation which is another key step in the carboxyl pathway on Pt (111) surfaces.

See more of this Session: Computational Catalysis II
See more of this Group/Topical: Catalysis and Reaction Engineering Division

(81c) Understanding of Environment-Dependent Mechanisms of Alkali Promotion In Heterogeneous Catalysis Using First-Principles Based Monte Carlo Simulation

Hongliang Xin and Suljo Linic, Chemical Engineering, University of Michigan, Ann Arbor, MI

Interactions among adsorbates on solid surfaces play a critical role in heterogeneous catalysis, electro-catalysis, self-assembly, nucleation, growth, and almost every other phenomenon governed by chemical or physical surface processes. For example, alkali promoters (Li, Na, K, Rb, Cs) enhance activity and/or selectivity of various catalytic processes, such as Fisher-Tropsch synthesis, water-gas shift reactions, ammonia synthesis, olefin epoxidation, and automotive three way converters. Although the impact of alkalis in heterogeneous catalysis has been recognized for decades, not much is known about the underlying physical mechanisms that govern the promotion.

We have utilized density-functional theory (DFT) calculations to develop a physically transparent model for the analysis of the interactions between adsorbates on metal surfaces. We have applied this model to investigate underlying mechanisms that govern the interaction between Cs promoters and oxygen adsorbates on the Ag(111) surface. We have also employed an atomistic thermodynamics approach and Monte Carlo simulations to extend the results of DFT calculations to relevant catalytic conditions (*T, p*). We found that the underlying mechanisms of chemical promotion by alkali promoters change significantly as a function of external conditions.

See more of this Session: Computational Catalysis II
See more of this Group/Topical: Catalysis and Reaction Engineering Division

(81d) Formic Acid Decomposition On Transition Metals: Trends Through Density Functional Theory Studies

Jessica Scaranto[1], Lars C. Grabow[2], Scott D. Tonelli[1], Jeffrey A. Herron[1], Suyash Singh[1], Brandon J. O'Neill[1], James A. Dumesic[1] and Manos Mavrikakis[1], (1)Chemical & Biological Engineering, University of Wisconsin-Madison, Madison, WI, (2)Chemical and Biomolecular Engineering, University of Houston, Houston, TX

Abstract

The investigation of formic acid (HCOOH) decomposition on transition metals is important in order to derive useful insights for vapor phase catalysis involving HCOOH and for direct HCOOH fuel cells. Here we present the main results obtained from periodic, self-consistent Density Functional Theory (DFT-GGA) calculations concerning the gas-phase decomposition of HCOOH on Pt(111), Pd(111) and Cu(111) surfaces. Accordingly, we analyzed the minimum energy paths for HCOOH decomposition into CO_2 and H_{-2} through the carboxyl (COOH) and formate (HCOO) intermediates, along with alternative bond-breaking possible steps in those intermediates. The DFT-derived parameters have been employed in a microkinetic model and then used to compare modeling results with experimental reaction kinetics data. This work will be discussed in the context of thermochemistry trends for the same reaction on a number of other late transition metal closed-packed surfaces.

Acknowledgement

This material is based upon work supported as part of the Institute for Atom-efficient Chemical Transformations (IACT), an Energy Frontier Research Center funded by the U.S. Department of Energy, Office of Science, Office of Basic Energy Sciences.

See more of this Session: Computational Catalysis II
See more of this Group/Topical: Catalysis and Reaction Engineering Division

(81e) Selective Activation and Conversion of Simple Carboxylic Acids and Esters

Ye Xu, Center for Nanophase Materials Sciences, Oak Ridge National Laboratory, Oak Ridge, TN and Lijun Xu, Center for Nanophase Materials Science, Oak Ridge National Laboratory, Oak Ridge, TN

Selective activation and conversion of simple carboxylicacids and esters

Ye Xu,* and Lijun Xu

Center for Nanophase Materials Sciences, Oak RidgeNational Laboratory, Oak Ridge, Tennessee 37831, USA

* xuy2@ornl.gov

The hydrogenation of carboxylic acids and esters to thecorresponding alcohols are industrially important processes, and will likely bean integral pathway in biorefineries for the production of valuable chemicalsand fuels from biomass feedstock because carboxylic acids and esters are commoncomponents of such feedstock. Acids, in particular, are more difficult toreduce than other classes of organic oxygenates. Ester hydrogenation is traditionallycatalyzed by Cu and Cu compounds, which requires high temperature andpressure. Pt group metals can decompose acids and esters at milder conditions butoffer very limited selectivity for alcohols in gas-phase experiments.

Recently Olcay et al. demonstrated that significantselectivity for ethanol can be obtained via the aqueous-phase hydrogenation(APH) of acetic acid on a series of metals, with Ru being the most active andselective at low temperatures.[1] Ru has also been used in the APH of other organic acids with good activity and selectivity.[2] We have performed density functional theory calculations to understand the role of different metal species and surface structure in the activation and conversion of simple carboxylic acids and esters on metal surfaces, using acetic acid and methyl acetate as representative compounds.

Our calculations for the activation and subsequenthydrogenation of acetic acid on several metals suggest that the activity of aceticacid conversion is dictated largely by the intrinsic reactivity of the metals.[1] In agreement with experimental results, the selectivity to ethanol is found to depend on reaction conditions. Dehydrogenation and decomposition are the dominant reaction pathway for acetic acid and methyl acetate at vacuum/low pressure conditions on Ru and Pd, which fundamentally limits the selectivity to ethanol.[3] The activity of C-H and C-O bond dissociations is different on close-packed, open, and under-coordinated Pd sites: Pd step edges and kinks promote both C-H and C-O bond dissociation steps primarily via electronic effects, while (100)-type square surface symmetry is more effective at stabilizing C-O bond dissociation transition states than (111)-type hexagonal symmetry does, thereby differentially promoting C-O relative to C-H bond dissociation.[4]

References

1 H. Olcay, L. Xu, Y. Xu, and G. W. Huber, *ChemCatChem* **2**, 1420 (2010).

2 Y. Q. Chen, D. J. Miller, and J. E. Jackson, *Ind.Eng. Chem. Res.* **46**, 3334 (2007).

3 L. Xu and Y. Xu, *Surf. Sci.* **604**,887 (2010).

4 L. Xu and Y. Xu, *Catal. Today*doi:10.1016/j.cattod.2010.12.021.

See more of this Session: Computational Catalysis II
See more of this Group/Topical: Catalysis and Reaction Engineering Division

(81f) Hydrogenation of Lactic Acid to Propylene Glycol and Propanoic Acid: A Selectivity Challenge

Suyash Singh and Manos Mavrikakis, Department of Chemical and Biological Engineering, University of Wisconsin-Madison, Madison, WI

Periodic, self-consistent Density Functional Theory (DFT) calculations are employed to analyze the hydrogenation of lactic acid[1, 2] to propylene glycol (desired product) and propanoic acid (undesired product). A comprehensive reaction network is proposed and key surface intermediates are identified for both reactions. Detailed adsorption thermochemistry and activation energy barrier calculations are performed to probe the mechanisms on Pt(111) and energetically most favorable pathways for the formation of both products are identified. A thorough investigation of the thermochemistry of all possible elementary steps over several close-packed transition metal surfaces (Rh, Ir, Ni, Pd, Cu, Ag, Au) is then performed to establish reactivity trends among metals. The results from these calculations point to a specific metal which might be an efficient monometallic catalyst.

These DFT calculations, in combination with microkinetic modeling and experiments, will be used to screen for improved (active and selective) alloy catalysts for lactic acid hydrogenation. Our analysis is expected to not only help with identifying improved ways for the manufacture of propylene glycol from biomass-derived lactic acid, but is also expected to capture the qualitative features of the vapor phase hydrogenation of oxygenated biomass-derived feed stocks.

References

1. Cortright RD, Sanchez-Castillo M, & Dumesic JA. Conversion of biomass to 1,2-propanediol by selective catalytic hydrogenation of lactic acid over silica-supported copper. Applied Catalysis B-Environmental **2002**; 39: 353-359.

2. Simonov MN, Simakova IL, & Parmon VN. Hydrogenation of lactic acid to propylene glycol over copper-containing catalysts. Reaction Kinetics and Catalysis Letters **2009**; 97: 157-162.

See more of this Session: Computational Catalysis II
See more of this Group/Topical: Catalysis and Reaction Engineering Division

(81g) Molecular Investigation of Catalytic Isomerization of Glucose to Fructose

Samir H. Mushrif, Nima Nikbin, Stavros Caratzoulas, Vinit Choudhary, Stanley Sandler, Vladimiros Nikolakis and Dionisios G. Vlachos, Chemical Engineering, University of Delaware, Newark, DE

Conversion of carbohydrates, the key components of biomass, to various intermediates, such as 5-hydroxymethyl-furfural (HMF), has gained significant attention in recent years. Even though some success has been achieved in the dehydration of fructose to HMF, there is a pressing need to selectively transform glucose, the most dominant sugar in biomass, to HMF. The pathway to HMF from glucose is believed to involve isomerization of glucose to fructose. In order to convert glucose to HMF, via fructose, in a single reactor system, it is desired to have a catalyst which actively converts glucose to fructose in an aqueous environment. Recent experimental studies at the Catalysis Center for Energy Innovation have demonstrated promising results using various catalysts, such as tin, chromium, titanium. Among them, the tin-beta zeolite catalyst can give considerably good performance. In the present work, we successfully adopt a first principles molecular modeling approach to investigate the fundamentals of the catalytic isomerization reaction, in explicit aqueous medium. Stepwise mechanistic details of the reaction and the dynamics and energetics of individual steps are computed. Results are compared to new experimental data obtained via combining a range of analytical, NMR, and other spectroscopic methods. Molecular details obtained from the present work serve as an input for developing and designing novel catalytic materials for the glucose isomerization reaction, thus providing for the first time a bottom-up approach of catalyst design for biomass conversion reactions.

See more of this Session: Computational Catalysis II
See more of this Group/Topical: Catalysis and Reaction Engineering Division

(103a) Coarse-Graining of Nanocoating Using MARTINI Force Field for Mechanical Property Evaluation

Rohan Uttarwar, Jeffrey Potoff and Yinlun Huang, Department of Chemical Engineering and Materials Science, Wayne State University, Detroit, MI

Nanocomposite coatings, either thermoplastics or thermosets, have been proven to possess significantly improved performance or novel functionalities, such as self cleaning, self healing, etc. Dramatically increased surface area of nanosize fillers inside the coating matrix is the main reason for achieving superior mechanical properties. Molecular simulation can be performed to identify the root cause behind enhancement in properties of nanocomposite coatings, which is increase in polymer number density around nanoparticles.[1] But these results of coarse-grained simulations can be difficult to validate with real experimental results due to loss of physical meaning and degrees of freedom of the system. Hence, effective coarse graining methodology needs to be developed to study molecular behavior of polymer systems and model with experimental validation.

Recently, a MARTINI based approach has been introduced to develop successfully coarse grained models for biomolecular systems.[2] A coarse grained model is generated from atomistic models, which should be more reliable in model accuracy and model-based prediction, and thus it can be compared with experimental data. In this paper, the MARTINI technique is adopted for coarse grained simulations of polymer nanocomposite material. Due to increasing environmental concerns, water based resin systems, such as acrylics, are used tremendously for nanocoating syntheses. Thus polymethyl methacrylate (PMMA) resin with spherical silica nanoparticles dispersed in it is modeled for simulating a nanocoating. The volume fraction of nanofillers is kept at 5%. All atomistic simulations are performed using the CHARMM force field. The verification to this model is obtained by obtaining density and glass transition temperature consistent with experimental data. The equilibrated structures of a polymer from all-atom model are used to generate the coarse grained model by mapping onto coarse-grained beads. The parameters from the MARTINI model are modified, and a new set of parameters are generated to obtain a coarse grained model with comparable densities and radius of gyration with that of atomistic structures. All molecular dynamics simulations are performed with NAMD, version 2.7.

The MARTINI models for pure polymer samples and nanocoating samples of different nanoparticle sizes are deformed with external force at a fixed strain rate in longitudinal and transverse directions to evaluate its mechanical properties. The Young's modulus of nanocomposite is observed to increase at lower temperatures. The relative change in stiffness of nanocomposite is lower below the glass transition (T_g) compared to above T_g due to increased mobility of polymer chains. It is also seen that the Young's modulus increases with decreases in the size of the nanoparticle. This justifies the increase in the surface area leading to improved mechanical properties. During the deformation testing, the externally applied stress reaches a peak and goes down, which dictates the onset of plastic flow of the nanocomposite. The behavior of polymer models under applied stress is in agreement with experimental results. This demonstrates the applicability of the use of coarse-grained potentials in the design of nanocoating formulations with superior mechanical properties.

References

1. Xiao J., Y. L. Huang, and C. Manke, "Computational Design of Thermoset Nanocomposite Coatings: Methodological Study on Coating Development and Testing," Chemical Engineering Science, 65, 753-771, 2010.
2. Marrink S. J., Risselada H. J., Yefimov S., Tieleman P. D. and Vries A. H., "The MARTINI Force Field: Coarse Grained Model for Biomolecular Simulations", Journal of Physical Chemistry B, 111, 7812-7824, 2007.

See more of this Session: Modeling and Simulation of Polymers I
See more of this Group/Topical: Materials Engineering and Sciences Division

(103b) Morphology Dependent Hydrophobic Drug Partitioning In PEO-PCL Micelles Investigated with Coarse-Grained Molecular Dynamics

Sharon M. Loverde[1], Michael L. Klein[2] and Dennis E. Discher[1], (1)Chemical & Biomolecular Engineering, University of Pennsylvania, Philadelphia, PA, (2)Institute for Computational Molecular Science and Department of Chemistry, Temple University, Philadelphia, PA

Atomistically accurate, coarse-grained (CG) polymer models have been developed and studied for poly (ethylene oxide) -poly(caprolactone) (PEO-PCL), with molecular dynamics simulations comparing well with experimental phase behavior. The anti-cancer drug paclitaxel is also being CG'd with intramolecular interactions again obtained from all-atomistic molecular dynamics. Solubility of paclitaxel indicates that partitioning of the drug is dependent on micellar morphology. Utilizing free energy techniques, it is found that the hydrophobic drug possesses a greater partitioning in a worm micelle morphology of the same diblock weight than a spherical micelle morphology. These findings are consistent with previously found experimental results from the Discher laboratory. Moreover, at larger loading concentrations, the taxol-PCL and taxol-taxol interactions significantly shift the micellar density profile, showing increased interactions with the PEO, which may be an explanation for the phenomena of 'burst release.'

See more of this Session: Modeling and Simulation of Polymers I
See more of this Group/Topical: Materials Engineering and Sciences Division

(103c) A Comparison Between Polymer Salt and Ionomer for Battery Applications

Kan-Ju Lin[1], Katherine Li[2] and Janna Maranas[1], (1)Chemical Engineering, Pennsylvania State University, University Park, PA, (2)Chemical and Biological Engineering, Princeton University

Salts solvated in a Polyethylene oxide (PEO) matrix are considered to be the most promising candidates for solid state polymer electrolytes in battery applications. Since the anions have high mobility, and do not participate in chemical reactions, they accumulate at the electrode/electrolyte interface, causing a concentration gradient. This degrades battery performance after several charge/discharge cycles. An electrolyte in which the cation is the only conducting species is desired. PEO based ionomers solvate the cations, and reduce anion motion because the anions are covalently bonded to the polymer backbone. Like their PEO-salt counterparts, these single ion conductors suffer from low conductivity.

In this study, we examine a PEO-ionomer system, and the corresponding PEO-salt system using molecular dynamics simulation. The PEO-ionomer system has been experimentally synthesized and studied at Penn State. To provide a clean comparison, we consider a hypothetical PEO-salt system, in which the cation and anion are the same, but the anion is no longer incorporated on the polymer backbone. Thus, the cation identity, the cation-anion, and the cation-PEO interactions are the same as in the PEO-ionomer system. Comparison of the two systems isolates the effect of attaching the anion to the polymer backbone by covalent bonding. A wide range of ion environments is observed in both systems. The number of ion crosslinks in the two two systems are similar, but 1.5 times more cations are solvated by the PEO matrix in the salt compared to the ionomer. As ion concentration increases in a PEO salt, the PEO chains slow significantly. In the ionomer, this mechanism is active, but ions also slow the polymer through ionic cross-linking. Thus, covalently bonding the anion to the PEO backbone slows its mobility in comparison to the PEO-salt system. This in turn impacts the mobility of the cations and anions. We also find that the cations in the PEO-salt system change their coordination more frequently than in the ionomer. We conclude that the use of single ion conductors requires a shift in mechanism, such as ion motion facilitated by string-like clusters. We show examples of such movement in the ionomer and suggest ways to increase this type of ion motion.

See more of this Session: Modeling and Simulation of Polymers I
See more of this Group/Topical: Materials Engineering and Sciences Division

(103d) Simulation of Gas Diffusion In Polymer Nanocomposites

Ben Hanson, Chemical Engineering, University of Texas, Austin, TX and Venkat Ganesan, Department of Chemical Engineering, The University of Texas at Austin, Austin, TX

ABSTRACT: The effectsof nanoparticles on the rates of gas diffusion through glassy polymers werestudied by a combination of molecular dynamics and Monte Carlo techniques. These techniques were used to overcome thelarge amount of computational time required to obtain gas trajectories in thediffusive regime for glassy systems.Polystyrene was simulated using molecular dynamics with a number of gasmolecules to obtain a probability profile of gas motion within the polymermatrix at a range of temperatures above and below glass transition temperature. These results were used in conjunction withMD simulations to examine the types of gas particle motion as a function of thesetemperatures. The use of the a timeextending kinetic Monte Carlo allowed simulation of a sufficient time frame tocalculate gas motion in the diffusive regime. We studied the effect of Fullerenenanoparticles added to the polymer upon the gas diffusivity in such polymers. These results will be discussed in this talkalong with a mechanistic explanation of our observations.

See more of this Session: Modeling and Simulation of Polymers I
See more of this Group/Topical: Materials Engineering and Sciences Division

(103e) Analysis of AFM Results with MD Calculations On Single-Molecule Stretching of Poly(ethylene oxide) In Water and Hexane

Mangesh Chaudhari, Department of Chemical and Biomolecular Engineering, Tulane University, New Orleans, LA and Lawrence R. Pratt, Chemical and Biomolecular Engineering, Tulane University, New Orleans, LA

Analysisof AFM results with MD calculations on single-molecule stretching of poly(ethylene oxide) in water and hexane

Mangesh I. Chaudhari and LawrenceR. Pratt

Department of Chemical and Biomolecular Engineering

Tulane University, New Orleans,LA, 70118, USA.

ABSTRACT :

Elasticityof a single polymer in solution is important for the modeling of foams/gels andcan be directly measurable by atomic force microscopy (AFM). The work comparesstretching behavior of poly (ethylene oxide) polymer in water and hexane.Parallel tempering molecular dynamic simulations on aqueous 11-, 21-, and31-mer poly (ethylene oxide) solutions differentiates low-, moderate-, andhigh- extension regions (FIG.1) and explains that water swells these polymers. The low extension (low force) behaviorobtained from experimental and simulation data is often complex. Our previousresults, which combined parallel tempering and window sampling, clarified thelow-extension behavior. Here we return to examine the measured high extension behaviorin detail. We compare simulation results for high polymer extensions withavailable experimental data, and we compare simulations for different solventswith one another. We analyzesimulation results on aqueous solutions from the perspective of cooperativeH-bonding of water molecules to the ether oxygens inorder to clarify the role of solvent H-bond bridges in the conformationalstructures of these polymers.

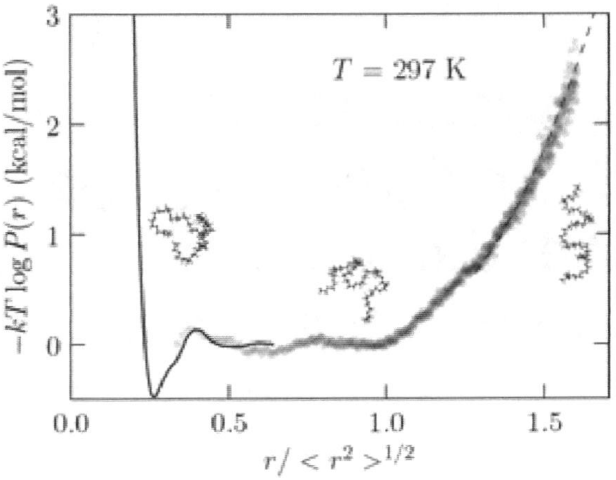

FIG. 1: Parallel tempering results forT=297 K distinguish low-, moderate-, and high-extension regimes. Thelow-extension regime includes the contact peak on the left of these graphs, andis associated with the analogous contact-minimum in the CH_4°CH_4potential-mean-force (pmf). That pmfhas been studied in numerous previous simulation calculations over severaldecades. The solid curve represents short-distance 'hydrophobic bonds' withwindow sampling that is consistent with the direct simulation results. X-axisis normalized with RMS end-to-end distance and emphasizes that the high extensionregion starts after this value. Parabola fit through high extensions iscomparable with experimental data available on stretching of these polymers.

See more of this Session: Modeling and Simulation of Polymers I
See more of this Group/Topical: Materials Engineering and Sciences Division

(103f) Molecular Dynamics Simulations of An Oxidized Vapor-Grown Carbon Nanofiber and Vinyl Ester Resin Interactions Leading to a Possible Interphase Formation In the Cured Nanocomposite

Changwoon Jang[1], **Sasan Nouranian**[2], Thomas E. Lacy[1], Steven R. Gwaltney[3], Hossein Toghiani[4] and Charles U. Pittman Jr.[3], (1)Department of Aerospace Engineering, Mississippi State University, Mississippi State, MS, (2)Center for Advanced Vehicular Systems (CAVS), Mississippi State University, Mississippi State, MS, (3)Department of Chemistry, Mississippi State University, Mississippi State, MS, (4)Dave C. Swalm School of Chemical Engineering, Mississippi State University, Mississippi State, MS

Molecular dynamics simulations were used to study the effect of vapor-grown carbon nanofiber (VGCNF) surface oxidation on VGCNF-vinyl ester (VE) resin monomer interactions. It was anticipated that the interfacial interactions between resin monomers and the oxidized carbon nanofiber surface would result in different local molar ratios of the resin monomers compared to the bulk resin. The resin was comprised of styrene and two VE monomers with either one or two bisphenol A groups in their backbones (designated as VE1 and VE2, respectively). Differences in local molar ratios of these resin monomers may result in the formation of an interphase region in the final cured nanocomposite. The time-averaged relative concentration profiles were used to monitor the temporal and spatial distributions of the resin monomers in a simulation cell of size 60×50×60 $Å^3$. Initially the cell was filled with monomers in a ratio consistent with a 33 wt% commercial epoxy vinyl ester resin (Derakane 441-400, Ashland Co.). The idealized VGCNF surface was represented by two overlapped (shingled) graphene sheets, which were in contact with the resin. Oxidation of the graphene sheets was represented by the introduction of functional groups on the edges and the surfaces of the graphene sheets. The edges were more highly oxidized than the surfaces, containing of hydroxyl, phenolic, lactone, quinone, hydroquinone, and anhydride functional groups. Hydroxyl, epoxide, aromatic, ketone and carboxylic acid groups were attached to the surfaces (e.g. basal planes). These functional groups have been previously identified in our laboratory [1]. The NVT (constant number of atoms, N; constant volume, V; constant temperature, T) ensemble was used for simulations. All simulations were conducted with a simulation time of 10 ns at 1000 K for monomer distribution equilibration followed by another 5 ns simulation at 300 K. The relative concentration profiles were generated for the direction roughly perpendicular to the graphene sheet surfaces. The VE1 monomer concentration was higher near the oxidized graphene surface, compared to a simulation performed using a pristine graphene [2], where VE1 was depleted near the graphene surface. Styrene accumulated in the interfacial region in both cases, but more so for the pristine surface. Overall, a ~5 Å thick interfacial region was observed near each surface of the graphene sheet. In these regions, the monomer molar ratios differed from those in the bulk resin. The increased concentration of VE1 near the oxidized carbon nanofiber surface may lead to a stiffer interphase compared to the bulk matrix. In addition, favorable VGCNF-matrix polar interfacial interactions may lead to increased interfacial adhesion and subsequently, increase the interfacial shear strength in the final cured nanocomposite. The results of this study can be extended to other resins and surfaces.

[1] Lakshminarayanan, PV, Toghiani, H, Pittman, CU. Nitric acid oxidation of vapor grown carbon nanofibers. Carbon 2004; 42: 2422-2433.

[2] Nouranian, S, Jang, C, Lacy, TE, Gwaltney, SR, Toghiani, H, Pittman Jr, CU. Molecular dynamics simulations of vinyl ester resin monomer interactions with a pristine vapor-grown carbon nanofiber and their implications for composite interphase formation. Carbon (In Press).

See more of this Session: Modeling and Simulation of Polymers I
See more of this Group/Topical: Materials Engineering and Sciences Division

(104a) Particle Based Multi-Scale Modeling of the Dynamic Response of Hexahydro-1,3,5-Trinitro-s-Triazine (RDX): Constant Energy Dissipative Particle Dynamics with Coarse-Grained Density Dependent Potentials

Joshua D. Moore[1], Sergei Izvekov[1], Martin Lisal[2] and John K. Brennan[1], (1)Weapons and Materials Research Directorate, U.S. Army Research Laboratory, Aberdeen Proving Ground, MD, (2)E. Hala Laboratory of Thermodynamics, Academy of Sciences of the Czech Republic, Prague, Czech Republic

Nanocomposites are inherently heterogeneous, containing several species types of disparate shapes and sizes. For example, energetic materials are often comprised of micro-sized crystallites (or grains) held together by a small amount of polymer binder, and in some instances contain metal nanoparticles. Accordingly, mechanical stimulation (e.g., shock or shear) of these materials can incite responses over a wide range of spatial and temporal scales. Localized regions of elevated thermal energy, or hot spots, often occur near microscale defects in the material such as voids and grain boundaries. However, some energy transfer processes are atomistically governed, for which field-based microscale, mesoscale, and continuum modeling techniques are currently unable to provide the required level of fidelity. Modeling the thermal and mechanical response of these materials atomistically, despite growing computational power and resources, remains a challenge due to the length and time scales required, as often billions of molecules are necessary to be modeled on micron-size scales.

To overcome some of these challenges, we have implemented multiscale techniques to bridge the atomistic and mesoscale levels of description. The multiscale coarse-graining method (MS-CG) [1,2], a *bottom-up* approach based on force-matching, has been utilized to derive particle-based coarse-grained potentials for hexahydro-1,3,5-trinitro-s-triazine (RDX), where one RDX molecule is mapped onto a single interaction site. To improve transferability of the MS-CG model (required to accurately capture thermodynamic responses during mechanical and thermal stimulation), explicit MS-CG mapping from the atomistic to the microscale is made from ambient to high pressures [2]. The resulting density dependent MS-CG model reproduces several atomistic properties within reasonable agreement from ambient to high pressures for the molecular crystal including the lattice structure, melting point, elastic and vibrational properties, and the phase diagram [3]. For the molten state, the 2-body and often multi-body structure are also well described with the model [3].

Despite the successes of the MS-CG model of RDX to describe static properties, it inevitably cannot account for accurate energy and momentum exchange in response to mechanical stimulation using traditional molecular dynamics. This is due to intramolecular degrees of freedom (i.e., intramolecular bonding) being coarse-grained out, and thus, the heat capacity is underestimated by nearly an order of magnitude [3]. To correct this deficiency, we account for momentum and energy transfer under mechanical shock by utilizing the constant-energy Dissipative Particle Dynamics method (DPD-E) [4-6]. Unlike traditional Dissipative Particle Dynamics [7-9], which conserves only momentum, the DPD-E method conserves both energy and momentum by assigning an internal energy to each particle, which for the MS-CG RDX case, helps to minimize the effect of coarse-graining the intramolecular degrees of freedom.

In this talk, we will present results for mechanical and thermal shock loading of our MS-CG model of RDX using DPD-E. Various modeling parameters have been investigated for sensitivity including spatial interaction cutoffs used to determine local densities for the density dependent potentials, the mesoparticle equation of state that accounts for the particle internal energy and the overall dynamics through energy dissipation. Results will be compared directly to atomistic simulation for both mechanical shock and the response due to thermal heating of the material.

This effort is supported by the recently established Institute for Multi-Scale Reactive Modeling of Insensitive Munitions (MSRM), which is a multi-team effort led by the U.S. Army Research Laboratory and the U.S. Army Armament Research, Development, and Engineering Center, involving various other national laboratories and academic groups totaling over 20 scientists. J.D.M. gratefully acknowledges support, in part, by an appointment to the Internship/Research Participation Program for the U.S. Army Research Laboratory administered by the Oak Ridge Institute of Science and Education through an agreement between the U.S. Department of Energy and U.S. ARL (Project ID: 201003211).

[1] S. Izvekov and G.A. Voth, *J. Chem. Phys.*, **123**, 134105 (2005).

[2] S. Izvekov, P.W. Chung, and B.M. Rice, *J. Chem. Phys.*, **133**, 064109 (2010).

[3] S. Izvekov, P.W. Chung, and B.M. Rice, *Submitted* (2011).

[4] J. Bonet Avalos and A.D. Mackie, *Europhys. Lett.*, **40**, 141 (1997).

[5] P. Español, *Europhys. Lett.*, **40**, 631 (1997).

[6] M. Lísal, J.K. Brennan, and J. Bonet Avalos, *Submitted* (2011).

[7] P.J. Hoogerbrugge and J.M.V.A. Koelman, *Europhys. Lett.*, **19**, 155 (1992).

[8] J.M.V.A. Koelman and P.J. Hoogerbrugge, *Europhys. Lett.*, **21**, 363 (1993).

[9] R.D. Groot and P.B. Warren, *J. Chem. Phys.*, **107**, 4423 (1997).

See more of this Session: Multiscale Modeling II

(104b) Temperature-Accelerated Molecular Dynamics Reveals That Insulin Can Undergo Large-Scale Conformational Reorganization On Binding to Its Receptor

Harish Vashisth, Chemistry and Biophysics, University of Michigan, Ann Arbor, MI and Cameron F. Abrams, Drexel University, Philadelphia, PA

Insulin regulates blood glucose levels in higher organisms by specifically binding to and activating a transmembrane glycoprotein known as the insulin receptor (IR). Detailed photo-crosslinking studies have suggested that insulin molecule reorganizes on binding to IR due to the displacement of its flexible B-chain C-terminus in the presence of a tandem hormone binding structural motif called CT-peptide. Although a detailed understanding of how insulin binding leads to receptor activation remains elusive, our recently proposed all-atom structural models of insulin/IR complexes provide information valuable for exploring structure-function repertoire of IR. However, the large-scale conformational change in the insulin was not studied in these structural models due to the absence of then-unresolved CT-peptide, and also the lack of adequate simulation techniques. We have proposed a new conformational sampling algorithm, temperature-accelerated molecular dynamics (TAMD) (2010, *PNAS*, **107**, 4961-4966), that provides an immediate opportunity to use all-atom simulation to directly address these important gaps in our understanding of insulin binding to IR. Using a combination of Metropolis Monte-Carlo (MC) and molecular dynamics (MD) simulations, we have constructed all-atom structural models of insulin/IR complexes in the presence of (now-resolved) CT-peptide. We further apply TAMD to the C-terminus of IR-bound (T and R) insulin molecules at a fictitious thermal energy of 6 kcal/mol, and observe that while insulin remains stably-bound to IR, it can undergo large-scale conformational change in the B-chain C-terminus at ~50-ns time-scale resulting in the exposure of hidden hydrophobic core of insulin. These results are significant because we demonstrate at the atomistic-scale for the first time that insulin molecules can undergo large-scale reorganization on binding to IR in presence of CT-peptide. Moreover, we suggest TAMD as a viable simulation technique to study such large-scale conformational changes in proteins.

(104c) Multiscale Modeling for Phospholipid Bilayer Simulations

Emily Curtis, Chemical & Biomolecular Engineering, North Carolina State University, Cary, NC

Multiscale Modeling for PhospholipidBilayer Simulations

A multiscale modelingapproach was used to develop an implicit-solvent intermediate-resolution modelto simulate the behavior of the lipid, 1,2-dipalmitoyl-*sn*-glycero-3-phosphocholine (DPPC) inwater. By reducing the DPPCrepresentation to 14 coarse-grained sites with unique properties, treatingsolvent implicitly, and employing discontinuous molecular dynamics, a very fastalternative to traditional molecular dynamics, we are able to simulate thespontaneous formation of a bilayer from a random solution of 512 DPPC moleculesin several hours on a fast workstation. Data used to calculate the coarse-grained model parameterswas obtained by running united-atom simulations with GROMACS. Radial distribution functions wereplotted for all pairs of non-bonded coarse-grained sites to estimate thehardsphere and square-well diameters. The interaction energy between each pair of non-bonded coarse-grainedsites was calculated using the Boltzmann inversion scheme. The relative stiffness of each lipid ismaintained by imposing pseudobonds, which limit the bond lengths to fluctuateby the same amount observed in the united atom simulations. In the model each coarse-grained sitehas its own realistic mass. Simulationresults show that the model accurately reproduces structural properties of theDPPC bilayer including the area per lipid, bilayer thickness, bond order andmass density profiles. Currentwork is focused on extending the model to lipid mixtures and simulating thephase separation that occurs as a bilayer composed of DPPC and 1,2-dipalmitoyl-*sn*-glycero-3-phosphate (PA) changes froman initially-homogeneous state (at neutral pH) to a phase-separated state atlow pH.

See more of this Session: Multiscale Modeling II
See more of this Group/Topical: Computational Molecular Science and Engineering Forum

(104d) A Coarse Grained Model of Inhibitors to Scavenger Receptor Uptake to LDL

Michael Tomasini and M. Silvina Tomassone, Chemical and Biochemical Engineering, Rutgers University, Piscataway, NJ

The buildup and oxidation of low-density lipoproteins (LDL) in the walls of arterial blood vessels results in atherosclerotic lesions and arterial damage. The immune response of this arterial damage is to recruit macrophages expressing scavenger receptors specific for the uptake of oxidized LDL. The result is the formation of foam cells and fatty streaks, the first signs of atherosclerotic plaques. One possible method to combat the formation of atherosclerotic plaques is through the use of inhibitors which prevent the uptake of oxidized LDL by the scavenger receptors. Recently, Tian et al. 2004 [1] developed a class of molecules termed amphiphilic scorpion-like macromolecules (AScM) comprised of a mucic acid head group with attached aliphatic chains and a long polyethylene glycol (PEG) tail used specifically to block uptake of oxidized LDL by scavenger receptors. In this work, we use computational simulations to explore the structure and interactions of AscM micelles with the specific scavenger receptor – scavenger receptor A (SR-A) in order to shed light on an optimal inhibitor of oxidized LDL uptake.

In our simulations, we use a coarse-grained model based on the MARTINI force field [2] to describe the AScM molecules as well as the collagenous portion of SR-A where binding is assumed to take place [3]. As the interaction of oxidized LDL is thought to be mediated by electrostatic interactions, we vary the number (0, 1, or 2) of carboxylate groups at different positions on the AScM (near the mucic acid head or at the PEG tail) to determine the effect on both AScM micellation (which it though necessary for SR-A binding) and AScM association with SR-A to determine the best inhibitor.

References:

[1] Tian L, Yam L, Zhou N, Tat H, and Uhrich KE. Amphiphilic Scorpion-like Macromolecules: Design, Synthesis, and Characterization. *Macromolecules*. 2004; **37**: 538-543.

[2] Monticelli L, Kandasamy SK, Periole X, Larson RG, Tieleman DP, and Marrink SJ. The MARTINI Coarse-Grained Force Field: Extension to Proteins. *J. Chem. Theory and Comput.* 2008; **4**(5): 819-834.

[3] Platt N and Gordon SJ. Is the Class A Macrophage Scavenger Receptor (SR-A) Multifunctional? – The Mouse's Tale. *J. Clin. Invest.* 2001; **108**(5): 649-654.

See more of this Session: Multiscale Modeling II
See more of this Group/Topical: Computational Molecular Science and Engineering Forum

(104e) Multiscale Molecular Modeling of Fullerol-Dendrimer Complexes

Seung Ha Kim and Monica H. Lamm, Department of Chemical and Biological Engineering, Iowa State University, Ames, IA

Dendrimers are large, synthetic macromolecules that have regular and highly branched architectures. The multivalency of dendrimers permits them to bind with other species and makes them attractive platforms for a wide range of applications, including medicine, surface science, catalysis, and water treatment. To better understand the physical mechanisms of dendrimer-guest assembly, we have developed a multiscale molecular model for simulating flexible macromolecules in solution, based on the force matching method for coarse graining. We illustrate this approach by simulating the complexation between poly(amidoamine) (PAMAM) dendrimers and $C_{60}(OH)_{20}$ (fullerol) in aqueous solution. Specifically, we systematically examine the effects of concentration, and of dendrimer molecular weight, on the resulting PAMAM-fullerol complexes and compare our model predictions to recent experimental measurements.

See more of this Session: Multiscale Modeling II
See more of this Group/Topical: Computational Molecular Science and Engineering Forum

(104f) Multiscale Modeling of Perfluoropolyether Lubricants with Functional Endgroups

Robert Smith, Pil Seung Chung, Lorenz T. Biegler and Myung S. Jhon, Chemical Engineering, Carnegie Mellon University, Pittsburgh, PA

Physical properties of nanoscale perfluoropolyether (PFPE) films, which are generally used as a perfluoro linear co-polymer lubricant in hard disk drives (HDDs) has been previously studied in disparate molecular/mesoscale simulations. The PFPE molecules must exhibit strong adhesion to the carbon overcoat surface as well as rapid self-healing properties; therefore, a fundamental understanding of nanoscale characteristics of lubricant structure which contribute to desired macroscale performance must be established for successful nanotechnology design.

Due to the hydroxyl functional groups at each chain end of well-known PFPE (e.g., Zdol and Ztetraol), PFPE on an amorphous carbon surface shows peculiar characteristics (i.e., layering structure and entanglement on the surface). Via coarse-grained (CG) molecular dynamics (MD), we qualitatively validated molecular scale phenomena related to nanostructured conformation and dynamic behavior of molecularly thin lubricant film [1, 2]. To validate this coarse-grained model, a procedure in multi-scale modeling [3] which links atomic scale phenomena to CG MD is established thereby providing a form of synergy between the atomistic and molecular scales. Accurate intra / intermolecular force field parameters for PFPE and carbon surfaces are implemented in CG MD to enable improvements in our molecular scale model.

In this study, we use *ab initio* methods to determine intramolecular force field parameters for PFPE molecules and their intermolecular interaction with other PFPEs as well as with various carbon overcoat surfaces. Calculations based on *ab initio* methods are used to determine the intramolecular (stretching, bending, and torsional) force field parameters via solving the *ab initio* Hessian matrix eigenvalue problem. The intermolecular interaction between endgroups in pure and blended PFPEs is quantified using the supermolecular approach to molecular interaction. In addition, we explore novel carbon overcoat technology in HDD by calculating PFPE interaction with graphene. The implementation of atomically thin graphene as a carbon overcoat has the potential for a factor of two improvement in areal recording density. We compare the PFPE interaction with the traditional carbon overcoat versus graphene. We also evaluate PFPE-PFPE and PFPE-carbon surface geometry as a function of the PFPE structures particularly for new-generation lubricants including DDPA. The force field is evaluated by comparing interaction energies and molecular geometries of equilibrium configurations obtained using *ab initio* calculations to those obtained via the classical force field analysis.

[1] Q. Guo, L. Li, Y.-T. Hsia, and M.S. Jhon, "A Spreading Study of Lubricant Film via Optical Surface Analyzer and Molecular Dynamics", *IEEE Trans. Magn.* 42(10), 2528-2530 (2006).

[2] P.S. Chung, H. Chen, and M.S. Jhon, "Molecular dynamics simulation of binary mixture lubricant films", *J. Appl. Phys.*, 103, 07F526 (2008).

[3] D. Kim, P.S. Chung, P. Jain, S.H. Vemuri, and M.S. Jhon, "Multiscale modeling of head disk interface", IEEE Trans. Magn., 46, 2401-2404 (2010).

See more of this Session: Multiscale Modeling II
See more of this Group/Topical: Computational Molecular Science and Engineering Forum

(104g) Nanoparticles At the Water-Decane Interface: Evidence of Emergent Behavior From Equilibrium Multi-Scale Simulations

Heng Fab, School of Chemical, Biological and Materials Engineering, The University of Oklahoma, Norman, OK and Alberto Striolo, School of Chemical Biological and Materials Engineering, The University of Oklahoma, Norman, OK

Pickering emulsions (emulsions stabilized by solid nanoparticles rather than by surfactants) are finding practical applications in several areas, including cosmetics, drug-delivery, and even in-situ upgrade of bio-oil to fungible bio-fuels [Science 327 (2010) 68]. It is known that solid particles adsorb at water-oil interfaces to reduce the contact area between the two immiscible phases. Stable emulsions are obtained when the particles strongly adsorb at the interfaces. We will report here equilibrium molecular dynamics simulation results employed to differentiate the behavior of silica-based nanoparticles used to stabilize either water-in-decane or decane-water emulsions. We will compare nanoparticles with similar surface properties, but different distribution of the surface groups. Particular attention will be given to Janus as opposed to homogeneous nanoparticles [Langmuir 27 (2011) 5264-5274]. The all-atom simulation results have been used to parameterize dissipative particle dynamics simulations. Using the latter tool, we have obtained interesting morphologies depending on nanoparticle concentration, and water-decane molar ratio in the various simulated systems. The simulation results will be discussed in relation to available experimental data.

See more of this Session: Multiscale Modeling II
See more of this Group/Topical: Computational Molecular Science and Engineering Forum

(119a) Gibbs Ensembe Monte Carlo Simulation of Phase Coexistence Involving Solids Using Periodic Retiling

Eric Grzelak[1], Amadeu K. Sum[1] and David T. Wu[2], (1)Chemical Engineering Department, Colorado School Of Mines, Golden, CO, (2)Chemical Engineering and Chemistry Departments, Colorado School Of Mines, Golden, CO

The determination of solid phase boundaries has long been a difficult task for simulation studies. Advanced methods typically employ difficult thermodynamic paths and long computational times to reach accurate equilibrium data. In this presentation we share our developments towards a new Monte Carlo algorithm that permits evaluation of phase coexistence properties involving solids. Called "Periodic Retiling," our technique allows for maintenance of solid phase order while particles are inserted and deleted from the phase. This is accomplished by altering the periodic boundaries but maintaining a standard minimum image convention. The complicated nature of the move is handled in the acceptance criteria through the maintenance of detailed balance. To prove the efficacy of the method the complete phase diagram, including solid-vapor and solid-liquid phase equilibria, of the Lennard-Jones fluid is reproduced and compared with literature values. When compared with more traditional methods of obtaining solid phase equilibria the Periodic Retiling method shows a remarkable decrease in computational effort and time.

See more of this Session: Recent Advances In Molecular Simulation Methods II
See more of this Group/Topical: Computational Molecular Science and Engineering Forum

(119b) Thermodynamics of a Small System In a μ T Reservoir

Thijs JH Vlugt[1], Sondre K Schnell[1], Jean-Marc Simon[2], Signe Kjelstrup[3] and Dick Bedeaux[3], (1)Delft University of Technology, Delft, Netherlands, (2)Chemistry - Laboratoire de Recherches sur la Ractivit des Solides (UMR 5613 CNRS), Universit de Bourgogne, Dijon, France, (3)NTNU Norway, Trondheim, Norway

To date, small systems have not received much attention in thermodynamics. The reason seems obvious: small systems have few particles, thus being far from the thermodynamic limit. However, the thermodynamics of small systems will become increasingly important in nanoscience. By using the formalism of Hill [1], we have derived the finite-size dependence of the enthalpy and thermodynamic factor in a μ T reservoir [2]. We find this scaling to be dependent on 1/L, where L is the linear size of the small system. The results are verified using molecular simulations in 2 and 3 dimensions. It turns out that one can compute the thermodynamic factor, adsorption enthalpy and activity coefficients from fluctuations at a very small scale, provided that the correct finite-size scaling is used. The method provides an efficient way to compute activity coefficients in liquid mixtures from MD/MC simulations in the NVE, NVT, or NPT ensemble. The methodology also applies to non-homogeneous systems like argon adsorbed in MFI-type zeolite. In this case the shape and size of the small system compared to the unit cell of the zeolite framework needs to be taken into account, it turns out that this should be in multiples of a quarter unit cell of MFI-type zeolite.

We consider the situation of a large system of size L_t in 3 dimensions, so that the total volume equals $(L_t)^3$. This large system can be in the microcanonical, canonical or grand-canonical ensemble. Inside this system, we construct an ensemble of smaller systems of size L and volume L^3. If L is much smaller than L_t, the small system can be considered to be in the grand-canonical ensemble. By investigating density and energy fluctuations of the small system of size L, we can obtain the molar enthalpy and thermodynamic factor of the small system. The formalism of Hill [1] can be used to show that the quantities obtained from the small system show a very large finite-size effect, and that there is a scaling with 1/L in sharp contrast to the more usual finite-size scaling of

$1/L^3$ [3].

We confirm this finite size scaling numerically for the WCA and Lennard-Jones systems, as well as for argon adsorbed in MFI-type zeolite. Clearly, by extrapolation of the linear regime, we obtain the correct quantities corresponding to the macroscopic system. At very small values of L, nook- and corner effects become dominant and deviations from the 1/L scaling are observed. If L is close to L_t, than the small subsystem will not be in the grand-canonical ensemble and deviations from the 1/L scaling will appear.

References

[1] T.L. Hill, Thermodynamics of Small Systems, Part 1, Benjamin, New York, 1963.

[2] S.K. Schnell, T.J.H. Vlugt, J.-M. Simon, D. Bedeaux, S. Kjelstrup, Chem. Phys. Lett., 504:199-201, 2011.

[3] D.P. Landau, K. Binder, A Guide to Monte Carlo Simulations in Statistical Physics, Cambridge University Press, 2000.

See more of this Session: Recent Advances In Molecular Simulation Methods II
See more of this Group/Topical: Computational Molecular Science and Engineering Forum

(119c) Monte Carlo Simulation Methods for Calculating the Interfacial Properties of Fluid Mixtures near a Solid Surface

Vaibhaw Kumar, Chemical and Biological Engineering, University at Buffalo, Buffalo, NY and Jeffrey Errington, Chemical & Biological Engineering, University of Buffalo, Buffalo, NY

Fluid mixtures are known to exhibit a rich variety of wetting behavior near a solid surface. In this presentation we describe our recent efforts to study this phenomenon with the help of molecular simulation. We focus on methodological developments that enable one to compute interfacial properties in an efficient manner. Our basic approach involves calculating the surface excess free energy (also known as the interface potential) as a function of the surface density of the fluid. The interface potential provides information regarding qualitative wetting features (such as wetting regime and order of wetting) as well as access to quantitative information, such as the contact angle and interfacial tensions. We outline several strategies for acquiring the components of the interface potential that are necessary to evaluate interfacial properties of interest. We demonstrate the utility of these methods by computing various interfacial properties of two Lennard-Jones mixtures, one of which shows surface-induced demixing.

See more of this Session: Recent Advances In Molecular Simulation Methods II
See more of this Group/Topical: Computational Molecular Science and Engineering Forum

(119d) Wang-Landau Simulations for Self-Assembling Systems

Katie A. Maerzke[1], Lili Gai[1], Peter T. Cummings[2] and Clare McCabe[1], (1)Chemical and Biomolecular Engineering, Vanderbilt University, Nashville, TN, (2)Center of Nanophase Materials Sciences, Oak Ridge National Laboratory, Oak Ridge, TN

Understanding the relative free energies of various conformations of macromolecules and the phase transitions in self-assembled systems is an extremely important area of study; however, limited progress has been made to date since the calculation of free energy remains a significant challenge for molecular simulation. A recent breakthrough in this area is the Wang-Landau method, which allows for the direct calculation of the density of states, and thus the free energy. Wang-Landau simulations for continuous molecular systems present several challenges regarding efficient implementation and sampling techniques that must be overcome to utililize the method broadly. We will discuss the ways in which we overcome these difficulties for a simple implicit solvent lipid model.

See more of this Session: Recent Advances In Molecular Simulation Methods II
See more of this Group/Topical: Computational Molecular Science and Engineering Forum

(119e) Incremental Gauge Cell Monte Carlo Simulation Method for Calculating Chemical Potential of Chain Molecules

Christopher J. Rasmussen, Aleksey Vishnyakov and Alexander V. Neimark, Chemical and Biochemical Engineering, Rutgers, The State University of New Jersey, Piscataway, NJ

Calculation of the chemical potential and free energy of polymer chains in atomistic simulations is a difficult problem. Many approaches have been proposed, all with their caveats. We suggest a new method, based on the chain-incrementing strategy of Kumar et al. [1] and its extension to the gauge cell method. The incremental chemical potential is calculating by adding a single monomer to a test chain and measuring the difference in the chain free energy. As such, the chain chemical potential represents the sum of these incremental values. Introduction of the gauge cell that contains monomers, which are added to or removed from the end of the test chain, provides three main benefits: first, it serves as a measure of the incremental chemical potential, second, it suppresses fluctuations and allows unstable states to be stabilized in the main cell, and third, exchange from main to gauge cell and back facilitates efficient sampling of state space. The method was applied to chains of Lennard-Jones beads with stiff harmonic bonds up to 500 monomers in length. We show that the suggested method quantitatively reproduces the modified Widom particle insertion method, and it is by the order of magnitude more efficient for long chains in terms of the computational time required for the same accuracy of chemical potential calculations. A special attention is paid to the studies of the behavior of single chains in confinements comparable in size with the free chain radius of inertia with and without adsorption potential and at various temperatures. At sufficiently low temperatures, the dependence of the incremental chemical potential on the chain length is superficially similar to adsorption and capillary condensation isotherms of simple Lennard Jones fluids. It reflects monolayer formation on the adsorbing wall with subsequent filling of the pore volume as the chain length increases. We find that the incremental gauge cell Monte Carlo method is an accurate and efficient technique to calculate the free energy of chain molecules that can be applied for bulk and confined systems that may find practical applications, as shown with several examples of modeling polymer partitioning on porous substrates and dynamics of chain translocation into nanopores.

[1] Kumar SK, Szleifer I, Panagiotopoulos AZ. Determination of the chemical potentials of polymeric systems from Monte Carlo simulations. Physical Review Letters. 1991;66(22):2935.

See more of this Session: Recent Advances In Molecular Simulation Methods II
See more of this Group/Topical: Computational Molecular Science and Engineering Forum

(119f) Accelerating Replica Exchange and Generalized Ensemble Simulations Through Gibbs Sampling In State Space

Michael R. Shirts, Chemical Engineering, University of Virginia, Charlottesville, VA and John D. Chodera, California Institute for Quantitative Biosciences, University of California, Berkeley, Berkeley, CA

The widespread popularity of replica exchange and generalized ensembles algorithms for simulating complex molecular systems in chemistry and chemical engineering has generated great interest in improving the efficiency of these protocols in ways that enhance phase space mixing and therefore increase sampling efficiency. We demonstrate how both classes of algorithms can be considered as special cases of a statistical technique called Gibbs sampling (not to be confused with Gibbs ensemble sampling) within a Markov chain Monte Carlo (MCMC) framework. Seen in this framework, new ways of updating of the thermodynamic states associated with these configurations can be easily identified and substantially increase mixing in a way that still sample from the desired distributions. We present some simple alternatives to standard protocols that improve mixing of the overall Markov chain, thus reducing simulation times required to converge. These improved schemes are demonstrated in several common applications, including expanded ensembles among Hamiltonians for estimation of free energies, parallel tempering, and multidimensional replica exchange umbrella sampling. In some standard cases, these simulation techniques can increase the rate of configurational sampling by four to ten times.

See more of this Session: Recent Advances In Molecular Simulation Methods II
See more of this Group/Topical: Computational Molecular Science and Engineering Forum

(119g) Optimization of CBMC Particle Swap Moves for GEMC Simulations

Peng Bai, Chemical Engineering and Materials Science, University of Minnesota, Minneapolis, MN and J. Ilja Siepmann, Depts. of Chemistry and of Chemical Engineering & Materials Science, University of Minnesota, Minneapolis, MN

Particle swap moves are usually the rate-limiting step for Gibbs ensemble Monte Carlo (GEMC) simulations of fluid phase equilibria for systems described by force fields with continuous potential functions. The acceptance probabilities for particle swap moves are often very low at low reduced temperatures and/or for molecules with articulated structures and strong orientation-dependent interactions. The configurational-bias Monte Carlo (CBMC) method can greatly aid the acceptance probabilities for articulated molecules and to a lesser extent for rigid or nearly rigid particles with a few interaction sites. The efficiency of the CBMC algorithm is influenced by multiple parameters and selection of the interactions considered in the Rosenbluth weights. In this work we assess the efficiency of different CBMC strategies for GEMC simulations using the SPC/E and TIP4P water models at 283, 343, and 473 K and demonstrate that much higher acceptance probabilities can be achieved than previously reported in the literature.

See more of this Session: Recent Advances In Molecular Simulation Methods II
See more of this Group/Topical: Computational Molecular Science and Engineering Forum

(134a) Pure Carbon-Dioxide Gas Absorption In a Counter-Current Foam-Bed Reactor

Amit A. Gaikwad, Department of Chemical Engineering, Indian Institute of Technology, Delhi, New Delhi, India and Ashok N. Bhaskarwar, Chemical Engineering, Indian Institute of Technology, Delhi, New Delhi, India

A foam-bed reactor is a gas-liquid contactor which offers low pressure drop and high gas hold-up. Thus it can be used for the treatment of large volumes of gaseous contaminants. Present work attempts to study the removal of carbon-dioxide gas, which could be obtained from power plant exhausts, by absorbing it in aqueous barium-sulfide using a counter-current foam-bed reactor.

The difficulty in scaling up and commercializing foam-bed reactors in spite of the advantages like very high interfacial area is the limitation of foam height. When the carbon-dioxide gas is sparged through the storage section i.e. a mixture of aqueous barium sulfide and a surfactant, the generated foam travels upwards into the foam section. The gas-liquid chemical reaction takes place in the foam section. The liquid hold-up in foam being small and chemical reaction being fast, after a particular height, the liquid is consumed by the chemical reaction, while unreacted gas is still left in the foam. This unreacted gas eventually leaves the reactor at the top where foam is broken. Thus, very tall foam columns cannot be used in practice because of the occurrence of this "depletion condition". This scale-up difficulty could be overcome by adding the solution of liquid-phase reactant from the top into foam, in addition to that introduced into the storage section.

Experiments were carried out for the carbonation of barium sulfide specifically to verify the anticipated advantages offered by such reactors and their effectiveness in removing carbon-dioxide gas. The parameters explored were the foam height, barium sulfide flow rate into the foam section, and number of inlets used for charging the barium sulfide solution into the foam section.

With the increase in foam height, the conversion of barium sulfide increased. An optimum value of barium sulfide flow rate into the foam section was obtained above and below which the conversion decreased. When two inlets were used at different locations to charge the same amount of the aqueous barium sulfide as in case of a single inlet, the conversion of barium sulfide increased marginally.

The experimental results were also compared with the pure-gas model of Gaikwad and Bhaskarwar (2008). The pure-gas absorption model predicted the experimental data to well within an accuracy of 20%. Comparison of conversions in semi-batch and counter-current foam-bed reactors showed that the latter perform better than the semi-batch reactors. Thus, with the counter-current mode of operation of the foam-bed reactors, there is a potential for large-scale treatment of carbon-dioxide gas.

(134b) Optimization of Carbon Capture Systems Using Surrogate Models of Simulated Processes

Alison Cozad[1], YoungJung Chang[1], Nick Sahinidis[1] and David C. Miller[2], (1)Chemical Engineering, Carnegie Mellon University, Pittsburgh, PA, (2)U.S. Department Of Energy, National Energy Technology Laboratory, Morgantown, WV

With increasing demand placed on power generation plants to reduce carbon dioxide (CO2) emissions, processes to separate and capture CO2 for eventual sequestration are highly sought after. Carbon capture processes impart a parasitic load on the power plants; it is estimated that this would increase the cost of electricity from existing pulverized coal plants anywhere from 71-85 percent [1]. The National Energy and Technology Lab (NETL) is working to lower this to below a 30 percent increase.

To reach this goal, work is being done not only to accurately simulate these processes, but also to leverage those accurate and detailed simulations to design optimal carbon capture processes. The major challenges include the lack of accurate algebraic models of the processes, computationally costly simulations, and insufficiently robust simulations. The first challenge bars the use of provable derivative-based optimization algorithms. The latter two can either lead to difficult or impossible direct derivative-free optimization.

To overcome these difficulties, we take a more indirect method to solving this problem by, first, generating an accurate set of algebraic surrogate models from the simulation then using derivative-based solvers to optimize the surrogate models. We developed a method that uses derivative-based and derivative-free optimization alongside machine learning and statistical techniques to generate the set of low-complexity surrogate models using data sampled from detailed simulations. The models are validated and improved through the use of derivative-free solvers to adaptively sample new simulation points. The resulting surrogate models can then be used in a superstructure-based process synthesis and solved using derivative-based methods to optimize carbon capture processes.

[1]. The Cost of Carbon Dioxide Capture and Storage in Geologic Formations. National Energy Technology Laboratory. http://www.netl.doe.gov/publications/factsheets/program/Prog065.pdf . April 2008. Accessed April 2010.

See more of this Session: Advances In CO2 Capture
See more of this Group/Topical: Engineering Sciences and Fundamentals

(134c) Performance Characterization of CO2 Capture Materials On Flue Gas Slip Streams

Feng Zheng, Kriston Brooks, Dale A. King, James Cabe and Paul Humble, Pacific Northwest National Laboratory, Richland, WA

A significant amount of U.S. research and development is currently focused on technologies for mitigating the rate of increase of atmospheric CO_2 concentration. The need for cost-effective separation of CO_2 from combustion exhaust is fundamental to achieving market penetration of CO_2 capture and storage technologies. These technologies will facilitate continued use of hydrocarbon energy resources such as coal and biomass while substantially reducing their contribution to atmospheric CO_2 concentration. Development and testing of CO_2 separation technologies and processes necessarily spans a broad range of scale requiring increasingly representative cycle configuration, gas composition, and process hardware throughout the development process. Accurately quantifying key material performance parameters and interactions as early as possible allows decisions on continued investment to be made with higher confidence and at lower cost. Flexible, continuous-scale test hardware and methods using actual flue gas streams play a key role after initial laboratory-scale batch testing and prior to large, costly engineering-scale testing.

This paper will present results of CO_2 capture material testing performed using an automated, dual-bed adsorption system for the evaluation of solid adsorbent materials. The mobile system was used to characterize material performance over a large number of loading and regeneration cycles in two configurations: Using bottled gases in a laboratory environment and with a flue gas slip stream from a 500 MW coal fired power plant. Baseline testing was performed on zeolite 13X using both thermal swing and vacuum swing regeneration. Automated adsorption and regeneration cycles, combined with an integrated data acquisition system, provide the ability to accumulate a large number of cycles and characterize material performance over long periods of operation. Performance trends and sensitivities will be discussed.

Having completed this baseline material testing, the test hardware and methods will be used to characterize performance of other sorbents with potential to reduce the incremental cost of electricity associated with flue gas CO_2 capture relative to zeolite 13X. A solvent test cart developed in parallel with the sorbent test cart will provide a similar characterization capability for liquid solvents.

See more of this Session: Advances In CO2 Capture
See more of this Group/Topical: Engineering Sciences and Fundamentals

Process Analysis of CO_2 capture with Ionic Liquid Solvent from Flue Gas

Y. J. Chen [a,b], Y. H. Yao[a,b], X. P. Zhang [a,*], S. J. Zhang[a], B. Z. Ren[b],

[a] *Institute of Process Engineering, Chinese Academy of Sciences, 100190, Beijing, China*

[b] *School of Chemical Engineering and Energy, Zhengzhou University, 450001,Zhengzhou, China*

Corresponding author: *xpzhang@home.ipe.ac.cn (X. P. Zhang)

Capturing CO_2 from the flue gas has been considered as one of the effective and possible options to eliminate the global warming effect now. As new solvents, ionic liquids have attracted great attentions by the academies and industries on account of their low volatility, high selectivity and high solubility. In this work , a rigorous simulation flow sheet including absorption and desorption units, and taking ionic liquids compound as solvent has been established which uses ELECNRTL as the global thermodynamic model, and the binary interaction parameters between ionic liquids and the other components in the gas mixture are regressed based on the experimental data (Zhao et al, 2010). Some key operation parameters, such as solvent ratio and reflux ratio, were determined by sensitivity analysis. The results showed that the new process could capture carbon dioxide from the flue gas as high as 95%; the energy of reboiler was estimated to be reduced more than 20% compared with the MEA method. Furthermore, a multi-objective optimization was proposed and performed considering both the energy consumption and environmental impact (green degree) of the new CO_2 capture process.

Acknowledgements

The authors sincerely appreciate the National Basic Research Program of China (973Program) (No.2009CB219907).General Program Youth of National Natural Science Foundation of China (Nos.20906096) for the financial support.

References

Zhao, Y. S., Zhang,X. P., Q., Dong, H. F, et al. (2010).Solubilities of gases in novel alcaminesionicliquid2-[2-hydroxyethyl(methyl)amino] ethanolchloride. *Fluid Phase Equilibria*., 302,60-64.

(134e) Application of Ionic Liquids In CO2 Capture

Suojiang Zhang, Institute of Process Engineering, Chinese Academy of Sciences, Beijing, China and Xiaochun Zhang, State Key Laboratory of Multiphase Complex Systems, Institute of Process Engineering, Chinese Academy of Sciences, Beijing, China

Recently, the capture of carbon dioxide (CO_2) has been paid much attention all over the world. Ionic liquids (ILs), as a new type of green medium and soft media materials, have been widely studied for its potential applications and prospective advantages. The unique properties of these liquids, and the ability to tailor properties by choice of cation, anion, and substituent, provide new opportunities to develop clean technology for CO_2 capture.

In this work, some research results about application of ILs in CO_2 capture are introduced, including ILs design and synthesis, engineering scaling up, process simulation and device design. **1) ILs design and synthesis**. Based on the database of ILs, and a quantitative relationship between multiscale structure and macro properties of ILs, a series of conventional (such as imidazolium-based ILs) and task-specific ILs (such as guanidinium-, alkanolamines- and amino acid based ILs), which are used to capture CO_2, were designed and synthesized. The physical/chemical mechanism between ILs and CO_2 were revealed. The good absorption capacities were obtained. **2) Engineering scaling up.** A set of novel equipment for in-situ measurement of the transport-reaction coupling performance in ILs was established based on the electric conductive and dynamic characteristics of ILs. The dynamic characteristics, such as formation, deformation, coalescence and movement of CO_2 and air bubbles in ionic liquids were directly observed, which provides scientific data for industrial scale-up of ILs-CO_2 systems. **3) Process simulation and device design.** System integration simulations were carried out for CO_2 capture process, and the energy conservation and economic were evaluated. Continuous absorption and desorption device for CO_2 capture process were developed.

See more of this Session: Advances In CO2 Capture
See more of this Group/Topical: Engineering Sciences and Fundamentals

(134f) Thermodynamics of Carbon Capture Through Reversible Ionic Liquids

Manish Talreja[1], Amy Rohan[2], Jackson Switzer[2], Mara Gonzlez Miquel[3], Swetha Sivaswamy[4], Kyle Flack[5], Elizabeth J. Biddinger[2], Manjusha Verma[6], Pamela Pollet[5], Francisco Rodriguez[3], Jose Palomar[7], Charles L. LIotta[8] and Charles Eckert[2], (1)Chemical and Biomolecular Engineering, Georgia Institute of Technology, Atlanta, GA, (2)School of Chemical & Biomolecular Engineering, Georgia Institute of Technology, Atlanta, GA, (3)Departamento de Ingeniera Qumica, Universidad Complutense de Madrid, Madrid, Spain, (4)Chemical and Biomolecular engineering, Georgia Institute of Technology, Atlanta, GA, (5)School of Chemistry and Biochemistry, Georgia Institute of Technology, Atlanta, GA, (6)Chemistry and Biochemistry, Georgia Institute of Technology, Atlanta, GA, (7)Seccin de Ingeniera Qumica (Departamento de Qumica Fsica Aplicada), Universidad Autnoma de Madrid,, Madrid, Spain, (8)Georgia Institute of Technology, Atlanta, GA

We have developed a novel class of reversible ionic liquids (RevIL) that offer numerous advantages over ethanolamine solvents for CO_2 capture applications. These novel RevIL are silylamine based compounds capable of dual mode of CO_2 capture through chemisorption (reaction of amine with CO_2 to form ammonium-carbamate ion pair) and physisorption (solubility of CO_2 in the ion pair.) The latter is particularly significant in the natural gas sweetening application. Subsequently, modest elevation in temperature reverses the reaction to regenerate the silylamine and CO_2 which can then be sent for sequestration. There are several key thermodynamic properties of the RevIL that have a significant impact on the economics of the carbon capture process. In this talk, we will discuss both theoretical and experimental approaches to determining these properties.

Thermodynamics of these systems is summed up by the following quantities of interest:

1. The reaction equilibrium constant K for the amine CO_2 reaction, which determines the extent of reaction.

2. Solubility of CO_2 in ionic liquid (Henry's Law Constant) and the associated enthalpy (Dsolution).

3. The heat required to raise the ionic liquid to the regeneration temperature: $Q = mC_pDp$, and (c) the regeneration temperature T_{reg} (Dreg - $T_{capture}$).

4. The enthalpy of reaction Drxn, which dictates the amount of heat required to regenerate the silylamine

For determining the first two properties (equilibrium constant and solubility), we have used *in situ* FTIR spectroscopy, while DSC/TGA is our primary tool for analyzing the latter two. We have also used COSMO-RS to supplement our experimental results. Based on these theoretical and experimental techniques, we will present a structure-property relationship based approach to finding a silylamine solvent with optimum set of thermodynamic properties. We will also make comparisons with non-silyl compounds to demonstrate the superiority of our compounds in CO_2 capture applications.

See more of this Session: Advances In CO2 Capture
See more of this Group/Topical: Engineering Sciences and Fundamentals

(134g) First Principles and Classical Simulations of Ionic Liquids for Carbon Dioxide Capture

Bo Zhang, US DOE-National Energy Technology Laboratory, Pittsburgh, PA, Adri C.T. van Duin, Department of Mechanical and Nuclear Engineering, Pennsylvania State University, University Park, PA and J. Karl Johnson, Dept. of Chemical and Petroleum Engineering, University of Pittsburgh, Pittsburgh, PA

Ionic liquids (IL) are promising materials for CO_2 capture due to their thermal and chemical stability. ILs have lower regeneration heat requirements compared with aqueous amine solutions because of their lower CO_2 absorption enthalpy (physical absorption). In addition, ILs have no measurable vapor pressure, and hence will not emit any volatile organic compounds. However, physisorption is not considered to be a viable approach, because the solubility and selectivity of CO_2 in ILs through physical absorption are too low at flue gas conditions. Functionalized ILs capable of specific reactions or interactions with CO_2 are designed to address the issues of solubility and selectivity at low partial pressures of CO_2. We are exploring ILs that can capture CO_2 through chemical interactions. Classical force fields can model the physical interaction between CO_2 and ILs fairly accurately, but they are not typically able to describe the bond breaking/forming events. Therefore, it is necessary to develop a new force field that can properly characterize the chemical interactions. ReaxFF is a very successful formalism for modeling chemical as well as physical interactions with an empirical approach. We are studying the IL $[P(C_4)_4][Gly]$ that has the potential to chemically react with CO_2. Our aim is to develop accurate ReaxFF parameters to allow large-scale simulations of the thermodynamics and transport properties of $CO_2/[P(C_4)_4][Gly]$ mixtures which likely involve chemical reactions. We use density functional theory (DFT) augmented with the van der Waals (vdW) interactions to explore the chemical and physical interactions between CO_2 and the functional group, $[Gly]^-$. Through first principles molecular dynamics, we have found there are different interaction sites for $CO_2/[Gly]^-$. One of these can be properly described by chemical binding, while another site could be classified as a quasi-chemical interaction. We also found that vdW interactions are important in determining the binding energies. We obtained several energy profiles along proposed reaction coordinates from our DFT calculations. We have also tested preliminary parameterizations of ReaxFF that are based on similar systems and have tested the validity of the parameterization against our DFT calculations.

See more of this Session: Advances In CO2 Capture
See more of this Group/Topical: Engineering Sciences and Fundamentals

(134h) Computational Design of Reactive Ionic Liquids for CO2 Capture

William F. Schneider, Bhabani Mallik and Chao Wu, Chemical and Biomolecular Engineering, University of Notre Dame, Notre Dame, IN

The physical characteristics and wide chemical tunability of ionic liquids make them an attractive alternative to aqueous amines for energy-related CO_2 separations. Most efforts to date to modify the solubility of CO_2 in ionic liquids have involved adding reactive amine functionalities to the anion or cation, with mixed success. Here we describe our recent work to use first-principles simulations to design a novel class of ionic liquids, aprotic heterocyclc anions, or "AHAs," that react selectively, reversibly, and controllably with CO_2. We examine trends in reactivity with AHA composition, and consider the electronics of AHA-CO2 binding that lead to the very wide range of accessible absorption energies and potentially interesting kinetic behavior. We consider the potential interferring effects of H_2O on CO_2 binding. Finally, we compare the CO2 chemistry of the anionic AHAs with isoelectronic, neutral carbenes, potentially suitable as absorbents in non-ionic-liquid applications. We demonstrate the feasibility of these heterocycles as a platform for a wide range of CO_2 separations applications.

See more of this Session: Advances In CO2 Capture
See more of this Group/Topical: Engineering Sciences and Fundamentals

(134i) Separation of CO2 and H2 Using Acetate and Amino Acid Ionic Liquids

Wei Shi, National Energy Technology Laboratory, U.S. Department of Energy, Pittsburgh, PA, David Luebke, US DOE/NETL, Pittsburgh, PA, Christina Myers, NETL/US DOE, Pittsburgh, PA and Dan C. Sorescu, Computational Chemistry, National Energy Technology Laboratory, Pittsburgh, PA

We will show theoretical and experimental results related to the use of acetate and amino acid ionic liquids to separate mixtures of H_2 and CO_2. Simulations show that H_2 solubility in [emim][CH_3COO] is very low, with Henry's law constant of about 10^4 bar and permeabilities in the range of 29-79 barrer at 313-373 K. Computational analysis also shows that CO_2 exhibits two absorption states in [emim][CH_3COO]. In State I, CO_2 molecules interact very strongly with the $[CH_3COO]^{-1}$ anion through strong complexation interactions leading to high CO_2 solubility. The C atom of CO_2 interacts with the O atoms of the $[CH3COO]^{-1}$ anion with an average distance of about 1.61 Å. CO_2 molecules in State I are significantly altered from isolated linear CO_2 molecules, with bond angles of about 138 degrees and C-O bond distances of 1.22 Å. Significant charge transfer occurs from the $[CH_3COO]^{-1}$ anion to CO_2 in State I, leading to 30-40% charge decrease on the $[CH_3COO]^{-1}$ anion. In State II, CO_2 interacts with [emim][CH_3COO] through simple van der Waals and electrostatic interactions, similar to CO_2 absorption in [emim][PF_6]. The predicted high CO_2 permeability and permeability selectivity for CO_2 over H_2 in [emim][CH_3COO] are verified by experiments. The experimental CO_2 permeability in [emim][CH_3COO] is in the range 1325-3701 barrer, and experimental CO_2/H_2 permeability selectivities of 21-37 at 313-373 K are observed. Simulations further demonstrate that $[emim]^{+1}$ cations compete with CO_2 to interact with the $[CH_3COO]^{-1}$ anion. Based on this observation, several possible cations are proposed to replace the $[emim]^{+1}$ cation and improve CO_2 solubility and permeability. The H_2 permeability in one of the proposed ILs is computed to be very low, suggesting that the proposed IL will exhibit high CO_2/H_2 separation performance.

Additionally, we have theoretically investigated CO_2 absorption mechanism in amino acid—based ionic liquids. One ionic liquid has been identified which is expected to show high CO_2 permeability and high permeability selectivity of CO_2 over H_2.

See more of this Session: Advances In CO2 Capture

(146a) Status of the ReaxFF Reactive Force Field Method and Applications to Combustion, Catalysis and Material Failure

Adri C.T. van Duin[1], Kaushik Joshi[2], Sung-Yup Kim[2] and Michael Russo[2], (1)Department of Mechanical and Nuclear Engineering, Pennsylvania State University, University Park, PA, (2) Mechanical and Nuclear Engineering, Pennsylvania State University, University Park, PA

The ReaxFFmethod provides a highly transferable simulation method for atomistic scalesimulations on chemical reactions at the nanosecond and nanometer scale. It combinesconcepts of bond-order based potentials with a polarizablecharge distribution.

Since it initial developmentfor hydrocarbons in 2001 [1], we have found this concept to be highlytransferable, leading to applications to elements all across the periodictable, including all first row elements, metals, ceramics and ionic materials(for example, [2-5]). For all these elements andassociated materials we have demonstrated that ReaxFFcan accurately reproduce quantum mechanics-based structures, reaction energiesand reaction barriers, enabling the method to predict reaction kinetics incomplicated, multi-material environments at a relatively modest computationalexpense.

In this presentation we willprovide an overview of recent developments of the ReaxFFmethod, including its availability in parallel simulation environments, andrecent application of this method to simulations on combustion, aqueous-phasereactions, catalysis and material failure.

[1] van Duin,A. C. T.; Dasgupta, S.; Lorant,F.; Goddard, W. A. Journal of Physical Chemistry A 2001, 105, 9396.

[2] Weismiller, M. R.; van Duin, A. C. T.; Lee, J.; Yetter,R. A. J. Phys. Chem. A 2010, 114, 5485.

[3] van Duin,A. C. T.; Bryantsev, V. S.; Diallo,M. S.; Goddard, W. A.; Rahaman, O.; Doren, D. J.; Raymand, D.; Hermansson, K. Journal of Physical Chemistry A 2010, 114,9507.

[4] Joshi, K.; van Duin, A. C. T.;Jacob, T. Journal of Materials Chemistry 2010, 20, 10431.

[5] Fogarty, J. C.; Aktulga, H.M.; Grama, A. Y.; van Duin,A. C. T.; Pandit, S. A. J. Chem. Phys. 2010, 132,174704/1.

See more of this Session: Combining Weak and Strong Forces: Reactive Forcefields and Van Der Waals Enabled-DFT

(146b) First-Principles Calculations of the Role of PVP In the Controlled Synthesis of Colloidal Ag Nanostructures

Kristen Fichthorn, Dept. of Chemical Engineering, Pennsylvania State University, University Park, PA, Wissam A. Al-Saidi, Dept.of Chemical and Petroleum Engineering, University of Pittsburgh, Pittsburgh, PA and Haijun Feng, School of Chemistry and Chemical Engineering, South China University of Technology, Guangzhou, China

We employ first-principles, density-functional theory (DFT) calculations to resolve the role of polyvinylpyrrolidone (PVP) in the controlled synthesis of colloidal Ag nanostructures. This polymer plays a central but poorly understood role in the selective formation of a large variety of metal nanostructures. One hypothesis is that PVP is an effective structure-directing agent because of its surface-sensitive binding. Examining the case of Ag, it has been proposed that PVP binds more strongly to Ag(100) than to Ag(111). To study this binding, we divide the PVP repeat unit into two logical submolecules: 2-pyrrolidone and ethane. While the interaction of ethane with Ag surfaces is almost purely van der Waals (vdW) in nature, the interaction of 2-pyrrolidone with Ag surfaces involves both direct bonding and vdW interactions. We compare two methods for including vdW interactions in DFT: Grimme's method and the Tkatchenko-Scheffler method and find the latter method to be the most suitable. We find that the binding of 2-pyrrolidone is stronger on Ag(100) than on Ag(111), consistent with experimental expectations. Our studies indicate that a rich interplay between direct bonding, vdW attraction, and Pauli repulsion is responsible for this surface sensitivity.

See more of this Session: Combining Weak and Strong Forces: Reactive Forcefields and Van Der Waals Enabled-DFT
See more of this Group/Topical: Computational Molecular Science and Engineering Forum

(146c) Molecular Model for Reactive Interactions of Water and Carbon Dioxide

Lukas Vlcek[1], Panchapakesan Ganesh[2], Paul R.C. Kent[2], Adri C.T. van Duin[3] and Ariel A. Chialvo[4], (1)Chemical Sciences Division, Oak Ridge National Laboratory, Oak Ridge, TN, (2)Oak Ridge National Laboratory, Oak Ridge, (3)Department of Mechanical and Nuclear Engineering, Pennsylvania State University, University Park, PA, (4)Chemical Science Division, Oak Ridge National Laboratory, Oak Ridge, TN

While the mixture of carbon dioxide with water has been a subject of countless experimental and theoretical investigations, explicit molecular models of the system are still limited to either non-reactive classical force fields or small-scale reactive systems studied by *ab initio* calculations [1]. Practical problems, however, often require simulations over extended time and length scales, such as for the study of dissolution and precipitation of carbonate minerals or of the effects of pH on surface reactions.

In the present study we explore an alternative to *ab initio* methods provided by a classical bond order-based reactive force field ReaxFF [2], and use it to capture the complex chemistry of aqueous CO_2. Using DFT *ab initio* methods, we first create a structure-energy database representing carbon dioxide reactions with water leading to the formation of carbonic acid as well as bicarbonate and carbonate ions, and then use the data to optimize ReaxFF parameters. In subsequent simulations, we test the capabilities and limits of the optimized model to reproduce available experimental data, such as the acidity constants of carbonic acid or rate constants for the main CO_2 + H_2O reactions. To describe reactions leading from molecular aqueous CO_2 to calcium carbonate, we also include calcium in our parameter optimization.

1) Nguyen, M. T.; Matus M. H.; Jackson V. E.; Ngan V T.; Rustad J. R.; Dixon D. A. J. Phys. Chem A 2008, 112, 10386.

2) van Duin, A. C. T.; Dasgupta, S.; Lorant, F.; Goddard, W. A. J. Phys. Chem. A 2001,105, 9396.

See more of this Session: Combining Weak and Strong Forces: Reactive Forcefields and Van Der Waals Enabled-DFT
See more of this Group/Topical: Computational Molecular Science and Engineering Forum

(146d) Noble Gases On Metal Surfaces: New Insights On Adsorption Site Preference

De-Li Chen[1], Wissam A. Al-Saidi[1] and J. Karl Johnson[2], (1)Dept.of Chemical and Petroleum Engineering, University of Pittsburgh, Pittsburgh, PA, (2)Dept. of Chemical and Petroleum Engineering, University of Pittsburgh, Pittsburgh, PA

Experiments have previously found that noble gases (Kr, Xe) adsorb on low-coordination atop sites on several different metal surfaces, rather than on high-coordination hollow sites. This unexpected preference for low-coordination sites has been previously ascribed to reduced Pauli repulsion, arising mostly from reduced exchange energy at the atop site. This explanation was based on density functional theory calculations within the local density approximation (LDA). In contrast, our calculations using he non-local van der Waals (vdW-DF) density functional show that site preference is due to a delicate balance between the electrostatics, which favor the hollow site, and kinetic energy which favors the atop site; exchange-correlation energies and, surprisingly, van der Waals interactions play a negligible role in site preference, although van der Waals interactions turn out to be critical in determining the magnitude of the adsorption energy and the correct equilibrium distance. Moreover, we find that the hollow site is a saddle point of index 1 or 2 (stationary points with 1 or 2 imaginary frequencies, respectively) on the 2-dimensional potential energy surface, while the atop site is the only true minimum. Therefore, the reason that hollow site occupation is not observed in experiments is that it is a transition state (see Figure 1) and so occupation of that site has a very short life-time. We show that this phenomenon holds for noble gas adsorption on close-packed Pt(111), Pd(111), Cu(111) surfaces, as well as the non-close-packed Cu(110) surface. Our results show that the inclusion of non-local vdW interactions is crucial for obtaining results in quantitative agreement with experiments for adsorption energies, equilibrium distances, and vibrational energies.

See more of this Session: Combining Weak and Strong Forces: Reactive Forcefields and Van Der Waals Enabled-DFT
See more of this Group/Topical: Computational Molecular Science and Engineering Forum

(146g) Screening and Design of Additives for Endothermic Fuels From First Principle Reactive Molecular Dynamics

Quan-De Wang[1], Xiao-Xiao Hua[2], Xue-Min Cheng[2], Juan-Qin Li[2] and Xiang-Yuan Li[2], (1)College of Chemistry, Sichuan University, Chengdu, China, (2)College of Chemical Engineering, Sichuan University, Chengdu, China

We report reactive molecular dynamics (MD) simulations on the high-temperature pyrolysis of *n*-decane and the pyrolysis of *n*-decane in presence of several additives employing the ReaxFF reactive force field. Apparent first-order kinetics is used to describe pyrolysis of *n*-decane, and the predicted apparent activation energy and pre-exponential factor show good agreement compared with experimental values. The thermal cracking of *n*-decane in presence of several additives is also investigated by using MD simulations employing the ReaxFF force field. The different chemical structures of the additives greatly affect the fitted apparent activation energy and pre-exponential factors. The selectivity of the main product distributions can be qualitatively described by ReaxFF reactive MD simulations. These results demonstrate that the ReaxFF can be used to unravel the complex reaction mechanisms of pyrolysis of hydrocarbons, and to screen and design of additives for hydrocarbon fuels for different purposes.

See more of this Session: Combining Weak and Strong Forces: Reactive Forcefields and Van Der Waals Enabled-DFT
See more of this Group/Topical: Computational Molecular Science and Engineering Forum

Beverly Brooks Hinojosa and Juan C. Nino, Department of Materials Science and Engineering, University of Florida, Gainesville, FL

Bismuth tri-iodide (BiI_3) is a semiconductor material for potential use as a room temperature gamma-ray detection, primarily due to its intermediate band gap, high density, and high effective atomic number. At room temperature, BiI_3 has a rhombohedral crystal structure (space group R-3, No. 148) with a layered structure of highly ionic Bi-I bonds in the layers and weak van der Waals bonding between layers. The band gap energy of the semiconductor material determines the number of electron-hole pairs generated under the radiation, and affects the energy resolution of the radiation detector. Despite the need to describe accurately the electronic structure of BiI_3, the experimentally and computationally determined band gap energy of BiI_3 has been reported with values ranging from 1.67 eV to 2.2 eV. More surprisingly, the band gap is overestimated by standard density functional theory (DFT), which is well known to underestimate band gaps. In this work, the Grimme's potential for van der Waals interactions is modified and incorporated within DFT software. The validity of the DFT-D3 model is discussed and comparisons of standard DFT and DFT-D3 results are made for various materials containing van der Waals interactions. The inclusion of van der Waals interactions on the predicted band gap in BiI_3 is also presented.

See more of this Session: Combining Weak and Strong Forces: Reactive Forcefields and Van Der Waals Enabled-DFT
See more of this Group/Topical: Computational Molecular Science and Engineering Forum

(147a) Density Functional Theory Research On Mechanisms and Rate-Determining States of Cyclohexanone Ammoximation Over Titanium Silicalite-1

Feng Xin and **Yanying QI**, School of Chemical Engineering and Technology, Tianjin University, Tianjin, China

Density functional theory research on mechanisms and rate-determiningstates of cyclohexanone ammoximation over Titanium Silicalite-1

Yanying Qi Feng Xin~

School ofChemical Engineering and Technology, Department of Chemical Engineering, TianjinUniversity, Tianjin 300072, China

Abstract

The concept of "rate-determining steps" hasbeen of great significance in deriving reaction kinetics from reactionmechanism for many decades. Nevertheless, more and more articles have got anagreement on identifying the rate-determining states more useful than therate-determining steps for both theoretical and practical researches. Therate-determining states provide the possibility of finding an appropriatecatalyst or improving the existed ones. Foran environmental friendly reaction of cyclohexanone ammoximation on titaniumsilicalite-1 (TS-1), the mechanisms should be confirmed from molecular simulation,and then its kinetics will be established by the rate-determining states andexperiments, which is useful for improving the catalyst performance.

In this paper, the pathways of catalyticammoximation of cyclohexanone in TS-1 and H_2O_2 system wassimulated in a cluster model of TS-1 containing a defect Ti site by using Dmol3 in Material Studios software. Densityfunctional theory (DFT) were utilized with exchange functional of Becke's 1988plus Lee-Yang-Parr's 1988 correlation energy function (BLYP), which was moresuitable to this system compared to other functions in Dmol3. The used atomicorbital basis sets were Double Numerical plus polarization (DNP), whichincluded a polarization p-function on all hydrogen atoms and a polarizationd-function on non-hydrogen atoms. All of the energies were calculated withzero-point energy (ZPE) correlation. For searching transition state, we chose themethod of complete linear synchronous transit and quadratic synchronous transit(complete LST/QST). For acquiring more accurate transition state, we used TSconfirmation tool, which began by approximating the Intrinsic Reaction Path (IRP)with QST and then performed subsequent refinements.

In the beginning,we compared our results of the formation of Ti-OOH with some experimental andcomputational data to correct the method and model. As a result, a novel activecenter was proposed, where TS-1 had Ti(1-OOH) defect active center with a ligand of NH_3.The transition states of all these reaction pathways were explored. A whole catalyticcycle was determined. In the imine mechanism, NH_3 and cyclohexanonereacted firstly to form an intermediate complex C_6H_{10}-OH-NH_2that was transformed to imine with a hydrogen transfer from the nitrogen to theoxygen. The formation of imine was rate-determining step in the above mechanism.While in the hydroxylamine mechanism, we compared the energies of differentreaction routes in the catalytic and non catalytic reactions of hydroxylamineand cyclohexanone. In the end, Campbell'sdegree of rate control was utilized to obtain the rate-determining transitionstates and intermediates.

Keywords:

degree of rate control£££

*Corresponding author. E-mail: xinf@tju.edu.cn

Tel: +86 22 27409533,   Fax: +86 22 27892359

See more of this Session: Computational Catalysis III
See more of this Group/Topical: Catalysis and Reaction Engineering Division

(147b) Understanding Hydrocarbon Reactivity In Zeolites with Molecular Dynamics Simulations

Paul M. Zimmerman[1], Martin Head-Gordon[1] and Alexis T. Bell[2], (1)Dept. of Chemistry, University of California at Berkeley, Berkeley, CA, (2)Department of Chemical Engineering, University of California Berkeley, Berkeley, CA

First principles simulations in zeolites reveal the principles of elementary catalytic reactions that are not captured by experiment. One such methodology for molecular dynamics simulations is quantum mechanical/molecular mechanical (QM/MM) simulation, which greatly improves computational efficiency. Although many such techniques exist, our method does so with little sacrifice in accuracy by systematic choice of MM parameters. This methodology is utilized to simulate short time kinetics involved in hydrocarbon conversions in zeolites. These trajectories can provide estimates of carbocation lifetimes, and we will describe dynamics studies of the tert-butyl carbocation in H-FER and H-MFI zeolites. The simulations provide estimates of the lifetime of this important intermediate, and these will be compared to lifetimes obtained from transition state theory.

See more of this Session: Computational Catalysis III
See more of this Group/Topical: Catalysis and Reaction Engineering Division

(147c) A Theoretical Investigation of the Methylation of Olefins by Surface Methyl Groups On Zeolites

Mark N. Mazar[1], Saleh Al Hashimi[2], Aditya Bhan[1] and Matteo Cococcioni[1], (1)Chemical Engineering and Materials Science, University of Minnesota, Minneapolis, MN, (2)Chemical Engineering, The Petroleum Institute, Abu Dhabi, United Arab Emirates

Emerging research in the conversion of methanol to hydrocarbons (MTH) over zeolitic acids suggests that observed catalytic rates and selectivity in the indirect "hydrocarbon pool" mechanism are primarily a result of olefin and arene methylation reaction sequences. In this research, cluster-based and periodic density functional theory studies were used to examine the kinetics and mechanism of olefin methylation by surface methyl groups on CHA, FER, BEA, MOR and MFI zeolite frameworks. The location of the Al atom was chosen such that the bridging O atom had access to the largest open volume of each framework. In each case, the transition state was found to consist of a nearly coplanar CH_3 species that was coordinated with both the original bridging O atom and the olefin suggesting that an sp^2 hybridized carbon species is involved in the transition state. Activation energies of 70, 86, and 98 kJ mol^{-1} for ethylene methylation were found for FER, CHA, and BEA, respectively; in each case, lower than that reported for a mechanism involving the co-adsorption of methanol and ethylene (104 kJ mol^{-1})(Svelle et al. *JACS* **2008**, *131*, 816-825). Observed trends in the activation energies and transition state geometries will be discussed with respect to the zeolite pore size and the local pore configurations.

See more of this Session: Computational Catalysis III
See more of this Group/Topical: Catalysis and Reaction Engineering Division

(147d) Density Functional Theory Study of Propane Ammoxidation Over Mo-V-Te-Nb-O M1 Catalyst

Junjun Yu[1], Muthukumar Kaliappan[1], Vadim V. Guliants[1] and Ye Xu[2], (1)Chemical Engineering and Energy and Materials Engineering, University of Cincinnati, Cincinnati, OH, (2)Center for Nanophase Materials Sciences, Oak Ridge National Laboratory, Oak Ridge, TN

The selective ammoxidation of propane into acrylonitrile catalyzed by the Mo-V-Te-Nb-O mixed metal oxide received considerable attention because of its environmentally friendly nature. This process consists of a series of steps involving propane oxidative dehydrogenation (ODH), ammonia and O_2 activation, and N-insertion into C_3 surface intermediates. However, the limited fundamental understanding of the catalyst has hindered the progress in improving their activity and selectivity required for commercial applications. In this study, density functional theory (DFT) calculations were performed to investigate the mechanism of propane ammoxidation over cluster models of the proposed selective and active sites present in the surface *ab* plane of the M1 phase, in order to understand the roles of the different metal cation species in the reaction. We have calculated the activation energy barriers for the oxidative dehydrogenation (ODH) of propane on vanadyl, molybdyl, and telluryl oxo groups (V=O, Mo=O, and Te=O). V^{5+}=O is more active than V^{4+}=O, in line with the current proposed reaction mechanism. However, Te=O is found to be the strongest adsorption site for H atoms and a significantly more active site for propane ODH than V=O regardless of the oxidation state of V. Radical C_3 species (isopropyl and allyl) energetically prefer to adsorb on V=O and Te=O. We also explored the adsorption of NH_x (x=0-3) species and the N-insertion steps into C_3 intermediates over proposed active sites present in the surface *ab* plane of Mo-V-Te-Nb-O M1 phase. The reactivity trends established for the constituent metal cations and the information about elementary reaction steps for propane ODH found in this study are highly useful for further improvement of the M1 catalyst.

See more of this Session: Computational Catalysis III
See more of this Group/Topical: Catalysis and Reaction Engineering Division

(147e) Active Sites for Olefin Metathesis On WO_3 Catalysts : A Density Functional Theory Study

Zhuo Cheng and Cynthia S. Lo, Department of Energy, Environmental and Chemical Engineering, Washington University, Saint Louis, MO

Industrial demand for polypropylene has spurred research on improved catalysts and mechanisms for propylene production. Recently, WO_3 was reported to exhibit high activity for the metathesis of ethylene and 2-butylene to form propylene. This study investigates active sites for this metathesis reaction on WO_3 surfaces, using first-principles calculations based on density functional theory. For the WO_3 orthorhombic crystal, the (001) surface has been found to be the most stable with sixfold coordinated tungsten sites W_{6c} and fivefold coordinated tungsten sites W_{5c}. The whole process is divided into two stages: 1. Initiation - the formation of active W-carbene species from the W-oxo precursor, and 2. Propagation - the metathesis reaction on these active sites to yield propylene. The oxidation states of W on the surface have a great influence on the adsorption of ethylene and 2-butylene. It is more favorable for W5c sites to adsorb ethylene while it is more favorable for W6c sites to adsorb 2-butylene, which were the compromising results of electrophilic and steric effects. The formations of the W-carbene species from ethylene and 2-butylene were found to follow different pathways. These study results reveal the relationship between WO_3 surface structure and metathesis activities, which is significant in guiding the preparation of more suitable WO3 catalysts for the production of propylene.

See more of this Session: Computational Catalysis III
See more of this Group/Topical: Catalysis and Reaction Engineering Division

(147f) Role of Ceria In Ti0.9Ce0.1O2 Materials for Adsorptive Desulfurization

Michael J. Janik[1], Siddharth Sitamraju[2], Jiahua Guo[2] and Chunshan Song[2], (1)Chemical Engineering, Pennsylvania State University, University Park, PA, (2)Energy and Mineral Engineering, Pennsylvania State University, University Park, PA

Fuel cell systems efficiently convert chemical energy to electrical energy. The reforming and water-gas shift catalysts used to produce hydrogen and the fuel cell electrodes are all sensitive to the sulfur compounds in the fuels. Conventional desulfurization methods, i.e. hydrodesulfurization (HDS), consume large amounts of energy to reach sulfur concentrations low enough for fuel cells. Adsorptive desulfurization (ADS) is a promising method being developed to produce ultra-low sulfur fuels. A $Ti_xCe_{1-x}O_2$ material has been reported by the Song group at Penn State to be a potential adsorbent for ADS due to its good selectivity and regenerability. The ratios of Ti:Ce were optimized to 9:1 for the maximum adsorption capacity. Comparing $Ti_{0.9}Ce_{0.1}O_2$ to TiO_2, the 10% CeO_2 addition significantly increases the capacity, despite only a slight surface area increase. We have applied density functional theory (DFT) calculations to examine the role of Ce addition to TiO_2 in promoting the adsorption of thiophenic molecules. A number of possible contributions of Ce addition were evaluated, including 1) stabilization of TiO_2 surfaces that adsorb thiophene strongly, 2) stronger adsorption to Ce-doped TiO_2 surfaces, 3) stabilization of reduced or over-oxidized sites on Ce-doped TiO_2, and 4) increased oxygen storage capacity to promote formation of adsorbed sulfone-like complexes. We find that Ce-doping of the surface can offer adsorption sites with stronger binding of thiophene, and can stabilize the anatase (100) surface which demonstrates strong adsorption affinity. Ceria may also serve as an oxygen reservoir to promote formation of adsorbed sulfone species to the TiO_2 surface. Thiophene adsorption energies are compared with cyclopentene to probe adsorption selectivity, and substitution of both thiophene and cyclopentene is considered to evaluate variations in selectivity. Calculations using DFT and dispersion corrected DFT, DFT+D, are compared.

See more of this Session: Computational Catalysis III
See more of this Group/Topical: Catalysis and Reaction Engineering Division

(147g) DFT Study of the Hg Oxidation Mechanism Over the V2O5-TiO2(001) Surface

Ana Suarez Negreira, Energy Resources Engineering, Stanford University, Stanford, CA and Jennifer Wilcox, Department of Energy Resources Engineering, Stanford University, Stanford, CA

We present a density functional theory study of Hg oxidation from the interaction between Hg and HCl with the V2O5(001) surface supported on TiO2(001). We investigate the intermediate products of the reaction and calculate the reaction energy barrier for each elementary reaction, taking into account several different adsorption geometries and intermediate products. We propose that Hg oxidation occurs through a Mars-van Krevelen type mechanism consisting of seven steps: adsorption and dissociation of HCl on a surface oxygen vacancy, combination of atomic chlorine with neighboring adsorbed Hg, combination of HgCl with adsorbed Cl, desorption of HgCl2, hydroxyl recombination, water desorption and dissociative oxygen adsorption. The rate-determining step in the reaction is calculated and the results are compared qualitatively to available experimental data.

See more of this Session: Computational Catalysis III
See more of this Group/Topical: Catalysis and Reaction Engineering Division

(183a) A New Theoretical Method for Rapid Prediction of Solvation Free Energy In Water

Shuangliang Zhao, UC Riverside, Riverside, CA and **Jianzhong Wu**, Chemical and Environmental Engineering, UC Riverside, Riverside, CA

We present a new theoretical method for rapid calculation of the solvation free energy in water by combining molecular simulation and the classical density functional theory (DFT). The DFT calculation is based on an accurate free-energy functional for water that incorporates the simulation results for long-range correlations and the fundamental measure theory for the molecular excluded-volume effects. The numerical performance of the theoretical method has been validated with simulation results and experimental data for the solvation free energies of halide (F^-, Cl^-, Br^- and I^-) and alkali (Li^+_-, $Na\text{--}^+$, K^+, Rb^+ and Cs^+) ions in water at ambient conditions. Because simulation is applied only to the particular thermodynamic condition of interest, the hybrid method is computationally much more efficient than conventional ways of solvation free energy calculations.

See more of this Session: Recent Advances In Molecular Simulation Methods III
See more of this Group/Topical: Computational Molecular Science and Engineering Forum

(183b) Solvent Dynamics In Ion Pair Dissociation

Ryan Gotchy Mullen[1], Joan-Emma Shea[2] and Baron Peters[1], (1)Chemical Engineering, University of California, Santa Barbara, CA, (2)Chemistry, University of California, Santa Barbara, CA

Having an accurate reaction coordinate provides mechanistic insight, facilitates rate constant calculations, and provides free energy landscapes that are kinetically meaningful. Reaction coordinates for reactions that involve the solvent are notoriously difficult to identify. The aqueous dissociation of NaCl would seem simple because there is only one solute coordinate, the ionic distance. However, many studies have shown that ionic distance is not a good coordinate. By using the aimless shooting version [1] of transition path sampling [2] and likelihood maximization [1], we have identified a reaction coordinate for this system. The coordinate combines a localized density of water molecules with the ionic distance. We also investigate the relationship between the transmission coefficient and reaction coordinate error. We discuss possible implications for simulations of charge recombination in photochemistry, ionic crystal growth by kink site insertion, and peptide aggregation.

1. B. Peters, G.T. Beckham, and B.L Trout, *J. Chem. Phys.* **127** 034109 (2007)
2. P.G. Bolhuis, D. Chandler, C. Dellago, and P.G. Geissler, Ann. Rev. Phys. Chem. **53**, 291 (2002)

See more of this Session: Recent Advances In Molecular Simulation Methods III
See more of this Group/Topical: Computational Molecular Science and Engineering Forum

(183c) Reducing the Statistical Error of Free Energy Calculations Through Functional Optimization of Pair Potential Pathways

Tri T. Pham, Department of Chemical Engineering, University of Virginia, Charlottesville, VA and Michael R. Shirts, Chemical Engineering, University of Virginia, Charlottesville, VA

Improving the efficiency of free energy calculations is important for many chemcial engineering and materials design applications. We show that for any pair potential, moderately accurate estimation of the radial distribution function for a solute molecule is sufficient to accurately estimate the statistical variance along a free energy pathway. This allows inexpensive analytical identification of low statistical error free energy pathways. We find that the computationally cheap two-body approximation for g(r) performs as well or better than 3D–RISM theory to identify low variance free energy pathways. We use this method to minimize the variance of solvation free energy over the entire parameter space of a generalized "soft core" potential. The free energy pathway arising from this optimization procedure reduces the total variance by at least 50% compared to thetraditional pathways. This free energy variance optimizationtechnique is generalizable to solvation in any homogeneous fluid and for any type of pairwise potential. We also compare this minimum pairwise statistical error to the theoretically lowest statistical error possible using less computationally tractable nonpairwise potentials.

See more of this Session: Recent Advances In Molecular Simulation Methods III
See more of this Group/Topical: Computational Molecular Science and Engineering Forum

(183d) Mechanisms and Free Energy Barriers for the Nucleation of Molecular Crystals From the Melt and From Solution

Erik E. Santiso, Manas R. Shah and Bernhardt L. Trout, Chemical Engineering, Massachusetts Institute of Technology, Cambridge, MA

Understanding crystallization is crucial for many applications in the food, chemical and pharmaceutical industries. Despite its widespread use, however, the design of crystalline materials is far from an easy task. With the exception of a few simple systems, the molecular-level details of the nucleation process are still poorly understood. On the one hand, the stochastic nature of nucleation, and the fact that it usually involves only a few molecules/ions, make it very hard to probe experimentally. On the other hand, the time scales involved are beyond the reach of traditional molecular simulation methods.

In this work, we combine a novel method to characterize the degree of order in molecular crystals [1] with the string method in collective variables [2] to obtain minimum free energy pathways for the nucleation of several systems from the melt and from solution. We then use milestoning with Voronoi tesselations [3] to obtain free energy barriers for nucleation. We compare some of the results from this approach with those obtained using transition path sampling. Finally, we show how this method can be extended to study aggregation in systems where the aggregates only exhibit partial ordering. These results are a first step toward a molecular-level understanding of nucleation in molecular systems.

[1] E.E. Santiso and B.L. Trout, *J. Chem. Phys.* **134**, 064109 (2011)

[2] L. Maragliano *et al.*, *J. Chem. Phys.* **125**, 024106 (2006)

[3] E. Vanden-Eijnden and M. Venturoli, *J. Chem. Phys.* **130** 194101 (2009)

See more of this Session: Recent Advances In Molecular Simulation Methods III
See more of this Group/Topical: Computational Molecular Science and Engineering Forum

(183e) Linking the COSMO-RS Model to Molecular Dynamics Simulations

Rolf E. Isele-Holder, Brooks D. Rabideau and Ahmed E. Ismail, Mechanical Engineering, RWTH Aachen University, Aachen, Germany

We describe how to link the results of the Conductor-like Screening Model for Real Solvents (COSMO-RS) model, a powerful tool for calculating thermodynamics equilibrium properties of organic compounds, with Molecular Dynamics (MD) simulations. In the COMSO-RS model, a liquid is regarded as a corresponding ensemble of pairwise-interacting surface segments whose charges are obtained from quantum mechanical COSMO calculations. A self-consistent solution of the partition function for calculating the chemical potential is enabled by applying the free segment approximation, in which a molecule's surface is divided into segments, which can be paired independently. Despite the loss of steric information of the system, COSMO-RS provides reliable results for small molecules, but is less reliable for more complex molecules.

To better understand the free segment approximation and interacting surface segments, various compounds widely used in chemical engineering, such as water, hexane, diethyl ether, acetone, toluene, and ethanol, were studied in MD simulations using the LAMMPS Molecular Dynamics Simulator and the OPLS-AA force field. By examining the liquid structure using weighted Voronoi tessellation, a convenient definition for surface contact in molecular simulations is established. According to the defined criteria, histograms describing the probability that the surfaces of certain atoms are in contact can be created. Comparing these results to histograms based on COSMO-RS calculations identifies a link between the two vastly different simulation approaches. This link will pave the way for future studies on systems in which COSMO-RS does not provide accurate results and should aid in identifying COMSO-RS's inadequacies as well as suggest possible improvements.

See more of this Session: Recent Advances In Molecular Simulation Methods III
See more of this Group/Topical: Computational Molecular Science and Engineering Forum

() Field Theoretic Simulations of the Interfacial Properties of Complex Coacervates

Robert Riggleman, Chemical and Biomolecular Engineering, University of Pennsylvania, Philadelphia, PA and Glenn H. Fredrickson, Chemical Engineering, University of California, Santa Barbara, Santa Barbara, CA

Many biological processes and emerging technologies, such as wet adhesives, biosensors, and responsive surfaces, rely on the association between oppositely charged polyelectrolytes. Such association is driven not only by the electrostatic interactions between the polyelectrolytes, but there is also a substantial entropy gain associated with counterion release upon complexation. In some cases, the association between oppositely charged polymers can lead to a solid precipitate while others can result in a fluid phase rich in polyelectrolytes (coacervate phase) coexisting with a polyelectrolyte-dilute solvent phase. For many of the applications seeking to exploit coacervation, characterization of the interface between the solvent phase and the coacervate is of paramount importance. In this talk, we will present the results of field-theoretic simulations for a coarse-grained polyelectrolyte model that exhibits complex coacervation. Our simulations sample the fully-fluctuating fields in three dimensions and provide a detailed characterization of the interface between the solvent and the coacervate phase for symmetric polyelectrolytes (where both the polycations and the polyanions carry identical charge densities) as a function of salt concentration and strength of the electrostatic fields. We will present two methods for calculating the interfacial tension from field-theoretic simulations, one based on the mechanical interpretation of the interfacial tension, and a second based on the thermodynamic interpretation. Both approaches agree quantitatively, and we will present scaling relationships between the surface tension and various parameters of our system.

See more of this Session: Recent Advances In Molecular Simulation Methods III
See more of this Group/Topical: Computational Molecular Science and Engineering Forum

(183g) The Stochastic Simulation of Isocyanate Amine Polymerization

Erdem Arslan, Computational Modeling Center, Air Products and Chemicals, Inc., Allentown, PA, Atteye H. Abdourazak, EHS&Q Functional Support, Air Products and Chemicals, Inc., Allentown, PA and Gamini A. Vedage, Chemicals R&D Technical Service, Air Products and Chemicals, Inc., Allentown, PA

The isocyanate amine reactions are used to produce various polyurea products which have important industrial uses. By varying isocyanate and amine types and ratios one could produce polyurea products which have range of functionalities. In this study, we developed a stochastic simulation algorithm to simulate isocynate amine polymerization reaction to predict molecular weight distribution of the final polyurea product. The developed algorithm accurately predicts the molecular weight distribution without requiring any experimental data and can help to synthesize new formulation by providing fast and accurate way of searching wide ranges of reactant ratios. The algorithm uses exact kinetic Monte Carlo simulation method where the reaction and the time between reaction events are sampled explicitly from calculated reaction probabilities. This method consider the time evalution of the species concentration as a random Markov Chain process where the propabilities of the reaction depends only the current state of the system. The new molecules are created and used-up molecules are destroyed dynamically so that large polymerization reaction network can be simulated efficiently both in terms of memory and computational time. We used the developed algorithm to simulate diamine/diisocyanate/monoamine polymerization. The linear polymerization, termination and cycle formation of dual functional molecules are included in the reaction network. The result of the algorithm showed good agreement with the matrix-assisted laser desorption/ionization (MALDI) measurements.

See more of this Session: Recent Advances In Molecular Simulation Methods III
See more of this Group/Topical: Computational Molecular Science and Engineering Forum

(190a) Molecular Modeling and Simulation for Closed-Loop Liquid-Liquid Immiscibility Behavior of Poly(ethylene oxide) In Aqueous Solution

Amir Vahid, The University of Akron, Akron, OH and **JRichard Elliott**, Chemical Engineering Dept, University of Akron, Akron, OH

The proposed force fields for poly(ethylene glycol) (PEG) or poly(ethylene oxide) (PEO) fail to predict the closed-loop liquid-liquid immiscibility diagrams in aqueous solutions. The proposed Step Potential Equilibria and Dynamics (SPEADMD/C) method implements the field biased discontinuous molecular dynamic and monte carlo (FBDMD/MC) technique using the Wertheim theory of hydrogen bonding for water and PEO. A new model for water has been proposed and the results have been compared to SAFT-VR molecular approach for different molecular weights. For calibration of our force field, various experimental data from macroscopic properties such as density and vapor pressure of glycols and glymes to microscopic properties such as radius of gyration used to obtain a universal transferable approach. Finally, the effect of hydrophobic and hydrophilic interactions on the excess entropy of mixing for PEO/water mixture has been evaluated.

See more of this Session: Poster Session: Computational Molecular Science and Engineering Forum
See more of this Group/Topical: Computational Molecular Science and Engineering Forum

(190aa) Polymorph Selection During the Crystallization of Iron Under the Conditions of Earth's Inner Core

John Persson, Jason M. Hicks, Caroline Desgranges and Jerome Delhommelle, Department of Chemistry, University of North Dakota, Grand Forks, ND

Using classical molecular dynamics simulations together with a many-body embedded atom model (EAM) potential, we study the crystallization of iron under conditions typical of the Earth's inner core. We simulate this process at a pressure of 330 GPa and at two different temperatures, 5680 K and 4970 K, corresponding to temperatures 20 % and 30 % below the melting point of pure iron, respectively. We show that, for both temperatures, crystal nucleation proceeds with the formation of a critical nucleus, in which all three main metallic structures, i.e. the body-centered cubic (BCC) polymorph, the hexagonal-close packed (HCP) polymorph and the face-centered cubic (FCC) polymorph, are present. While we find that the critical nucleus is predominantly BCC, we also identify a significant HCP signature and, to a lesser extent, a FCC signature. During the crystal growth at the higher temperature, we observe that the BCC fraction in the nucleus constantly rises at the expense of both close-packed structures, leading to post-critical nuclei largely dominated by the BCC polymorph. However, we show that the crystallization mechanisms and the polymorph selection process strongly depend on temperature since crystal growth at the lower temperature yields a highly polycrystalline, phase composed of blocks of the BCC, FCC and HCP structure. Depending on the actual temperature of the Earth's inner core, our results suggest that the Earth's inner core could be either predominantly BCC or a polycrystalline mixture of BCC, HCP and FCC.

See more of this Session: Poster Session: Computational Molecular Science and Engineering Forum
See more of this Group/Topical: Computational Molecular Science and Engineering Forum

(190ab) Predicting the Solubility of Pharmaceutical Solids by Molecular Simulation

Andrew S. Paluch, Department of Chemical and Biomolecular Engineering, University of Notre Dame, Notre Dame, IN and Edward J. Maginn, Chemical & Biomolecular Engineering, University of Notre Dame, Notre Dame, IN

Given the time and expense required for current rational drug design methods, molecular simulation holds a promising role as a drug design tool. We will present results of two unique approaches to predict the solubility limit of solids. The solubility limit of pharmaceutical solids is a critical thermodynamic property that is related to their bioavailability, and is also necessary for their synthesis and processing. In the first approach, free energy calculations are conducted for both the solid and solution phase of the solute to satisfy the phase co-existence criteria. In this manner, estimates of the solubility limit may be found in the complete absence of experimental data.

The second approach is a simplified method to estimate the solubility of sparingly soluble solids. The method assumes that the activity coefficient of the solute is relatively insensitive to the solute composition, resulting in a simplified thermodynamic model that involves the calculation of the residual chemical potential of a single solute molecule in solution at a finite concentration using an appropriate free energy simulation technique. With knowledge of a single experimental solubility data point, difficult simulations of the solid phase and use of analytical reference states are avoided.

Both methods have advantages over more empirical descriptor-based methods in that the simulations enable insight into the underlying molecular driving forces responsible for solubility trends. In this fashion, we will present simulation results that supplement solution theory to predict the existence of solubility enhancement in multi-component solvents. Results are presented for the solubility of paracetamol (acetaminophen) in pure and mixed solvents. We find that the methodology gives reasonable agreement with available experimental results, and is capable of predicting the existence of solubility enhancement in mixed solvents.

See more of this Session: Poster Session: Computational Molecular Science and Engineering Forum
See more of this Group/Topical: Computational Molecular Science and Engineering Forum

(190ac) Measurements of Solvents Properties for CO2 Capture Applications

Mohamed S. Elanany, Research and Development, Saudi Aramco Oil Company, Dhahran, Saudi Arabia

Post combustion capture technology using chemical solvents has the greatest potential to reduce CO_2 emissions in near future. A cost effective process is required to capture CO_2 from flue gas streams which have CO_2 partial pressure less than 15%. For achieving this goal, developing novel solvents that have outstanding performance in terms of absorption capacity, absorption rate, and regeneration energy is a crucial step. In the current study, Intelligent Gravimetric Analyzer (IGA) is used to investigate the absorption capacity of different solvents. Moreover, computational molecular modeling employing Density Functional Theory (DFT) is used to calculate the basicity of alkanolamine solvents. The computational method showed good agreement with the obtained experimental data.

See more of this Session: Poster Session: Computational Molecular Science and Engineering Forum
See more of this Group/Topical: Computational Molecular Science and Engineering Forum

(190ad) A Comparative Study On the Enhanced Sampling of Small Peptides

Michael A. Deighan and Jim Pfaendtner, Chemical Engineering, University of Washington, Seattle, WA

Molecular dynamics (MD) provides a useful way to investigate the underlying thermodynamics of a system that are difficult to resolve by experiment alone. However, MD as it stands is insufficient when the subject of interest involves large-scale conformational changes in soft-matter systems (e.g., proteins or polymers). Such events tend to occur on the order of microseconds or longer, a timescale far beyond the range of MD. In response to this, numerous enhanced sampling algorithms have been developed to effectively explore conformational space in a reasonable amount of time. Two common classes of methods for overcoming energy barriers are: 1) those that manipulate a system's energy by adding an external bias (e.g., umbrella sampling or metadynamics) and 2) those that periodically expose a system to temperatures above the melting point (e.g., parallel tempering).

The work we present focuses on the convergence of conformational sampling and thermodynamic energy landscapes of an explicitly solvated 20-residue tryptophan-cage protein (Trp-cage). Three types of simulations are compared: classical parallel tempering (PT) [1], parallel tempering and well-tempered metadynamics (PT-WTM) [2], and a new method that uses metadynamics within the well-tempered ensemble of Bonomi and Parrinello [3]. This new method allows for a broad temperature range to be sampled using many fewer replicas than PT or PT-WTM. Dubbed Well-Tempered Exchange (WTEx), not only does this approach significantly reduce computational cost – compared to the aforementioned methods – but it also enables the possibility of scaling to systems larger than what was previously considered practical for PT simulations.

As a proof of concept, initial simulations of an N-methylalanylacetamide (alanine dipeptide) toy system are presented. The underlying free-energy surface (FES) agrees well with umbrella sampling calculations and previous estimates. Following subsequent optimizations of the WTEx parameter space, we extend the method to larger systems of practical interest. The overall convergence and efficiencies of obtaining the FES for PT (100 replica), PT-WTM (100 replica), and WTEx (10 replica) are presented. Finally, the approach is used to examine the folding of a more complicated structure – bovine pancreatic trypsin inhibitor.

[1] Sugita & Okamoto, *Chem. Phys. Let.* **314**, 141-151 (1999)

[2] Barducci et al., *Biophys. Jour.* **98**, L44 (2010)

[3] Bonomi & Parrinello, *Phys. Rev. Let.* **104**, 190601 (2010)

(190ae) Substrate Binding Affinity and Proton Translocation Studies for Lactose Permease of E. Coli

Pushkar Y. Pendse, Department of Chemical and Biomolecular Engineering, University of Maryland, College Park, College Park, MD and Jeffery Klauda, Department of Chemical and Biomolecular Engineering, University of Maryland, College Park, MD

See more of this Session: Poster Session: Computational Molecular Science and Engineering Forum

See more of this Group/Topical: Computational Molecular Science and Engineering Forum

(190af) Homology Modeling and Molecular Dynamics Simulations of a Bi-Directional Membrane Bound Hydrogenase From R. Eutropha H16

Shawn A. Buckley, Chemical Engineering, University of Tennessee, Knoxville, TN and Paul Frymier, Chemical and Biomolecular Engineering, University of Tennessee, Knoxville, Knoxville, TN

Homology modeling was used to create a three dimensional structure for the bi-directional membrane bound hydrogenase from *R. eutropha.* H16, which has not been solved using any method. The crystal structure was based on the membrane bound hydrogenase from *A. Vinosum*; the only crystallized membrane bound hydrogenase to date. The predicted structure of the hydrogenase contains two subunits, HoxK with three iron-sulfur centers, and HoxG which contains the nickel iron reaction center. HoxK appears to contain two distinct domains. The first is the iron-sulfur cluster containing domain, which is involved in the transportation of electrons to the reaction center. The second domain is not well defined in the homology model. However, molecular dynamics simulations were used to predict that this domain is a transmembrane alpha helix. The predicted structures of the subunits of the hydrogenase agree reasonably well with experimentally determined structures, whit the exception of the C-terminal alpha helix on HoxK. This homology model will be later used in fusion models of hydrogenase from *R. eutropha* and photo system I from *Thermosynechococcus elongates*, to facilite the biological production of hydrogen.

See more of this Session: Poster Session: Computational Molecular Science and Engineering Forum
See more of this Group/Topical: Computational Molecular Science and Engineering Forum

(190ag) Functionalized Calixarene Based Surfaces for Highly Selective Protein Capture

Samrish Variyath, Chemical Engineering, University of Cincinnati, Cincinnati, OH and **Vadim Guliants**, Department of Chemical and Materials Engineering, University of Cincinnati, Cincinnati, OH

Dialysis related amyloidosis (DRA) is a significant disease associated with patients with chronic renal failure. Current hemodialysis techniques are unable to remove β2microglobulin (β2m) from the blood of dialysis patients, resulting in β2m aggregates and protein deposits in their joints that may lead to carpal tunnel syndrome, spondyloarthropathies, hemarthrosis, joint pain and immobility. We thereby propose the development of functionalized calixarene based surfaces that target specific areas, like charged amino acid clusters and cavities on the surface of proteins to enhance selective protein capture, using molecular dynamics (MD) simulations. Such surfaces can also find applications in pharmaceutical manufacturing, biosensor technology, medicine and immobilized enzyme technology

The molecular-level understanding of protein interactions with calixarene-functionalized surfaces is currently very limited. We therefore use MD as a cost effective way to determine optimal functionality, surface grafting strategy for calixarenes. We have modeled diverse calixarenes with functional groups targeting charged, polar uncharged and hydrophobic patches on β2m, lysozyme and cytochrome C. We rank these affinities in terms of Gibbs free energy using protein – ligand docking simulations. We confirm the predictions of docking simulations by performing all atom simulations of proteins with free functionalized calixarenes. Finally, we predict the selectivity of calixarene based surfaces from the results of the above simulations.

The interactions of the carboxylic acid groups on calixarenes with the ε-amino groups of lysine residues of cytochrome C have been previously demonstrated experimentally by others to be responsible for functionalized calixarene selectivity between cytochrome C and lysozyme. The electrostatic potential maps of the cytochrome C and lysozyme from our investigations have provided different charged amino acid clusters on these proteins, which otherwise have similar molecular weights and isoelectric points. We demonstrate that functionalized calixarenes provide a unique architecture for protein recognition and capture. Their central cavity provides for an extended contact with large target molecules, such as proteins. Their upper or lower rims can be functionalized to maximize the interactions with target molecules.

See more of this Session: Poster Session: Computational Molecular Science and Engineering Forum
See more of this Group/Topical: Computational Molecular Science and Engineering Forum

(190ah) Molecular and Coarse-Grained Analysis of Flap Motion In Lipase Enzymes

Patrick R. Burney and Jim Pfaendtner, Chemical Engineering, University of Washington, Seattle, WA

Lipase enzymes are a common component ofmost pathways for metabolizing lipids. They catalyze the breakdown of largelipid structures by means of hydrolyzing ester bonds. Their importance in this role explainswhy human lipases have been implicated in a variety of health conditions,including heart disease and diabetes. Industrial utilization of lipases(purified from various microorganisms) has predominantly been for foodproduction and in detergents, yet in the past decade for the production ofbiodiesel and other sustainable fuels from plant-derived substrates.

Despite their importance in bothtechnological application and human health, little is known about the dynamicsof lipases at the atomic level. Previous experimental investigations have showna common structural motif among lipases is a flap-like domain that protects theactive site from the solvent until in the enzyme contacts a solvent-lipidinterface. The *C. rugosa*lipase (CRL) structure has been resolved for both the flap-open and flap-closedstates. The two states have almost identical structure, except for the 26amino-acid flap, which differs by 17 Angstrom betweenthe open and closed states. This similarity has led to the conclusion that theflap-opening mechanism is a hinge-like motion. Given the importance of thisconformational change in regulating lipase activity, the molecular details of theflap mechanism are essential knowledge for engineering improved lipases. However, direct observation of this processis difficult or impossible via experiment alone. Therefore, we have used molecular andcoarse-grained simulations to investigate the behavior of CRL in explicitwater. The major challenge of usingclassical MD in investigating conformational rearrangement and rare events(e.g., flap opening/closing) lies in sufficient sampling. As we demonstrate, even microsecond-longsimulations of the all-atom (AA) system cannot provide enough sampling tocalculate the equilibrium probability distribution of this process. We use theAA simulation trajectories to develop coarse-grain (CG) models based on therecently developed ED-CG [1] method. The ED-CG models help identify dynamically similar domains within theopen and closed states. Finally, weuse the metadynamics [2] algorithm to calculate the free-energy landscape andidentify key structural transitions in the opening and closing process.

[1] Z. Zhang, L. Lu, W.G. Noid, V. Krishna, J. Pfaendtner,and G.A. Voth (2008). "A Systematic Methodology forDefining Course-Grained Sites in Large Biomolecules." Biophysical Journal95: 5073-5083.

[2] A. Laio and M. Parrinello(2002). "Escaping free-energy minima." PNAS 99(20): 12562-12566.

See more of this Session: Poster Session: Computational Molecular Science and Engineering Forum
See more of this Group/Topical: Computational Molecular Science and Engineering Forum

(190ai) A Molecular Simulation Study of the Adsorption of Polycyclic Aromatic Hydrocarbons and Ozone On Atmospheric Ice Films

Thilanga P. Liyana-Arachchi, Kalliat T Valsaraj and Francisco R. Hung, Cain Department of Chemical Engineering, Louisiana State University, Baton Rouge, LA

A molecular simulation study of the adsorption of polycyclic aromatic hydrocarbons and ozone on atmospheric ice films

Thilanga P. Liyana-Arachchi, Kalliat T. Valsaraj and Francisco R. Hung

Cain Department of Chemical Engineering, Louisiana State University, Baton Rouge, LA 70803

Polycyclic aromatic hydrocarbons (PAHs) consist of two or more carbon-hydrogen rings in which at least one ring has an aromatic structure. . PAHs are known to have important carcinogenic and mutagenic effects. Furthermore, these compounds can undergo photo chemically induced oxidation and nitration reactions with reactive oxygen species (ROSs) [e.g., ozone (O_3) and radicals such as singlet oxygen, hydroperoxy (HO_2), hydroxyl (OH) and nitrate (NO_3)]. Reactions of PAHs with ROSs produce oxy- and nitro-PAHs that are even more toxic than PAHs. PAHs can be adsorbed at the surfaces of water droplets, atmospheric aerosols, fog and mist, and ice and snow. This process consists of adsorption to the air/ice interface and dissolution in the bulk quasi-liquid layer (QLL). The processes taking place at atmospheric air/ice therefore have a profound impact on the fate and transport of PAHs and other trace gases in the atmosphere.

However, and despite their relevance, a fundamental understanding of the adsorption and heterogeneous reactions of PAHs and ROSs at air/ice interface is still lacking. Here we report a molecular simulation study where we attempt to elucidate molecular-level details of the adsorption mechanism of naphthalene, phenanthrene and ozone molecules, we also study the interactions between naphthalene and ozone molecules at the air/ice interface, as compared to similar processes at the air/water interface. From our results, Surface adsorption was found to be the predominant mechanism for the uptake of naphthalene, phenanthrene and ozone on ice films. Also, the favorability of surface adsorption against bulk phase dissolution increases as the temperature decrease for naphthalene and phenanthrene on ice. Naphthalene and phenanthrene was not observed to be incorporated in the structure of solid ice; the uptake of PAHs on ice therefore only consists of adsorption at the interface and dissolution in the bulk of the QLL. Also PAHs preferred to have a flat orientation on the air/ice interface. Analysis of the number of contacts between naphthalene and ozone molecules (which could lead to oxidation rate of the PAHs) at the air/ice interfaces indicate that, when the number of ozone molecules was held constant, the number of contacts showed a linear relationship to the number of naphthalene molecules for all systems. However, when the naphthalene concentration was held constant, for all systems we observed a linear relationship at low ozone concentrations and a plateau in the number of contacts is reached at high ozone concentrations. This is consistent with the reaction trends that were observed for experimental studies. We also discuss the structural and dynamical properties of the naphthalene films on air/ice surface as the naphthalene concentration varies, and how ozone adsorption affects these properties.

See more of this Session: Poster Session: Computational Molecular Science and Engineering Forum
See more of this Group/Topical: Computational Molecular Science and Engineering Forum

(190aj) Characterization of Bimetallic Electrocatalytic Nanoparticles On Boron-Doped Carbon Supports

Zhongtao Zhang and C. Heath Turner, Chemical and Biological Engineering, University of Alabama, Tuscaloosa, AL

Currently, Pt-M bimetallic alloy nanoparticles like PtMo, PtRu, PtW and PtNi have shown promisefor proton exchange membrane fuel cells (PEMFCs) applications, due to their high electrocatalytic activity and CO tolerance on pristine carbon supports. Meanwhile, our previous research has indicated that Pt nanoparticles are significantly stabilized on boron-doped carbon supports. In this work, we use computational tools to design new catalysts, which benefit from both good stability and CO tolerance, in order to improve the performance of fuel cell electrodes. The electronic structures, geometries, and interactions between Pt-M (M=Mo, Ru, W, and Ni) bimetallic alloy nanoparticles and pristine or B-doped single-wall carbon nanotubes (SWCNT) are investigated using density functional theory (DFT) calculations. Moreover, the adsorption energies and geometries of CO on SWCNT-supported (pristine and B-doped) and unsupported Pt are calculated by DFT to shed light on their potential use as electrocatalysts.

See more of this Session: Poster Session: Computational Molecular Science and Engineering Forum
See more of this Group/Topical: Computational Molecular Science and Engineering Forum

(190ak) Evaluation of Density Functional Theory (DFT+U) to Describe Doped Cerium Oxide Surfaces

Matthew D. Krcha[1], Micheal J. Janik[1] and Kerry.M. Dooley[2], (1)Chemical Engineering, Pennsylvania State University, University Park, PA, (2)Chemical Engineering, Louisiana State University, Baton Rouge, LA

Density functional theory (DFT) has trouble representing systems which contain cerium atoms, evidenced by improper delocalization of 4*f* electrons in reduced ceria. The use of DFT+U, with the Hubbard U correction term, localizes the 4*f* electrons in systems which contain reduced ceria. Though, this correction has been widely applied to investigate ceria surface chemistry, producing results in agreement with experiment. However, the applicability to metal-doped ceria surface, with a surface Ce atom replaced with another metal, has not been established. Here, we investigate the electronic structure of doped ceria systems, specifically investigating whether ceria or the doped atoms reduce upon oxygen vacancy formation or hydrocarbon adsorption. To determine which atoms reduce a number of different techniques can be used including Bader charge analysis, site projected atomic charges, density and partial density of states plots, and orbital imaging. We investigate a series of transition metal dopants, and report periodic trends in reducibility of doped ceria surfaces. We also investigate whether dopants serve as a reduction center or alter the reducibility of nearby ceria atoms, and test the sensitivity of surface reducibility to the U parameter. Comparisons are made with X-ray adsorption spectroscopy experiments.

See more of this Session: Poster Session: Computational Molecular Science and Engineering Forum
See more of this Group/Topical: Computational Molecular Science and Engineering Forum

(190al) Parameterization of ReaxFF Force Field for Pd-Ceria From First Principles DFT Calculations

Thomas Patrick Senftle, Department of Chemical Engineering, Penn State University, State College, PA, Adri C.T. van Duin, Department of Mechanical and Nuclear Engineering, Pennsylvania State University, University Park, PA and Michael Janik, Department of Chemical Engineering, Pennsylvania State University-University Park, University Park, PA

Recent Density Functional Theory (DFT) studies of Pd/ceria have demonstrated the stability and activity of the Pd-ceria system. However, the computational intensity inherent to quantum-level calculations prevents DFT studies from reaching the necessary length and time scales required to characterize the structural rearrangement and reaction dynamics of the catalyst surface. For this reason, a Reactive Force Field (ReaxFF) model, parameterized with quantum mechanical data obtained from DFT, that is capable of reaching the required length and time scales will be used to study the dynamic restructuring of the catalyst surface during reaction. DFT calculations are performed to identify the optimal structure and energy of various Pd/O/C/H species. Climbing Image-Nudged Elastic Band (CI-NEB) calculations are used to identify transition states and activation energies. The resultant DFT data is then used to train the ReaxFF force field for use in reactive molecular dynamics (RMD) simulations of the Pd-ceria catalyst.

See more of this Session: Poster Session: Computational Molecular Science and Engineering Forum
See more of this Group/Topical: Computational Molecular Science and Engineering Forum

(190an) Determination of Vapor-Liquid Equilibria Using Hybrid Monte Carlo Wang-Landau Simulations In the Isothermal-Isobaric Ensemble

Tsvetan Aleksandrov, Erica A. Kastl, Caroline Desgranges and Jerome Delhommelle, Department of Chemistry, University of North Dakota, Grand Forks, ND

Because of the large temperatures and pressures involved, the experimental determination of the vapor-liquid equilibria and of the critical properties of metals is fraught with difficulties. We show in this work how we determine these properties for a metal using Hybrid Monte Carlo Wang-Landau simulations in the isothermal-isobaric ensemble on the example of copper. We use a many-body potential, known as the quantum corrected Sutton-Chen embedded atom model, to model the interactions between Cu atoms. We also discuss recent applications of this method to the determination of the vapor-liquid equilibria of molecular fluids, such as e.g. branched alkanes.

See more of this Session: Poster Session: Computational Molecular Science and Engineering Forum
See more of this Group/Topical: Computational Molecular Science and Engineering Forum

(190ao) Phase Equilibria of Polyaromatic Hydrocarbons by Hybrid Monte Carlo Wang-Landau Simulations

Jason M. Hicks, Andrew Magness, Caroline Desgranges and Jerome Delhommelle, Department of Chemistry, University of North Dakota, Grand Forks, ND

Using a combination of Hybrid Monte Carlo simulations in the isothermal-isobaric ensemble with a Wang-Landau sampling, we parametrize a force field for polyaromatic hydrocarbons (PAHs). The proposed force field gives an accurate description of the vapour-liquid equilibria of naphthalene, phenanthrene and anthracene. In particular, the model yields a better account of the dependence of the vapour pressure on temperature than existing models.

See more of this Session: Poster Session: Computational Molecular Science and Engineering Forum
See more of this Group/Topical: Computational Molecular Science and Engineering Forum

(190ap) Computational Screening of Metal-Organic Frameworks: Application for Flue Gas Separation

Taku Watanabe and David S. Sholl, School of Chemical and Biomolecular Engineering, Georgia Institute of Technology, Atlanta, GA

Metal-organic frameworks (MOFs) have been extensively studied over the last decade especially for gas adsorption and separation applications. Due to the intense research effort, there are more than 5000 MOFs known today and material selection has become a daunting task. We have applied multi-level computational screening process to identify promising candidate MOFs to be used in a mixed-matrix-membrane. Of all the structures extracted from the database, we examined over 300 structures with detailed molecular simulations. We found some MOFs with exceptionally high diffusion selectivities (>100). Adsorption isotherms were also computed to obtain the sorption selectivity of CO_2 and N_2. Finally the ideal selectivities were calculated, and materials were ranked based on their performance.

See more of this Session: Poster Session: Computational Molecular Science and Engineering Forum
See more of this Group/Topical: Computational Molecular Science and Engineering Forum

(190aq) Accelerating Simulations In the Gibbs and Canonical Ensembles with GPUs

Jason R. Mick, Chemical Engineering and Materials Science, Wayne State University, Detroit, MI and Jeffrey J. Potoff, Chemcial Engineering and Materials Science, Wayne State University, Detroit, MI

The use of molecular simulation to study complex physical phenomena at the atomic level has grown exponentially over the last decade with increasing CPU power and the development of parallel molecular dynamics codes that scale efficiently over thousands of processors. As a result, the simulation of 100,000 atom systems has become routine, state of the art calculations are now possible on systems of over 1 million atoms, examples of which include the tobacco mosaic virus[1] and the ribosome[2]. Despite our current ability to simulate such systems, there is a demand to simulate increasingly larger systems for longer timescales.

Graphics processors (GPUs) have begun to play a crucial role in extending molecular simulation to even larger length and time scales [4-6]. In comparison to CPUs, GPUs possess a highly parallel structure allowing for more effective processing of large data sets. For this reason, intrinsic parallel traits prevalent in particle/molecular simulations may be exploited for more efficient computation on the GPU [7,8]. Each multiprocessor in the CUDA programming model has the ability of executing a significant number of threads in parallel. These threads, organized into individual blocks, work cooperatively to efficiently share and synchronize operations.

In this work a GPU-accelerated Monte Carlo simulation engine, capable of performing simulations in the canonical, isobaric-isothermal and Gibbs ensembles[6], is presented. The code uses a modular architecture and is written in C++. Sections of the code designed to run on the GPU, such as the calculation of pairwise interactions, utilize the Compute Unified Device Architecture (CUDA) from NVIDIA Corp. In comparison to optimized serial, CPU-bound code, our GPU-accelerated Monte Carlo simulation engine provides an order of magnitude reduction in wall-clock time for a system containing 65,536 particles interacting via Lennard-Jones potentials. Similar performance increases are observed for simulations in the Gibbs ensemble.

1. Freddolino, P.L., et al., *Molecular dynamics simulations of the complete satellite tobacco mosaic virus.* Structure, 2006. **14**(3): p. 437-49.

2. Sanbonmatsu, K.Y., S. Joseph, and C.S. Tung, *Simulating movement of tRNA into the ribosome during decoding.* Proc Natl Acad Sci U S A, 2005. **102**(44): p. 15854-9.

3. Stone, J.E., et al. *Accelerating molecular modeling applications with graphics processors.* Journal of Computational Chemistry, 2007. **28**(16): p. 2618-40.

4. *AMBER 11 NVIDIA GPU ACCELERATION SUPPORT*. http://ambermd.org/gpus/ (accessed April 29, 2011)

5. Brown, W.M., *Porting LAMMPS to GPUs*. SOS 14 Conference.

6. Panagiotopoulos, A.Z., *Direct determination of phase coexistence properties of fluids by Monte Carlo simulation in a new ensemble*. Molecular Physics. 1987, **61**(4):p. 813-26.

7. Anderson, J.A., et al., *General purpose molecular dynamics simulations fully implemented on graphics processing units*. Journal of Computational Physics, 2008. **227**(10): p. 5342-59.

8. Liu, W., et al., *Molecular dynamics simulations on commodity GPUs with CUDA*. Proceedings of the 14th International Conference on High Performance Computing, 2007. p. 185-96.

See more of this Session: Poster Session: Computational Molecular Science and Engineering Forum
See more of this Group/Topical: Computational Molecular Science and Engineering Forum

(190ar) Nanoarchitecture Conformational Deformities, Stability and Energy Predictions: Supramolecular Deoxyribose Nucleic Acid and Functionalized Carbon Nanotube In Silico Conjugates

Riaz A. Khan, Medicinal Chemistry Department, Qassim University, Qassim, Saudi Arabia

Nanoarchitecture Conformational Deformities, Stability and Energy Predictions: Supramolecular Deoxyribose Nucleic Acid and Functionalized Carbon Nanotube *in silico* Conjugates

Riaz A. Khan*

Department of Medicinal Chemistry, College of Pharmacy, Qassim University, Qassim 51452, KSA

Abstracts: Functionalized single-walled carbon nanotube (f-CNT) and supramolecular deoxyribose nucleic acid (DNA) conjugate models were *in silico* developed using molecular modeling software programs. The physico-chemical characteristics and conjugate geometries for different CNT-DNA conjugates with varying masses and componential ratios in empirical and non-empirical settings as well as binding patterns and feasibility of conjugation at different sites were evaluated. The physico-chemical properties, configurational deformity in conjugate geometries and their congenial components, stability of specific conjugate in terms of energy levels and their hydrophilicity were also predicted. The characteristics of CNT-DNA conjugates with componential ratios of 1:2 were found optimal with molecular weight ratios ranging between *single* f-CNT of 2300 amu with two units of DNA of either 75 bps or, of 150 bps length, later for heavy uploading as molecular transporter and other similar engagements for CNT-DNA entity in a bio-medium environment.

Keywords: Functionalized carbon nanotubes (f-CNT), CNT-DNA Conjugate, Molecular modeling, Energy Minimizations, Conformational Deformities, Molecular Transporters, Bio-vehicle.

See more of this Session: Poster Session: Computational Molecular Science and Engineering Forum
See more of this Group/Topical: Computational Molecular Science and Engineering Forum

(190b) EOMIP-CCSD for Ground and Excited States Energies of Open-Shell Systems with Spin-Orbital Coupling

Zheyan Tu and **Fan Wang**, School of Chemistry, Sichuan University, Chengdu, China

A novel two-component Equation-Of-Motion CCSD method for Ionization-Potential (EOMIP) based on a previously developed CC method with spin-orbital coupling (SOC)[1-4] included in post-SCF part based on CCSD is implemented in the CFOUR program package[5]. This approach could be use to calculate both ground and some excited states energies for open-shell systems with SOC. The EOMIP-CC method is computationally more efficient than methods that include SOC in SCF part[6,7] and free from spin contaminant due to the closed-shell nature of reference states. SOC in the present approach is introduced either in the CC part or in the EOMIP part. The later implementation is particularly efficient since the computational effort for the EOMIP part is much smaller than that of the CCSD calculation without SOC. This approach has been used to calculate some open-shell systems. The splitting value of lowest ^2P term for Kr^+, Xe^+, Rn^+, Br, I and At, as well as the splitting value of lowest ^2D term for Zn^+, Cd^+, Hg^+, Cu, Ag, and Au. Meanwhile, the ionization energy contributed from s^2p^6 to s^2p^5 for Kr, Xe, Rn and the ionization energy contributed from $d^{10}s^2$ to d^9s^2 for Zn, Cd, Hg are calculated. Taking I_2^+ as a typical molecule, the equilibrium bond length and harmonic frequency for five lowest state of $^2\Pi_{g,3/2}$, $^2\Pi_{g,1/2}$, $^2\Pi_{u,3/2}$, $^2\Pi_{u,1/2}$ and $^2\Sigma_g^+$ are also calculated. The computational results show that the method with SOC included in EOMIP can provide reliable results for atoms and molecules containing even sixth row elements.

[1] F. Wang, J. Gauss and C. van Wüllen, *J. Chem. Phys.*, 129, 064113(2008)

[2] F. Wang and J. Gauss, *J. Chem. Phys.*, 129, 174110(2008)

[3] F. Wang and J. Gauss, *J. Chem. Phys.*, 131, 164113(2009)

[4] Z. Y. Tu, F. Wang and J. Gauss, unpublished

[5] http://slater.chemie.uni-mainz.de/cfour/

[6] L. Visscher, E. Eliav and U. Kaldor, *J. Chem. Phys.*, 115, 9720(2001)

[7] S. Hirata, T. Yana, R. J. Harrison, M. Kamiya, and P. D. Fan, *J. Chem. Phys.*, 126, 024104(2007)

(190c) Recovering Single-Molecule Energetics and Kinetics From Force Spectroscopy and Steered Simulations

Gaurav Arya, Arijit Maitra and Dario Meluzzi, Department of Nanoengineering, University of California, San Diego, La Jolla, CA

Understanding how biological molecules and polymers fold into complex 3D structures, bind to each other, and undergo conformational transitions is important from the point of view of designing drugs, dissecting disease mechanisms, designing sensors, and DNA sequencing. Dynamic single molecule force spectroscopy provides a powerful approach for probing the underlying energy landscape governing such molecular processes. These sophisticated experiments operate by imposing gradually increasing forces on the molecular system being probed and recording its force-extension behavior until eventual rupture. An outstanding question in this field is how to recover the intrinsic energy landscape of the molecular system from such force measurements. In this talk I will describe the development of new theoretical models for extracting the height and location of activation energy barriers and intrinsic transition rates from single-molecule force measurements [1,2]. The models go beyond the current state-of-the-art by accounting for both the finite stiffness of the pulling device and the non-linear stretching of the polymeric handles often used for connecting the molecule of interest to the device. I will also discuss our recent efforts in combining such theoretical models with steered molecular dynamics simulations for computing the stability of secondary structures in proteins and the binding free energies of drugs molecules [3].

[1] Maitra and Arya, "Model accounting for the effects of pulling-device stiffness in the analyses of single-molecule force measurements," *Phys. Rev. Lett.*, 104, 108301, 2010

[2] Maitra and Arya, "Influence of pulling handles and device stiffness in single-molecule force spectroscopy," *Phys. Chem. Chem. Phys.*, 13, 1836, 2011

[3] Meluzzi, Maitra, and Arya, in preparation.

See more of this Session: Poster Session: Computational Molecular Science and Engineering Forum
See more of this Group/Topical: Computational Molecular Science and Engineering Forum

(190d) Accurate Prediction of the n-Hexane/Water and 1-Octanol/Water Partition Coefficients of Complex Chemicals Using Molecular Simulation

Ioannis G. Economou[1], Nuno M. Garrido[2], Antnio J. Queimada[2], Miguel Jorge[2] and Eugenia A. Macedo[3], (1)Department of Chemical Engineering, The Petroleum Institute, Abu Dhabi, PO Box 2533, United Arab Emirates, (2)Departamento de Engenharia Qumica, Universidade do Porto, Porto, Portugal, (3)Departamento de Engenharia Qumica, Faculdade de Engenharia da Universidade do Porto, Porto, Portugal

In recent years molecular simulation has emerged as a useful tool to predict physical properties of complex chemical systems. A methodology to estimate the *n*-hexane / water and 1-octanol / water partition coefficients of environmentally relevant solutes, namely substituted alkyl-aromatic molecules, chlorobenzenes, polychlorinated biphenyls (PCBs) and polychlorinated diphenyl ethers (PCDEs) using molecular simulation is elucidated here. The partition coefficients are calculated based on the absolute solvation Gibbs energies in each phase which are estimated from molecular dynamics simulations employing the thermodynamic integration approach. Very encouraging results, with average absolute deviations of 0.4 log P units are presented. Consequently, our molecular-based approach with a strong physical background can provide reliable prediction of the partition coefficients in different solvent pairs without the *a priori* knowledge of experimental data.

See more of this Session: Poster Session: Computational Molecular Science and Engineering Forum
See more of this Group/Topical: Computational Molecular Science and Engineering Forum

(190e) Obtaining Protein Solvent Accessible Surface Area (SASA) When Structural Data Is Unavailable Using Osmotic Pressure

Devin W. McBride and Victor G. J. Rodgers, Bioengineering, University of California, Riverside, Riverside, CA

A major problem in analyzing protein complexes is the inability to know the protein solvent accessible surface area (SASA) when their structures are unavailable. Here we provide an algorithm, via the free-solvent osmotic pressure model, that predicts SASA using only concentrated solution osmotic pressure data. Sheep hemoglobin monomer and bovine β-lactoglobulin are used to verify this method. In addition, the SASA for the structurally-unknown calf lens α-crystallin aggregate is predicted. Using osmotic pressure data, the predicted SASA value for sheep hemoglobin, 22,398 ± 1,244 $Å^2$, was in excellent agreement with the computational model predictions from the x-ray diffraction structure (24,304 $Å^2$-26,100 $Å^2$). Similarly, the SASA values for bovine β-lactoglobulin, determined from osmotic pressure data of pH solutions 5.1, 6.0, and 8.0, were 5,765 ± 1,031 $Å^2$, 6,656 ± 1,082 $Å^2$, and 9,141 ± 1,060 $Å^2$, respectively, were in good agreement with the computationally determined SASA value (7,500 $Å^2$-8,628 $Å^2$). In addition, the SASA for the aggregate of calf lens α-crystallin (800 kDa) was found to be 417,691 ± 16,790 $Å^2$. These results illustrate that this novel method can provide an important experimental alternative in estimating SASA for proteins and, possibly, their complexes in solution.

See more of this Session: Poster Session: Computational Molecular Science and Engineering Forum
See more of this Group/Topical: Computational Molecular Science and Engineering Forum

(190g) Development of *N*-Functionalized Imidazoles for Industrial Gas Capture

Jerrad A. Thomas[1], **Zhongtao Zhang**[1], Matthew S. Shannon[2], Jason E. Bara[2] and C. Heath Turner[1], (1) Chemical and Biological Engineering, University of Alabama, Tuscaloosa, AL, (2)Chemical & Biological Engineering, University of Alabama, Tuscaloosa, AL

The efficient capture of CO2 from industrial power generation sources is a major challenge. Current studies are investigating the adsorption of CO2 into ionic liquids (ILs). Imidazolium-based ILs have received a great deal of attention for CO2 capture due to their unique physical properties. However, it has recently been reported that imidazoles – neutral counterparts to imidazolium-based ILs – can offer enhancements in viscosity, cost, and CO2 capacity. However, the lack of thermophysical property data for neutral imidazole substitutions and chemical permutations makes it difficult to identify and develop future candidates for specific gas processing applications without having to obtain extensive experimental datasets. Using *ab initio* and molecular mechanics methods, we elucidate thermodynamic and structural properties of a variety of *N*-functionalized imidazole compounds. *Ab initio* calculations are used to determine parameters (such as partial charges) for the all-atom Optimized Potentials for Liquid Simulations (OPLS-AA) force field, followed by simulations using the molecular dynamics package Gromacs. We make comparisons with our experimentally-synthesized compounds, providing guidance towards new synthesis targets.

See more of this Session: Poster Session: Computational Molecular Science and Engineering Forum
See more of this Group/Topical: Computational Molecular Science and Engineering Forum

(190h) Water Molecule Behaviors In TiO2 Nano-Slits with Heterogeneous Modification by Carbon

Ming-Jie Wei, Xiaohua Lu and Linghong Lu, State Key Laboratory of Materials-oriented Chemical Engineering, Nanjing University of Technology, Nanjing, China

It is well known that Titanium dioxide (TiO_2) is biocompatible and environmentally friendly. Consequently, TiO_2 is widely applied in many fields, such as implant materials, photocatalysis, pigment, even cosmetic additive, etc. Recently, TiO_2 is used as catalyst support in industrial catalysis, because of its activation on catalyst deposit and chemical reaction., For the concentration of reaction and product near catalysts will break the reaction equilibria, the transport of reactants (absorption of reactant and desorption of product) is also critical to the circle of whole catalysis, besides the reaction part. A kind of mesoporous TiO_2, synthesized by our group [1], is used to be catalyst support for desulfurization from gasoline. The conversion of organic sulfur into hydrogen sulfide was always unsatisfactory until the surface of TiO_2 modified by carbon heterogeneously. Moreover, our theoretical study [2] on mass transport at solid-liquid interface shows that the transport coefficient is the key point in many processes like dissolution and sorption. Since our previous work of simulation approves that water molecule behaviors on carbon and TiO_2 surfaces are widely divergent [3], we preferred to simulating the behaviors of water molecules on heterogeneous carbon- TiO_2 surface.

In this work, we study the static structure properties of water confined to nano-slits, made by slab geometry of rutile (110) covered by graphite (0001). The model contains two slabs of rutile(110), covered by a piece of graphite(0001), opposite to each other with water molecules between them, and their distance is ranging from 0.4 to 2.0 nm by NVT-ensemble (T=300K) molecular dynamics(MD) simulations. In the models with coverage of carbon being 7%, the 1×1 nm^2 piece of graphite(0001) is in the center of the rutile (110) surface(about 3.6×3.9 nm^2), while the one being 93%, 1×1 nm^2 carbon atoms is dug from the whole one. The 0% and 100% model is no graphite covered and wholly graphite covered, respectively.

We are interested in diffusion of the water molecules in the nano-slits. Results from MD simulations showed that the covering of carbon indeed increased the diffusion of water molecules, especially at the center of the slits, and the one at slits center is increased while the coverage enlarged. However, the diffusion near surfaces did not follow the rules above. Due to our previous work, the molecules of the first layer near the rutile (110) surface are immobile. When carbon covers the TiO_2 surface, the water molecules near surfaces show an interesting result. At the coverage of 7%, the water of first layer near the surface is a little mobile, particularly near the edge of carbon piece. On the other side, at the coverage of 93%, the one near the surface is still immobile, and the diffusion of second layer the water is reduced. The results conclude that the edge of the carbon covering could enhance desorption of water from TiO_2 surfaces. Hence, the small amount coverage and well dispersed of carbon would be the better choice. Such conclusion could also be grafted into many oxides models.

References

[1] M. He, X. H. Lu, X. Feng, L. Yu and Z. H. Yang, A simple approach to mesoporous fibrous titania from potassium dititanate, Chem. Commun., 2004, 19, 2202-2203.

[2] (a)Y. H. Ji, X. H. Lu, Z. H. Yang, and X. Feng, Thermodynamic analysis of the theoretical energy consumption in the removal of organic contaminants by physical methods, Sci. China Chem., 2010, 53 (3), 671-676.(b) Y. H. Ji, X. Y. Ji, C. Liu, X. Feng and X. H. Lu, Modelling of Mass Transfer Coupling with Crystallization Kinetics in Microscale, Chem. Eng. Sci., 2010, 65(9), 2649-2655.

[3] Wei M, Zhou J, Lu X, Zhu Y, Liu W, Lu L, Zhang L. Diffusion of Water Molecules Confined in Slits of Rutile TiO_2(110) and Graphite(0001). Fluid Phase Equilib., 2011, 302, 316–32

See more of this Session: Poster Session: Computational Molecular Science and Engineering Forum
See more of this Group/Topical: Computational Molecular Science and Engineering Forum

(190i) Influence of Pore Geometry On the Structural and Dynamic Properties of Confined Ionic Liquids

Joshua D. Monk, **Ramesh Singh**, Harsha Dissanayake, Garrett Dupre, Minhtrang Do and Francisco R. Hung, Cain Department of Chemical Engineering, Louisiana State University, Baton Rouge, LA

Ionic liquids (ILs) confined in nm-sized pores are of interest for their potential use in applications such as electrochemical double layer capacitors. In these devices, interactions exist between the nanoporous electrodes and the ionic liquids and the behavior of the confined ionic liquids is not fully understood. This project focuses on the influence of the carbon geometry on the confined ILs. We investigate the behavior of the ionic liquid, 1-*n*-butyl,3-methylimidazolium hexafluorophosphate, [bmim][PF_6] confined inside three carbon models: (1) a slit-like graphitic pore model, (2) a multi-walled carbon nanotube, and (3) a model CMK-3 material, which consist of amorphous carbon rods arranged in a hexagonal array. Molecular dynamics simulations were performed using the non-polarizable potential for [bmim][PF_6] developed by Bhargava and Balasubramanian [1] to understand the influence of pore size and geometry on the structural and dynamic properties of the confined IL. We observe that the IL exhibits similar layering effects near the carbon surfaces of the three pore models. It has also been found that the dynamics of the ions depend strongly on their location with respect to the carbon surface, with slower dynamics generally observed near the pore surfaces. To identify the behavior of the ILs in these simulations we performed analysis on the density profiles, orientation of the ions, and the dynamics at specific locations within the carbon pore models.

[1] B. L. Bhargava and S. Balasubramanian, *J. Chem. Phys.* **127**, 114510 (2007)

See more of this Session: Poster Session: Computational Molecular Science and Engineering Forum
See more of this Group/Topical: Computational Molecular Science and Engineering Forum

(190j) Structural and Dynamical Properties of the Ionic Liquid [BMIM+][PF6-] Confined Inside a Graphitic Slit Pore

Ramesh Singh, Joshua D. Monk and Francisco R. Hung, Cain Department of Chemical Engineering, Louisiana State University, Baton Rouge, LA

A fundamental understanding of the behavior of ionic liquids (ILs) confined in nanopores is necessary to explore and optimize applications in electrochemical double-layer capacitors (EDLCs) and dye-sensitized solar cells (DSSCs).

In this work we have used molecular simulation to investigate the structural and dynamical properties of the IL [$BMIM^+$][PF_6^-] when confined inside a slit-like graphitic pore. Different structural properties (local density profiles, order parameters) and dynamical properties [mean squared displacements (MSDs) in different directions, van Hove correlation functions and intermediate scattering functions] were determined for the confined IL. Our results suggest that the local dynamics of the ions depend strongly on their position inside the pore. Close to the pore walls, the typical relaxation times are larger than those for the ions in the center of the pore. Cations have faster relaxation times than anions everywhere in the system. Important spatial heterogeneities in the dynamics of the confined IL are observed, and those are compared against results observed for the same IL in bulk systems at the same temperatures. The dynamics of the confined ions is one important factor that determines the internal resistance in an EDLC, which ultimately affects its specific power (i.e., how quickly an EDLC can deliver energy).

See more of this Session: Poster Session: Computational Molecular Science and Engineering Forum
See more of this Group/Topical: Computational Molecular Science and Engineering Forum

(190k) Theoretical Investigation of the Mechanism for [Emim]Ac Dissolve Cellulose

Zhen-Dong Ding and Hai-Jun Wang, Jiangnan University, Wuxi, China

Despite ILs are capable of dissolving cellulose, the mechanism of this dissolution process is not well understood. To date, about the mechanism of ILs dissolving cellulose, speculation, ^{13}C NMR analysis and molecular dynamics simulations have been used to interpret it by some researchers. However, it has not been qualitatively studied by the methods of quantum chemical (QM) calculations and atom in molecules (AIM) theory. In this work, QM calculations and AIM theory have been used to investigate the mechanism of ionic liquid 1-ethyl-3-methylimidazolium acetate ([emim]Ac) dissolving cellulose. It was found that the absolute value of interaction energies of [emim]Ac with cellulose monomer is bigger than that of cellulose dimer. [emim]Ac interacts with cellulose or water molecules mainly via H-bonds. Moreover, the anionic part of [emim]Ac plays a major role in the interaction with them. Further research suggests that the H-bonds between cellulose and [emim]Ac are weaken or even destroyed when the water are added to the [emim]Ac-cellulose system which indicates that water molecules prefer to form H-bonds with [emim]Ac. Consequently the H-bonds between cellulose monomers connected again, and cellulose would precipitate out from the [emim]Ac-cellulose liquor.

See more of this Session: Poster Session: Computational Molecular Science and Engineering Forum
See more of this Group/Topical: Computational Molecular Science and Engineering Forum

(190l) Molecular Modeling of the Ionic Liquid [EMIM+][TFMSI-] Inside a Slit Nanoporous Electrode

Nav Nidhi Rajput[1], Joshua D. Monk[2], Ramesh Singh[2] and Francisco R. Hung[2], (1)Cain Department of Chemical Enginering, Louisiana State University, Baton Rouge, LA, (2)Cain Department of Chemical Engineering, Louisiana State University, Baton Rouge, LA

Considering the relevance of ionic liquids as alternative electrolytes in energy-related devices such as electrochemical double-layer capacitors (EDLCs) and dye-sensitized solar cells (DSSCs), it is essential to explore more about their behavior in the vicinity of the charged walls of a nanoporous electrode. Structural properties of an IL confined inside a nanoporous electrode affect macroscopic properties such as the capacitance in an EDLC; and the dynamical properties of the confined IL is one of the factors that determines the macroscopic electrical resistance in electrochemical devices.

Model electrodes represented as slit-like graphitic nanopores and the IL 1-ethyl-3-methylimidazolium bis (trifluoromethylsulfonyl)imide [$EMIM^+$][$TFMSI^-$] are studied using molecular dynamics (MD) simulations. Our goal is to understand the effect of pore size and surface charge density on the structural, dynamical and electric properties of ILs confined inside nanoporous materials. As previous studies have largely concentrated on the structure and electric properties of the confined ILs, we have especially focused on the dynamics of the system, which were studied in detail by measuring mean squared displacements (MSDs), van Hove self correlation functions (VHSCFs) and self intermediate scattering function (SISFs).

Our results indicate that structural properties such as the local density profiles, the local orientation, and the radial distribution functions of the cations and anions vary strongly with pore size and charge density in the pore walls. These variables also affect the dynamical properties of the confined IL. Our results suggest that the dynamics of the ions inside the slit pore are highly heterogeneous and depend strongly on their position with respect to the pore walls, as well as on the surface charge density of the walls. For example, our results show that the layer of cations that is closer to a positively charged wall, as well as the layer of anions closer to a negatively charged wall, exhibit extremely slow dynamics when compared to bulk systems.

See more of this Session: Poster Session: Computational Molecular Science and Engineering Forum
See more of this Group/Topical: Computational Molecular Science and Engineering Forum

(190m) OpenTox: In Silico Toxicity Prediction for Chemical Safety

David Gallagher, CAChe Research, Beaverton, OR, Sunil Chawla, Seascape Learning LLC, CA and Barry Hardy, Douglas Connect, Switzerland

The European Union (EU) chemical legislation REACH* requires that chemicals imported into or manufactured in the EU must be registered with safety information, from December 2010. It has been estimated that over 9 million additional test animals will be required to address the registration requirements of REACH, if no alternative methods are accepted. In particular, chronic and reproductive toxicity, *in vivo* mutagenicity, and carcinogenicity endpoints require the largest number of test animals within REACH, as no alternative *in vitro* assays are available. However, the number of test animals could be significantly reduced by utilizing existing experimental data in conjunction with Quantitative Structure Activity Relationship (QSAR) models, as part of the weight of evidence.

To address the challenge, the European Commission has funded the OpenTox (www.OpenTox.org) project to develop an open source framework, that provides *in silico* models including QSAR, validation/ reporting procedures, and unified access to experimental toxicity data. Also, end-user applications are being developed for model development/validation, toxicity prediction, reporting and analyses, and unified data access, for use by Toxicologists, Biologists, Chemists and Computational Scientists.

An overview of the progress and accomplishments of this 3-year European-funded OpenTox project will be presented.

*http://ec.europa.eu/environment/chemicals/reach/reach_intro.htm

See more of this Session: Poster Session: Computational Molecular Science and Engineering Forum
See more of this Group/Topical: Computational Molecular Science and Engineering Forum

(190n) MD Simulation of CO2/CH4 Gas Separation with Hybrid Membrane

Zhenxing Wang, Chemical Engineering, Virginia Polytechnic Institute and State University, Blacksburg, VA and **Luke E. K. Achenie**, Department of Chemical Engineering, Virginia Polytechnic Institute and State University, Blacksburg, VA

Membrane isone of the most widely used materials for gas separation. Many efforts havebeen made to improve the performance of membrane and hybrid organic-inorganicmembrane is one novel type with plenty of advantages. Thus we studied thishybrid membrane to understand the mechanism and further to find new researchdirection of experiment.

Regularshaped pores are created in silicon-dioxide crystal structure. The pores areblocked by different number of phenyl groups which are randomly bonded tosilicon atoms on the wall of the pores. Gas molecules are disposed in one sideof membrane structure and the other side is left as vacuum. Gas molecules are restrictedfrom to move to the other side by a repulsive wall. The simulation is under375K and 20MPa in 1 to 4 ns, which is depending on the permeanceresults. The effect of the pore's radius on the selectivity and permeance is studied. The permeanceis of the order of 10^{-4} $^{mol}m^{-2}$ s^{-1} Pa^{-1} . The selectivity changes from0.31 to 1.15. We also researched how the number of phenyl groups in the channelaffects the performance and the related mechanism. The orientation of thephenyl groups can cause significant difference. Also the distances between theatoms of the phenyl group and nearby atoms from the frame lead to differentresult for the separation. Based on the simulation results, one membranestructure is constructed which yield to the selectivity of 68 at thelevel of permeance.

See more of this Session: Poster Session: Computational Molecular Science and Engineering Forum
See more of this Group/Topical: Computational Molecular Science and Engineering Forum

(190o) Computation Design of Ionic Liquids for Thermal Energy Storage

Amirhossein Mehrkesh and **Arunprakash T. Karunanithi**, Department of Civil Engineering, University of Colorado Denver, Denver, CO

We present a computational design approach for designing novel ionic liquids with targeted properties for thermal energy storage applications. We combine group contribution approach and computer-aided optimization methods to solve this design problem.

See more of this Session: Poster Session: Computational Molecular Science and Engineering Forum
See more of this Group/Topical: Computational Molecular Science and Engineering Forum

(190p) Investigation of the Hyperthermal Collisions of Atomic Oxygen with Graphene Using ReaxFF Reactive Force Field

Sriram Goverapet Srinivasan, Department of Mechanical and Nuclear Engineering, The Pennsylvania State University, State College, PA and Adri C.T. van Duin, Department of Mechanical and Nuclear Engineering, Pennsylvania State University, University Park, PA

We have investigated the hyperthermal collisions of atomic oxygen with a graphene sheet through molecular dynamic simulations using the ReaxFF reactive force field. A pristine sheet and a sheet with a single vacancy defect were considered in the study. Two different graphene sheet sizes, one with 24 atoms and the other with 600 atoms were studied in the simulations. The impinging oxygen atoms had a translational energy of 5eV. Since 5eV oxygen atoms are not sufficiently energetic to sputter a carbon atom from graphene, the sheet was functionalized with epoxides in order to observe various possible reactions. Interesting reactions observed include O_2 formation, epoxide formation, sheet damage and inelastic scattering of the impinging oxygen. The results of our simulations, in terms of the number of reactive events and reaction mechanisms, compared well with the DFT based results of Paci et al (*J.Phys. Chem.A,2009,vol 113,pp 4677-4685)* for the small sheet. In case of the larger sheet, the percentage of trajectories yielding oxygen atom scattering and epoxide formation were much higher while O_2 formation was much lower than that for the smaller sheet. This can be attributed to excessive artificial symmetry that makes the smaller sheet too stiff and may prohibit certain phonon modes. Thus size effects are important in studying the dynamics of these events.

See more of this Session: Poster Session: Computational Molecular Science and Engineering Forum
See more of this Group/Topical: Computational Molecular Science and Engineering Forum

(190q) Multiscale Molecular Modeling of the Host-Guest Interactions of PAMAM Dendrimers with Organic Species

Seung Ha Kim and Monica H. Lamm, Department of Chemical and Biological Engineering, Iowa State University, Ames, IA

Dendrimers are large, synthetic macromolecules that have regular and highly branched architectures. The multivalency of dendrimers permits them to bind with other species and makes them attractive platforms for a wide range of applications, including medicine, surface science, catalysis, and water treatment. To better understand the physical mechanisms of dendrimer-guest assembly, we have developed a multiscale molecular model for simulating flexible macromolecules in solution, based on the force matching method for coarse graining[1]. This modeling approach facilitates the simulation of systems that are large enough to study competing effects between host-guest, host-host, and guest-guest interactions, rather than being limited to modeling a single host-guest complex in solution. To illustrate the effectiveness of our approach, we have simulated the complexation of poly(amidoamine) (PAMAM) dendrimers with a diverse set of guest species: phenanthrene[2], human serum albumin, and fullerol.

[1] S. H. Kim and M. H. Lamm, submitted.

[2] M. Lard, S. H. Kim, S. Lin, P. Bhattacharya, P. C. Ke, and M. H. Lamm. *Phys. Chem. Chem. Phys.*, **12**, 9285-9291 (2010).

See more of this Session: Poster Session: Computational Molecular Science and Engineering Forum
See more of this Group/Topical: Computational Molecular Science and Engineering Forum

(190r) Molecular Simulations of Network Polymers of Intrinsic Microporosity: Structure Generation by a Simulated Polymerization Algorithm and Gas Adsorption Studies

Lauren J. Abbott[1], Kyle E. Hart[1], Ping Lin[2] and Coray M. Colina[1], (1)Department of Materials Science and Engineering, Pennsylvania State University, University Park, PA, (2)The Materials Simulation Center, Pennsylvania State University, State College, PA

Polymers of intrinsic microporosity (PIMs) are a new class of amorphous, microporous materials that derive their porosity by the inclusion of rigid and nonlinear units, which prevent space-efficient packing. Network PIMs provide an enhanced rigidity, which results in greater microporosity. These microporous materials, with interconnected pores of sizes smaller than 2 nm, are of interest for a variety of applications including separations, adsorption, gas storage, and catalysis.

The performance of microporous materials is dependent on their porosity, which is difficult to characterize experimentally. Therefore, molecular simulations offer a unique perspective by allowing examination of the structures on the molecular level. This detail provided by the simulations, such as geometric measures of the pore structures, offers additional insight into understanding the nature of PIMs and assists in the design and tailoring of these materials for specific applications.

In this work, we apply a recently developed structure generation procedure to network PIMs, including those derived from hexaazatrinaphthylene (HATN) and cyclotricatechylene (CTC). This approach utilizes a simulated polymerization algorithm to form networks by connecting repeat units close together in a periodic box. The resulting simulated structures are characterized by densities, surface areas, pore volumes, pore size distributions, and structure factors. Additionally, gas adsorption isotherms are calculated using grand canonical Monte Carlo (GCMC) simulations to further study their adsorptive properties. The results show qualitative agreement to available experimental data.

See more of this Session: Poster Session: Computational Molecular Science and Engineering Forum
See more of this Group/Topical: Computational Molecular Science and Engineering Forum

(190s) A Coarse-Grained Model for Polyethylene Glycol (PEG) Polymer

Qifei Wang[1], David J Keffer[1] and Donald M. Nicholson[2], (1)Chemical and Biomolecular Engineering, University of Tennessee Knoxville, Knoxville, TN, (2)Computer Science and Mathematics Division, Oak Ridge National Laboratory

A coarse-grained (CG) model of Polyethylene Glycol (PEG) was developed and implemented in CG Molecular Dynamics (MD) simulations of PEG chains with degree of polymerization (DP) 20 and 40. In the model, two repeat units of PEG are grouped as one CG bead. Atomistic MD simulation of PEG chains with DP = 20 was first conducted to obtain the bonded structural probability distribution functions (PDFs) and nonbonded pair correlation function (PCF) of the CG beads. The bonded CG potentials are obtained by simple inversion of the corresponding PDFs. The CG nonbonded potential is parameterized to the PCF using both an inversion procedure based on the Ornstein-Zernike equation with the Percus Yevick approximation ($OZPY^{-1}$) and a combination of $OZPY^{-1}$ with the iterative Boltzmann inversion (IBI) method ($OZPY^{-1}$+IBI). As a simple one step method, the $OZPY^{-1}$ method possesses an advantage in computational efficiency. Using the potential from $OZPY^{-1}$ as an initial guess, the IBI method shows fast convergence. The CGMD simulations of PEG chains with DP = 20 using potentials from both methods satisfactorily reproduce the structural properties from atomistic MD simulation of the same systems. The $OZPY^{-1}$+IBI method yields better agreement than the $OZPY^{-1}$ method alone. The new CG model and CG potentials from $OZPY^{-1}$+IBI method was further tested through CGMD simulation of PEG with DP = 40 system. No significant changes are observed in the comparison of PCFs from CGMD simulations of PEG with DP = 20 and 40 systems, indicating that the potential is independent of chain length.

See more of this Session: Poster Session: Computational Molecular Science and Engineering Forum
See more of this Group/Topical: Computational Molecular Science and Engineering Forum

(190t) Effective Potential Between Nanoparticles In Suspension

Gary S. Grest[1], **Qifei Wang**[2], Pieter in 'T. Veld[3] and David J. Keffer[2], (1)Sandia National Laboratories, New Mexico, (2)Chemical and Biomolecular Engineering, University of Tennessee Knoxville, Knoxville, TN, (3)Polymer Research,, BASF, Ludwigshafen,, Germany

Abstract

Results of Molecular Dynamics simulations are presented for the pair distribution function between nanoparticles in an explicit solvent as a function of nanoparticle diameter and interaction strength between the nanoparticle and solvent. Effect of including the solvent explicitly is demonstrated by comparing the pair correlation function of nanoparticles to that in an implicit solvent. The nanoparticles are modeled as a uniform distribution of Lennard-Jones particles, while the solvent is represented by standard Lennard-Jones particles. The diameter of the nanoparticle is varied from 10 to 25 times that of the solvent for a range of nanoparticle volume fractions. As the strength of the interactions between nanoparticles and the solvent increases, the solvent layer surrounding the nanoparticle is formed which increases the effective radii of the nanoparticles. The pair correlation functions are inverted using the Ornstein-Zernike integral equations to determine an effective pair potential between the nanoparticles mediated by the explicit introduction of a solvent.

See more of this Session: Poster Session: Computational Molecular Science and Engineering Forum
See more of this Group/Topical: Computational Molecular Science and Engineering Forum

(190u) Multi-Scale Modeling of a Cathode/ Electrolyte Interface In Proton Exchange Membrane Fuel Cells (PEMFCs)

Kuan-Yu Yeh, Michael V. Glasspool, Michael J. Janik and Janna K. Maranas, Chemical Engineering, Pennsylvania State University, University Park, PA

The slow oxygen reduction reaction (ORR) at the cathode represents the largest performance loss of a PEMFC. Few studies have examined the impact of the cathode/electrolyte interfacial structure on surface reaction kinetics. We employed a solvated Pt(111) DFT model system to probe the dependence of ORR energetics on water density which represents interfacial structure variation. The variable electrode potential is modeled using double reference method (C. D. Taylor et al., 2006), which allows us to consider the coupled effect of solvent reorganization and electrode potential on reaction energetic. For comparison, a homogeneous electric field model perpendicular to the surface was applied on a vacuum slab model to evaluate trends in oxygen reduction under the influence of electric fields. Although the solvated model predicts a potential dependent rate limiting step of ORR, in agreement with experiment, the DFT conclusions are limited by the use of single water structure. Variations in DFT results due to the model choice emphasize the importance of using molecular dynamics (MD) to provide most probable interfacial structures. A reactive force field, based on the Central Force Model (F. H. Stillinger and A. Rahman, 1978), was used to model the electrolyte layer. A polarizable and potential controlled electrode model, termed electrode charge dynamic (C. G. Guymon et al., 2005), was used to represent the electrode surface The MD simulations quantify the interfacial water and ion structure as a function of electrode potential and acid concentration. Significant structural difference of interfacial water between MD results and the bilayer water used in DFT methods are found. Representative interfacial structures from MD simulations will be used as the starting structure for DFT methods to study the influence of interfacial water structure on oxygen reduction kinetics.

See more of this Session: Poster Session: Computational Molecular Science and Engineering Forum
See more of this Group/Topical: Computational Molecular Science and Engineering Forum

(190v) Computer Simulation of Carbon Nanotubes In Liquid Crystalline Solvents

Shivkumar Bale and Francisco R. Hung, Cain Department of Chemical Engineering, Louisiana State University, Baton Rouge, LA

Computer simulation of carbon nanotubes in liquid crystalline solvents

Shivkumar Bale and Francisco R. Hung

Cain Department of Chemical Engineering, Louisiana State University, Baton Rouge, LA 70803

Liquid crystals (LCs) have been used in the past to control the orientation of carbon nanotubes (CNTs). If the CNTs are adequately dispersed in the LC, they will tend to align with their long axes parallel to the director field **n**(**r**) of the LC. Furthermore, and since the direction of **n**(**r**) can be easily modified by using electric and magnetic fields, flows and/or external surfaces, dispersion of CNTs in LCs provides an excellent way to manipulate and align nanotubes. However, one of the main challenges in using LCs to manipulate and align CNTs is on how to efficiently disperse the nanotubes in the LC. Suspensions of CNTs in LCs tend to be very unstable, and most of the nanotubes promptly aggregate. The key point to obtain high-quality dispersions of CNTs in LCs is to understand the interactions between the nanotubes and the LC molecules. Molecular simulations are uniquely positioned to investigate interactions at the molecular level of detail, as well as their effects on the macroscopic properties of the systems. Results from this research are relevant for potential applications of these systems in displays, nanoscale electronics, optical sensors, and in the formulations of composites with unique mechanical, thermal and/or electronic properties.

We performed molecular dynamics simulations of systems of CNTs in different LCs and determined potentials of mean force (PMF) in order to understand the molecular-level interactions between the CNTs and the LCs. The PMF measures the free energy required to keep the two nanotubes separated by a distance d, when they are immersed in a given LC solvent. In our research, we considered several representative LC molecules, such as two members of the 4-cyano-4'-n-alkylbiphenyl series, 5CB and 8CB, as well as 5CF (4'-(3,4-difluoro-phenyl)-4-pentyl-bicyclohexyl). We used single walled CNTs with armchair (n, n) configuration in our calculations. Other properties, such as local densities, radial distribution functions and scalar order parameters, are also reported. Our results indicate that the PMF minima for all systems is at the state where the two CNTs are very close (i.e., no LC molecules in between the CNTs). It is also observed that the features of the PMF strongly depend on the molecular structure of the LC molecules.

(190w) Sintering Rate and Mechanism of TiO2 Nanoparticles by Molecular Dynamics

Beat Buesser, Arto J. Grhn and Sotiris E. Pratsinis, Department of Mechanical and Process Engineering, ETH Zurich, Zurich, Switzerland

Sintering Rate and Mechanism of TiO_2Nanoparticles by Molecular Dynamics

B. Buesser, A.J. Grhn and S.E. Pratsinis

Particle Technology Laboratory, Institute of Process Engineering,

Department of Mechanical and Process Engineering, ETHZrich, 8092 Zrich, Switzerland

Titania (TiO_2)nanoparticles have many attractive applications in photovoltaic[1] and photocatalytic[2] processes, to name a few.The performance of TiO_2 nanoparticles depends considerably on theirsize and composition which are determined by the sintering rate during their synthesis.The sintering rate is crucial for designing of gas-phase processes withcontrolled product particle size, structure, composition and eventualperformance in a number of applications[3].

Kobata *etal.*[4] and Seto *et al.*[5] have proposed sinteringrates for TiO_2 and validated them by accounting for the detailedfluid mechanics of their hot wall reactors forming rather large TiO_2nanoparticles (d_p = 10 - 100 nm). Little is known, however,for the sintering rate of small TiO_2 nanoparticles (d_p< 10 nm) as it is difficult to reliably measure it.

On the otherhand molecular dynamics (MD) simulations have been used to study the sinteringto full coalescence of metallic and metalloid nanoparticles[6,7] though much less has beendone for ceramic ones as their force fields and potentials are difficult todetermine. Koparde and Cummings[8] investigated thesintering of two TiO_2 nanoparticles up to t = 0.5 ns bytracking the shrinkage of the center-to-center distance and the growth of thesintering neck using the force field of Matsui and Akaogi[9]. Theycompared MD with phenomenological sintering models and investigated the meltingof TiO_2 nanoparticles[10].

The above MD ofTiO_2 sintering have reached up to 0.5 ns residence time, a durationthat is not sufficient for complete coalescence. So most of the surface areareduction of such small particles has taken place by adhesion and neck growththat limited the detailed understanding of sintering mechanisms and, mostimportantly, the extraction of quantitative sintering rates that are needed inprocess design of nanoparticle manufacturing.

Here,sintering of rutile TiO_2 nanoparticles is investigated by graphicalprocessing unit (GPU) accelerated MD[11] from adhesion and neckgrowth to finally full coalescence up to several hundred nanoseconds. Thisallows determining the sintering rate of very small TiO_2nanoparticles ($d_p < 5$ nm) by monitoring the evolution oftheir surface area (Figure 1). For the smallest particle diameters, theMD-obtained sintering rates were smaller than that predicted by theorydeveloped for larger particles4,5. Ions on the particlesurface exhibited higher net displacement than bulk ones revealing that surfacediffusion is the dominant sintering mechanism of TiO_2 nanoparticles.An expression for the sintering rate of rutile TiO_2 nanoparticleshas been extracted from MD, bridging the gap of knowledge from a few moleculesto several nanometers, the key size range for nanoparticle properties andperformance. This MD-derived sintering rate facilitates the use of phenomenologicalmodels[12] in design of processesfor large scale manufacture and processing of small nanoparticles[13].

Financial support from the Swiss National ScienceFoundation (SNF) grant # 200021-119946/1 and European Research Council isgratefully acknowledged.

References

1. Gratzel M. Photoelectrochemicalcells. *Nature.* 2001; 414, (6861), 338-344.

2. Formenti M, Juillet F, MeriaudeauP, Teichner SJ, Vergnon P. Preparation in a hydrogen-oxygen flame of ultrafinemetal oxide particles. Oxidative properties toward hydrocarbons in the presenceof ultraviolet radiation. *J. Colloid Interface Sci.* 1972; 39, (1),79-89.

3. Strobel R, Pratsinis SE. Flameaerosol synthesis of smart nanostructured materials. *J. Mater. Chem.* 2007;17, 4743-4756.

4. Kobata A, Kusakabe K, Morooka S.Growth and transformation of TiO_2 crystallites in aerosol reactor. *AIChEJ.* 1991; 37, (3), 347-359.

5. Seto T, Shimada M, Okuyama K.Evaluation of sintering of nanometer-sized titania using aerosol method. *AerosolSci. Technol.* 1995; 23, (2), 183-200.

6. Arcidiacono S, Bieri NR, PoulikakosD, Grigoropoulos CP. On the coalescence of gold nanoparticles. *Int. J.Multiphas. Flow.* 2004; 30, (7-8), 979-994.

7. Zachariah MR, Carrier MJ. Moleculardynamics computation of gas-phase nanoparticle sintering: a comparison withphenomenological models. *J. Aerosol Sci.* 1999; 30, (9), 1139-1151.

8. Koparde VN, Cummings PT. Moleculardynamics simulation of titanium dioxide nanoparticle sintering. *J. Phys.Chem. B.* 2005; 109, (51), 24280-24287.

9. Matsui M, Akaogi M. Moleculardynamics simulation of the strucutral and physical properties of the fourpolymorphs of TiO_2. *Mol. Simulat.* 1991; 6, 239-244.

10. Koparde VN, Cummings PT. Sintering oftitanium dioxide nanoparticles: a comparison between molecular dynamics andphenomenological modeling. *J. Nanopart. Res.* 2008; 10, (7), 1169-1182.

11. Garland M, Le Grand S, Nickolls J,Anderson J, Hardwick J, Morton S, Phillips E, Yao Z, Volkov V. ParallelComputing Experiences with CUDA. *Micro IEEE.* 2008; 28, (4), 13-27.

12. Koch W, Friedlander SK. The effect ofparticle coalescence on the surface area of a coagulating aerosol. *J.Colloid Interface Sci.* 1990; 140, (2), 419-427.

13. Pratsinis SE. Aerosol-basedtechnologies in nanoscale manufacturing: from functional materials to devicesthrough core chemical engineering. *AIChE J.* 2010; 56, (12), 3028-3035.

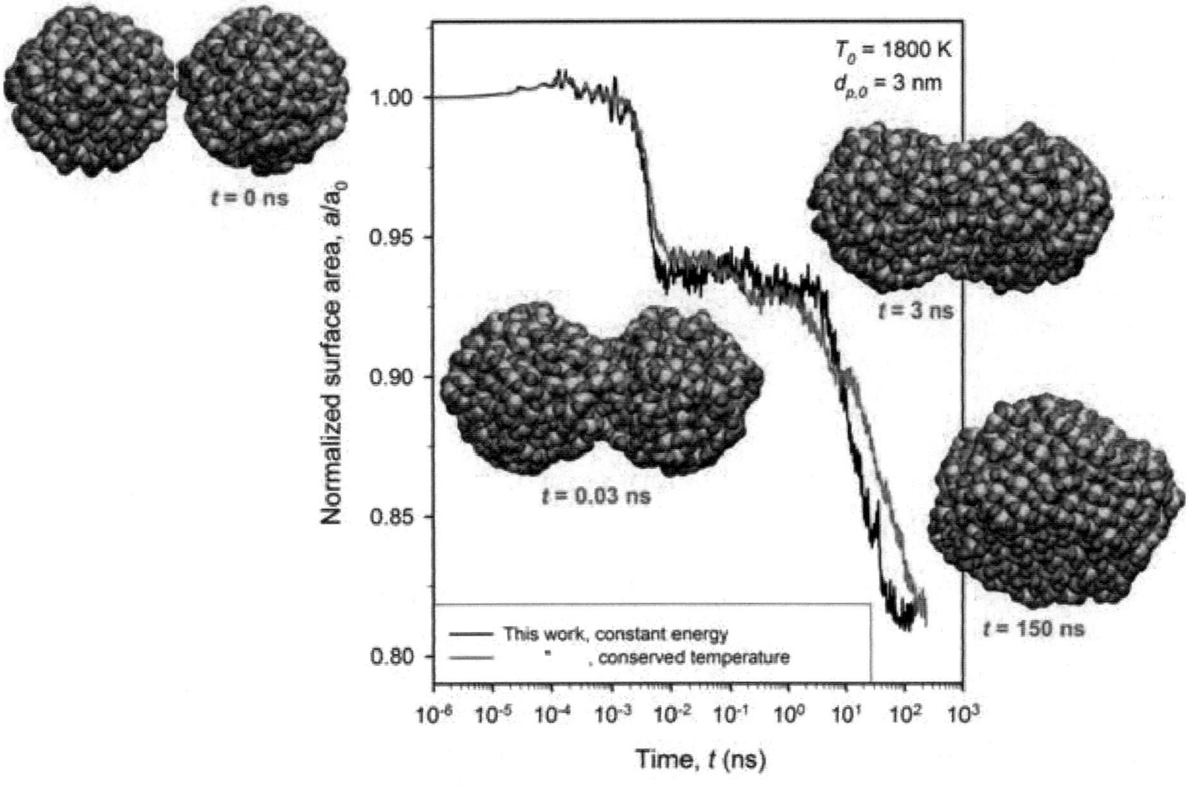

Figure1 Evolution ofsurface area of two TiO_2 nanoparticles (d_0 = 3 nm)at 1800 K.

(190x) Enantioselective Crystallization Process Design From Predictive COSMO Calculations

Ronald Zinke and Matthias Stein, Max Planck Institute for Dynamics of Complex Technical Systems, 39106 Magdeburg, Germany

The separation of enantiomers from a racemic mixture is of great interest to the chemical and the pharmaceutical industry. Enantiomers are non-superimposable mirror images of each other and thus have identical physical and chemical properties but may have different biological effects. Enantioselective crystallization is an attractive strategy since only one enatiomer crystallizes in its pure form whereas the other remains in solution. The rationalization of such a process is a challenge both to theoretical and experimental approaches.

The process design relies on an accurate evaluation of binary and ternary solubilities. COnductor-like Screening models, i.e. COSMO-RS [1] and (refined-) COSMO-SAC [2, 3] are efficient methods to calculate solubilities from a molecular perspective of both the solute and the solvent. They give the possibility to calculate chemical potentials via the determination of activity coefficients of molecules in solution and thus are further able to compute different equilibrium thermodynamic properties of mixtures. We are using these methods to design and rationalize enatioselective crystallization processes.

In [4] the authors describe a crystallization scheme to isolate and thus purify single enantiomers from an asymmetric mixture of stereoisomers. The corresponding process was adapted to demonstrate its enantioselective separation potential for the example of bicalutamide, the pharmaceutical ingredient of the drug Casodex(TM). The followed process is based on an appropriate choice of a pair of solvent and antisolvent. After identification of suitable candidates, the chiral separation of enantiomers of (R,S)-bicalutamide was performed. The experimental implementation of the enantioselective crystallisation was supported theoretically by using COSMO-SAC and refined COSMO-SAC calculations. COSMO-SAC was used to identify suitable antisolvent candidates and both COSMO-SAC approaches were used to calculate binary solubilities. The aim was to test the accuracy of calculated solubilities of these methods. In [4] COSMO calculations were only of supporting character and all experimentally relevant process steps were made on the basis of experimental results only.

We here report an enantioselective crystallization of bicalutamide based on *in silico* predictions from molecular solution. We used different implementations of COSMO models, and evaluated their accuracies in particular with regard to the possibility of designing the process. It is our aim to assess the accuracy of current COSMO implementations for the enantioselective crystallization of bicalutamide. Therefore the method COSMO-RS and the derived versions (refined-) COSMO-SAC of COSMO-based models were evaluated and compared in view of critical crystallization specific tasks. First, the search for a suitable antisolvent with reduced bicalutamide solubility was performed using COSMO methods. In a second step, model-based estimates of relevant thermodynamic properties required to perform the selective crystallization of single enantiomers were conducted. Finally, our *in silico* process design was verified in the lab. Bicaluatmide was used as a benchmark case for a pharmaceutical substance but also well-characterized chiral chemical compounds, such as mandelic acid, were tested. Limitations and inadequacies in the underlying theory, the accuracy of calculated molecular properties and their sensitivities to the choice of COSMO implementation were analyzed.

[1] A. Klamt and G. Schüürmann, COSMO: A new approach to dielectric screening in solvents with explicit expressions for the screening energy and its gradient, J. Chem. Soc. Perkin Trans., 2 (1993) 799-805.

[2] S.-T. Lin and S. I. Sandler, A priori phase equilibrium prediction from a segment contribution solvation model, Ind. Eng. Chem. Res. 46, (2002) 899-913.

[3] S. Wang and S. I. Sandler, Refinement of COSMO-SAC and the applications, Ind. Eng. Chem. Res. 46, (2007) 7275-7288.

[4] H. Kaemmerer, M. J. Jones, H. Lorenz and A. Seidel-Morgenstern, Selective crystallization of a chiral compound-forming system-Solvent Screening, SLE determination and process design, Fluid Phase Equilibria 296, (2010) 92-205.

See more of this Session: Poster Session: Computational Molecular Science and Engineering Forum
See more of this Group/Topical: Computational Molecular Science and Engineering Forum

(190y) Molecular Simulation of the Nucleation and Growth of C60 Nanoparticles From the Supersaturated Vapor and From the Undercooled Liquid

Kenneth N. Ngale, Caroline Desgranges and Jerome Delhommelle, Department of Chemistry, University of North Dakota, Grand Forks, ND

Using molecular simulation, we study the molecular mechanisms underlying the nucleation and growth of C60 nanoparticles from the supersaturated vapor and from the undercooled liquid. We show that in both cases, nucleation proceeds through the formation of small clusters composed of the metastable hexagonal close-packed (HCP) polymorph. Finally, we observe two different types of growth leading either to crystallites dominated either by the metastable HCP polymorph or by the stable face-centered cubic (FCC) polymorph.

See more of this Session: Poster Session: Computational Molecular Science and Engineering Forum
See more of this Group/Topical: Computational Molecular Science and Engineering Forum

(190z) Molecular Simulation of the Nucleation and Growth of Pd-Ni Nanoparticles

Serges E. Tatsinkou, Caroline Desgranges and Jerome Delhommelle, Department of Chemistry, University of North Dakota, Grand Forks, ND

Using molecular simulation methods, we determine the microscopic mechanisms underlying crystal nucleation and growth (up to crystallites of a 2.5nm radius) in various Pd-Ni alloys under ambient pressure. Our results show that the structure of the critical nucleus strongly depends on the composition. Simulations of the growth of critical nuclei into large post-critical nuclei show that these trends also govern crystal growth of Pd-Ni.

See more of this Session: Poster Session: Computational Molecular Science and Engineering Forum
See more of this Group/Topical: Computational Molecular Science and Engineering Forum

Ethanol Synthesis From Syn-Gas: How Surface Diffusion of Intermediates Impacts the Product Distributions Predicted for Bimetallic Catalysts

Ming He, James McAliley and ***David A. Bruce****, Chemical and Biomolecular Engineering, Clemson University, Clemson, SC*

Introduction

Ethanol has been receiving great interest for its use as a potential alternative fuel, a potential source of hydrogen, as well as a raw material for the synthesis of many chemicals and polymers.

A major challenge associated with the synthesis of ethanol from syn-gas is an inability to find a low-cost catalyst that promotes the proper combination of CO dissociation and CO insertion steps, so as to yield ethanol as the primary reaction product and inhibit the formation of methane, longer chain alkanes, and other coking reaction products. For this purpose, sub-nanometer size, bimetallic cobalt-palladium (Co-Pd) particles are studied for the desired ethanol reaction. On Co catalysts CO molecules are readily dissociated, followed by C-C coupling reactions to form higher hydrocarbons. On Pd-based catalysts CO molecules adsorb in the molecular state, followed by subsequent hydrogenations to methanol.

DFT simulations and Bronsted-Evans-Polanyi (BEP) relations were used to map out the full reaction mechanism from syn-gas to ethanol. Microkinetic models were built, considering the reaction steps as well as the diffusion of intermediate species between different metal surface sites: Co_3, Pd_3, and mixed CoPd sites.

Result and Discussion

By DFT (B3LYP, LACVP**) calculations, we identified 37 surface reactions in our proposed mechanism for ethanol formation from syn-gas. Based on the most stable morphology of the CoPd cluster, the surface sites can be classified into three categories: Co_3 sites (Co sites), Pd_3 sites (Pd sites) and mixed Co_2Pd or $CoPd_2$ sites (CoPd sites). With three different possible reaction sites and the complex reaction mechanism, there are more than 100 possible surface reaction steps from which to calculate rate constants. To solve this computationally intensive problem, the implementation of scaling methods is critical.

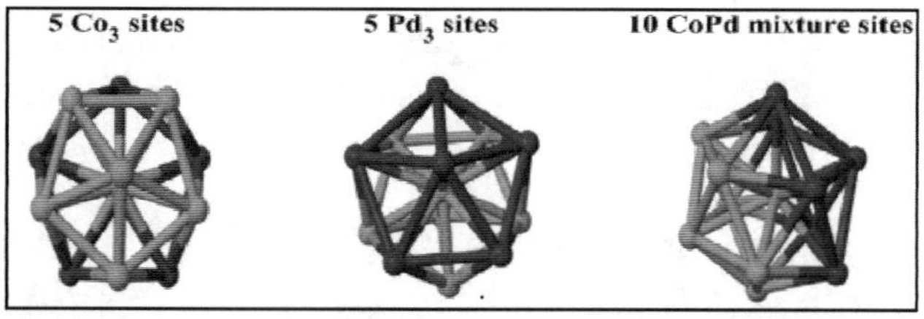

Figure 1. Different sites on the most stable conformation of a Co7Pd6 cluster (Co, blue; Pd, pink)

A widely used scaling method for estimating activation energies is the Bronsted-Evans-Polanyi (BEP) relation. The BEP relation linearly correlates the transition state energy of an elementary step to the reaction enthalpy of that step. It has been successfully applied to many catalytic systems for efficiently locating activation barriers.

To examine the intrinsic nature of the three distinct catalytic sites, the tool of microkinetic modeling was implemented. In this application, the preferred product is ethanol. Other byproducts generated along the process are methane, methanol, and acetaldehyde.

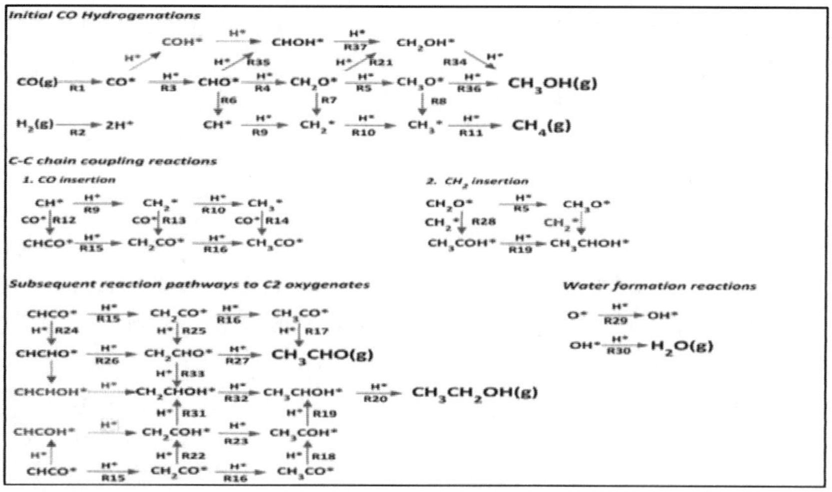

Figure 2. Full reaction mechanism of ethanol formation reacion from syn-gas

The pseudo-steady state hypothesis (PSSH) is applied to calculate surface concentrations of intermediate species. Initially, separate microkinetic models were built, targeting reaction mechanisms on isolated Co sites, CoPd sites, and Pd sites.

On the pure Co and pure Pa catalyst sites, we find good agreement of reaction selectivity between our model and experiments. Methane is the major product on the Co sides, with a selectivity of 98%. The selectivity of methane on a Co catalyst in experiment at similar conditions was 97%. On the Pd side, the microkinetic model predicts that methanol is the major product. Experiments have also shown methanol is the major product on Pd catalysts. It is interesting to note that the C2 oxygenate is formed on the interface sites (CoPd sites). Acetaldehyde is the major product with a selectivity of 58%.

For a cluster of such small scale, surface diffusion of intermediates between different reaction sites should be fairly common during the reaction. Also, because of the existence of heterogeneous reaction profiles, species may take part in distinct reactions on different sites. Thus, a proper description of the intermediates' diffusion behaviors is critical to reveal the overall reaction mechanism on the cluster.

The incorporation of surface diffusion into microkinetic models is very rare due to the complication it brings to the evaluation of kinetic parameters and the extra simulation effort it requires. Surface diffusion can be seen as a reaction between the diffusing species and an empty site. By this meaning, the treatment of diffusion is identical to that of a reaction. Ten species diffusion behaviors are studied here. They are HCO^*, CH_2O^*, CH_3O^*, CH_3CO^*, CH_2CHO^*, CH_3COH^*, CH_3CHOH^*, OH^*, CH_2^* and CH_3^*.

In the 10 species studied for surface diffusion, only the CH_3^* and CH_3CO^* species' diffusions have major effect on overall reactions. The other species have no impact on the reaction mechanism, since the diffusion rates between different sites are much faster than the reaction rates.

Diffusion has two significant effects on the reaction system. First, the overall reaction rate is much faster on Co sites than the other two sites. There is almost no products formation on CoPd sites or Pd sites. Moreover, on Co sites the selectivity of methane is 85%. The ethanol selectivity is 13% after accounting for surface diffusion. These results imply a synergetic effect between Co and Pd favoring the production of ethanol on the bimetallic cluster that would not occur on either of the individual metals.

CO insertion into the metal-carbon bonds of adsorbed methyl groups is the primary pathway to produce C_2 oxygenates. We observed that overall reaction rates are limited by the hydrogenation of CO to methoxy species. In part, this is due to the fact that the CoPd and Pd sites are primarily covered with CO, due to strong CO binding strengths on these two sites. Thus, a significant fraction of product forming reactions occur on the Co sites, which make methane the major product. Some ethanol can also form on Co sites via the diffusion of CH_3 and CH_3CO species between Co sites and CoPd sites, thus facilitating CO insertion reactions (Figure 3).

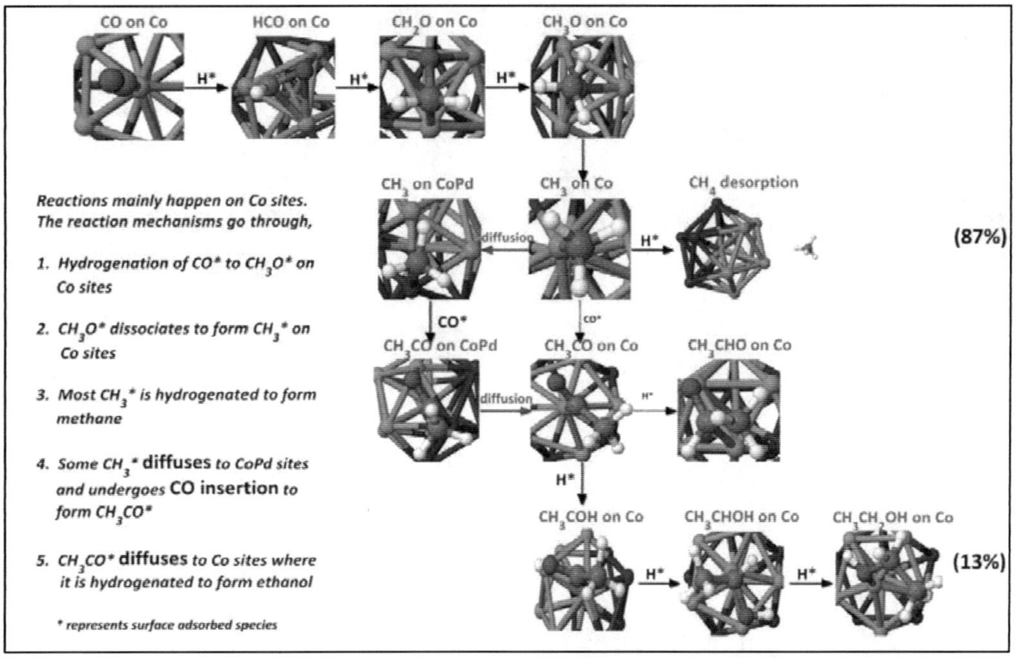

Figure 3. The overall reaction mechanism of ethanol formation from syn-gas on CoPd cluster

Conclusions

By using DFT simulations coupled with BEP methods, we calculate more than 100 intermediate reaction steps. The synergetic effect between Co and Pd is seen on the interface CoPd sites, where C2 oxygenates are the major products.

For this system, CO insertion into the metal-carbon bonds of adsorbed methyl groups is the primary pathway for C_2 oxygenate production. Due to most CoPd and Pd sites being blocked by adsorbed CO, a significant fraction of product forming reactions occur on the Co sites, which leads to methane being the major product (90%). However, the diffusion of CH_3 and CH_3CO between Co and CoPd sites facilitates CO insertion reactions, which ultimately leads to some ethanol (10%) being produced on the Co sites. These results indicate the key for ethanol formation on bimetallic clusters is the presence of intermixture sites, which can effectively reduce the reaction barrier differences of CO insertion reactions and hydrogenation reactions on the CH_3* species. This feature has to be coupled with fast diffusions of intermediates to accomplish ethanol production.

References

1. Spivey, J. J.; Egbebi, A. *Chemical Society Reviews* 2007, *36*, 1514-1528.
2. Subramani, V.; Gangwal, S. K. *Energy & Fuels* **2008**, *22*, 814-839.
3. Lahtinen, J.; Vaari, J.; Kauraala, K. *Surface Science* **1998**, *418*, 502-510.
4. Cheng, J.; Hu, P.; Ellis, P.; French, S.; Kelly, G.; Lok, C. M. *Journal of Physical Chemistry C* **2008**, *112*, 9464-9473.
5. Neurock, M. *Topics in Catalysis* **1999**, *9*, 135-152.
6. Conrad, H.; Ertl, G.; Latta, E. E. *Surface Science* **1974**, *41*, 435-446.

7. Bligaard, T.; Nørskov, J. K.; Dahl, S.; Matthiesen, J.; Christensen, C. H.; Sehested, J. *Journal of Catalysis* **2004**, *224*, 206-217.
8. Nørskov, J. K.; Bligaard, T.; Logadottir, A.; Bahn, S.; Hansen, L. B.; Bollinger, M.; Bengaard, H.; Hammer, B.; Sljivancanin, Z.; Mavrikakis, M.; Xu, Y.; Dahl, S.; Jacobsen, C. J. H. *Journal of Catalysis* **2002**, *209*, 275-278.
9. Liu, P.; Logadottir, A.; Nørskov, J. K. *Electrochimica Acta* **2003**, *48*, 3731-3742.
10. Wintterlin, J.; Zambelli, T.; Trost, J.; Greeley, J.; Mavrikakis, M. *Angewandte Chemie International Edition* **2003**, *42*, 2850-2853.

(211b) Development and Application of a ReaxFF Description for Formation of Hydrocarbons On Iron and Iron Carbide Fisher-Tropsch Catalysts

Chenyu Zou[1], Adri van Duin[1] and Dan C. Sorescu[2], (1)Mechanical Engineering, Pennsylvania State University, State College, PA, (2)Computational Chemistry, National Energy Technology Laboratory, Pittsburgh, PA

A ReaxFF reactive force field description for iron and iron carbide bulk phases, surfaces and their interactions with water, carbon monoxide and hydrocarbon species has been developed, which puts the foundations together for large scale, fully reactive molecular dynamics simulations of Fisher-Tropsch (FT) catalysis on iron catalysts.

The ReaxFF description in this project will be tested and trained against a suite of ab initio data, and will be validated using dynamical simulations on relatively small systems(50-2500 atoms), enabling us to compare ReaxFF reaction kinetic and diffusion data with experimental observations.

This project aims to close the gap between atomistic-scale simulations and experiment, and the methods developed here will greatly enhance the ability to design and test materials on an atomistic scale. Specifically, the large system sizes (>>1000 atoms) and large simulation time (> 1ns) will enable us to study the ageing, poisoning and regeneration of iron-based FT-catalysts.

See more of this Session: Computational Catalysis IV
See more of this Group/Topical: Catalysis and Reaction Engineering Division

(211c) Investigation of Structure Sensitivity for the CO Oxidation Chemistry On Pt and Au

Michail Stamatakis, Chemical Engineering, University of Delaware, Newark, DE and Dionisios G. Vlachos, University of Delaware, Newark, DE

The CO oxidation is a prototypic catalytic reaction for the study of structure sensitivity on metal surfaces. Yet, experimental results from different groups appear contradictory: studies on single crystal Pt, Ir, Rh and Pd surfaces have shown that CO oxidation is structure insensitive and attribute this effect on the surface being predominantly covered by CO, resulting in the rate being controlled by the CO desorption. Other studies, however, show that steps and terraces exhibit different activity, with the latter being much more active than the former, due to the preferential poisoning of the steps by CO. Thus, it still remains an open question whether the CO oxidation chemistry is structure sensitive and under what conditions.

In this study, we perform multiscale simulations that combine kinetic Monte Carlo with density functional theory based input on two different metals, namely Pt which exhibits strong binding for CO and O, and Au which exhibits weak binding for these species. We consider several surfaces with different ratios of step/terrace site numbers under various conditions, and investigate whether CO oxidation is structure sensitive. We show that the steps in the Au surfaces are much more active than the terraces. However, for Pt surfaces, poisoning effects by CO and/or O are important and may render the steps inactive, thereby resulting in the CO oxidation chemistry appearing structure insensitive under certain conditions.

See more of this Session: Computational Catalysis IV
See more of this Group/Topical: Catalysis and Reaction Engineering Division

(211d) Selective Methane-to-Methanol Oxidation On Bimetallic Transition Metal Surfaces

Kathryn Bjorkman and Linda J. Broadbelt, Department of Chemical and Biological Engineering, Northwestern University, Evanston, IL

Direct methane conversion to useful chemicals is a catalysis problem that has challenged researchers for many years. Methanol is one such desired product because of its potential use in a wide range of industrial applications, including methanol-to-olefin (MTO) processes to form ethylene and propylene, which then may be used as raw material in production of other chemicals. However, part of the challenge in converting methane to more useful products is that the products of those reactions are often more reactive than the methane itself. This results in selectivity problems and higher extents of oxidation. Nanoscale bimetallic catalytic particles have the potential to improve upon existing monometallic micro- and macroscale catalysts by introducing unique electronic and geometric structure effects to selectively convert methane to methanol while avoiding overreaction to complete oxidation products.

Quantum mechanical calculation methods are used to study the binding energies of methane-derived intermediates on monometallic and bimetallic transition metal surfaces in an effort to select the best catalytic sites for selective methane-to-methanol oxidation. Linear scaling relationships between binding energies and calculated activation energy barriers are established to showcase trends for bimetallic surfaces and provide for detailed reaction pathway analysis.

See more of this Session: Computational Catalysis IV
See more of this Group/Topical: Catalysis and Reaction Engineering Division

(211e) Mechanistic Studies of Oxygen Reduction by Hydrogen On PdAg(110)

Carrie A. Farberow[1], Andres Godinez-Garcia[2], Juan Francisco Perez-Robles[2], Omar Solorza-Fera[3] and Manos Mavrikakis[1], (1)Department of Chemical and Biological Engineering, University of Wisconsin - Madison, Madison, WI, (2)Departamento de Materiales, CINVESTAV-IPN, Quertaro, Mexico, (3)Departamento de Qumica, CINVESTAV-IPN, Ciudad de Mxico, Mexico

The reduction ofoxygen by hydrogen is a metal-catalyzed reaction important in the direct synthesisof hydrogen peroxide (HOOH) and at the cathode in proton exchange membrane fuelcells (PEMFCs). Due to its technological applicability and widespread historicalinterest within the catalysis community, many detailed mechanistic studies ofthis reaction have been conducted on monometallic and alloy surfaces. Most recently, bimetallic catalysts containing Pt[1, 2] have garnered much attention forfuel cell applications and PdAu alloys[3] have been the subject ofresearch for hydrogen peroxide synthesis applications. In this study,first principles electronic structure calculations based on periodic,self-consistent, density functional theory (DFT) were utilized to study themechanism for oxygen reduction by hydrogen on an alternative Pd-noble metalalloy surface, namely PdAg(110). Our results demonstrate that the minimumenergy pathway involves the initial formation of a peroxyl (OOH) intermediatefollowed by O-O bond scission, consistent with the minimum energy pathway shownfor oxygen reduction by hydrogen on monometallic Pd and Ag (111) surfaces.[4] The lower activation energybarrier for O-O bond scission in OOH versus hydrogenation of OOH to form HOOHsuggests that PdAg is not an effective catalyst for the direct synthesis ofhydrogen peroxide. The detailed thermochemistry and activation energy barriersof important elementary steps and intermediates in oxygen reduction by hydrogenon PdAg(110) are compared and contrasted with the analogous results recentlyreported for the monometallic surfaces.[4]

1. Stamenkovic, V.R.; Mun, B. S.; Arenz, M.; Mayrhofer, K. J. J.; Lucas, C. A.; Wang, G. F.;Ross, P. N.; Markovic, N. M., Trends in electrocatalysis on extended andnanoscale Pt-bimetallic alloy surfaces. *Nature Materials* **2007,** *6*(3), 241-247.

2. Zhang, J. L.;Vukmirovic, M. B.; Xu, Y.; Mavrikakis, M.; Adzic, R. R., Controlling thecatalytic activity of platinum-monolayer electrocatalysts for oxygen reductionwith different substrates. *Angewandte Chemie-International Edition* **2005,***44* (14), 2132-2135.

3. Edwards, J. K.; Solsona,B.; N, E. N.; Carley, A. F.; Herzing, A. A.; Kiely, C. J.; Hutchings, G. J., Switching Off Hydrogen Peroxide Hydrogenation in the Direct Synthesis Process. *Science***2009,** *323* (5917), 1037-1041.

4. Ford, D. C.;Nilekar, A. U.; Xu, Y.; Mavrikakis, M., Partial and complete reduction of O-2by

hydrogen on transition metal surfaces. *Surface Science* **2010,** *604*(19-20), 1565-1575.

See more of this Session: Computational Catalysis IV
See more of this Group/Topical: Catalysis and Reaction Engineering Division

(211f) A First Principles Analysis of the Selective Hydrogenation of Unsaturated Ketones: Influence of Hydrocarbon Structure

Bing Hao and Matthew Neurock, Department of Chemical Engineering, University of Virginia, Charlottesville, VA

The selective hydrogenation of carbonyl groups in unsaturated ketones (UK) is important in the catalytic synthesis of various pharmaceutical and fine chemical intermediates as well as in biomass conversion strategies [1]. The ability to selectively hydrogenation the C=O group over that of the C=C bond presents a significant challenge as the hydrogenation of the C=C is thermodynamically favored over that of the C=O bond [2,3]. Recent experimental observations reveal that the adsorption and selectivity for the hydrogenation of unsaturated ketones depends strongly on the substituents on the C=C bond [4,5]. One successful example is the selective hydrogenation of the sterically-hindered C=O group in ketoisophorone (KIP) over Pd/Al_2O_3 [6]. Very little, however, is known about the specific role that substitutions play in the elementary hydrogenation mechanisms of unsaturated ketones. To probe this effect, density functional theory (DFT) calculations were carried out to explore the elementary hydrogenation steps for a series of model ketones over different metal surfaces, beginning with the simplest unsaturated ketone, methyl vinyl ketone. Methyl substituents were then added one by one to the C=C bond to establish their influence on the kinetics.

Increasing degree of substitution at the C=C bond was found to weaken the adsorption strength of UK reactants and alter the energetics of hydrogenation steps by increasing the hydrogenation barrier at the substituted carbon, while lowering the barrier to hydrogenate at its nearby carbon. Increasing the degree of substitution notably decreases the activation barrier for the first hydrogenation step of UK thus increasing the likelihood of selectively hydrogenating the C=O bond over the C=C bond.

We have also examined the influence of substituents on the hydrogenation of KIP over Pd. KIP is a much more complex ketone as it is comprised of three reducible groups and as such offers a number of different products that can form. Increasing the degree of substitution on the C=C bond was again found to increase the hydrogenation barrier for C=C while inducing a decrease in barrier to hydrogenate the sterically hindered C=O bond. This is in agreement with the experimental results which show enhanced selectivity to 4-HIP.

Reference

(1) R.A. Sheldon, H. van Bekkum, Fine Chemicals through Heterogeneous Catalysis, Weinheim: WILEY-VCH, (2001).

(2) C. Mohr, P. Claus , Science Progress, 84 (2001) 311-334.

(3) P. Gallezot, D. Richard, Catalysis Reviews-Science and Engineering, 40 (1998) 81-126

(4) C. Milone, R. Ingoglia, A. Pistone, G. Neri, F. Frusteri, S. Galvagno, Journal of Catalysis, 222 (2004) 348-356.

(5) C. Milone, R. Ingoglia, M.L. Tropeano, G. Neri, S. Galvagno, Chemical Communications, (2003) 868-869.

(6) M. von Arx, T. Mallat, A. Baiker, Journal of Molecular Catalysis a-Chemical, 148 (1999) 275-283.

See more of this Session: Computational Catalysis IV
See more of this Group/Topical: Catalysis and Reaction Engineering Division

(211g) Reaction Free Energies In Solutions for Glucose Conversion Into Biomass Intermediates

Vinit Choudhary, Dionisios G. Vlachos and Stanley I. Sandler, Chemical Engineering, University of Delaware, Newark, DE

Sugar dehydration reactions are of great importance in the conversion of biomass into fuels and chemicals. In this study, we have computed reaction free energies in solution for glucose conversion into levulinic acid and levoglucosan (LG). We computed vapor phase energies using the G4 ab initio method in Gaussian and estimated solvation free energies using COSMO-SAC (conductor-like screening model-segment activity coefficient) model. We used water and dimethyl sulfoxide (DMSO) as solvents. Our calculations predict glucose isomerization into fructose and mannose is reversible in both water and DMSO, whereas fructose dehydration into 5-(hydroxymethyl)furfural (HMF) is practically irreversible reaction. We predict that glucose dehydration into LG can be a reversible reaction depending on the reaction conditions as well as the solvent choice. In the vapor phase, at higher temperature, the entropic effect dominates and LG formation becomes favorable. It is found that glucose dehydration into levoglucosan is not thermodynamically feasible below 200°C in aqueous media but can become feasible in organic solvents. These calculations provide guidance into how to selectively control sugar dehydration by changing the reaction conditions. Experiments conducted in our lab and from the literature are contrasted with model predictions and good overall agreement is found.

See more of this Session: Computational Catalysis IV
See more of this Group/Topical: Catalysis and Reaction Engineering Division

(212a) Time-Dependent, Thermal-Capillary Analyses of the Micro-Pull-Down Crystal Growth System

Gaurab Samanta, Andrew Yeckel and Jeffrey J. Derby, Chemical Engineering and Materials Science, University of Minnesota, Minneapolis, MN

The micro-pull-down crystal growth technique is becoming increasingly popular to grow small-diameter, fiber crystals. Recently, it is also being developed for quick screening of promising scintillator crystals for radiation detectors in the crystal growth facility of E.D. Bourret-Courchesne at Lawrence Berkeley National Laboratory (LBNL). In this use, time delays are often associated with establishing suitable growth conditions for micro-pull-down experiments for new materials that have never been grown before.

The goal of our research is to develop computational models that can address the determination of suitable growth conditions in the micro-pull-down systems using a minimal amount of information. In keeping with this goal, we first developed a two- dimensional, quasi-steady-state (QSS), thermal-capillary model of the micro-pull-down system to grow sapphire crystals. We used this model to identify growth windows with respect to process parameters representing heater input, ambient temperature, pull rate, and melt static head. However, we did not establish the stability of the QSS solution branches.

In this presentation, we focus on results from transient simulations of the thermal- capillary model used to establish the stability behavior of those solution branches. For this purpose, starting with a quasi-steady-state solution, we provide step changes to any of the process parameters stated above and evolve the thermal-capillary system in time. By following the crystal radius evolution over time, we conclude whether a particular solution branch is stable or not. We also conduct transient simulations of a batchwise set-up of the micro-pull-down process and, for the first time, present temporal predictions of system evolution.

See more of this Session: Computational Studies of Electronic and Photonic Materials I
See more of this Group/Topical: Materials Engineering and Sciences Division

(212b) Simulation of Heat Transfer and Convection During Sapphire Crystal Growth by the Heat Exchanger Method

Hyun Gyoon Park[1], **Nan Zhang**[2] and Jeffrey J. Derby[2], (1)Advanced Materials Engineering, Korea Polytechnic University, Shiheung, Gyeonggi, South Korea, (2)Chemical Engineering and Materials Science, University of Minnesota, Minneapolis, MN

Sapphire single crystals are an important substrate for the fabrication of gallium-nitride light-emitting-diodes (LEDs) that emit white light. Theses solid-state devices promise great energy savings over both incandescent and compact fluorescent lighting technologies, and their market is expanding at double-digit growth rates. Key to continuing cost reduction for such LED-based lighting is reducing the cost of sapphire without compromising quality. The heat exchanger method (HEM) represents a possible yield improvement over existing growth technologies (primarily Kyropoulos growth) and is the subject of the study presented here.

We have performed quasi-steady state analysis using the commercial code CrysMAS to investigate the effects of furnace geometry and processing parameters on the temperature distribution, convection field, and melt-crystal interface shape during the crystal growth of sapphire by heat exchanger method (HEM). The temperature profile in the system is calculated to have lower temperatures both at the base of crucible and at the center top of the melt; however, the presence of convection flow arising from Marangoni force and buoyancy can ameliorate the latter effect. The interface shape is also predicted to change from flat to convex as crystal growth proceeds. Changes in geometry of furnace, such as segmentation of heater, changing the crucible position relative to heater center, and altering the width of crucible support, are shown to affect the temperature profile significantly. Based upon such results, optimization of furnace design will be presented.

See more of this Session: Computational Studies of Electronic and Photonic Materials I
See more of this Group/Topical: Materials Engineering and Sciences Division

(212c) Analysis of Current-Driven Morphological Evolution of Monolayer-Thick Coherently Strained Heteroepitaxial Islands On Substrates

Georgios I. Sfyris, Dwaipayan Dasgupta, M. Rauf Gungor and Dimitrios Maroudas, Chemical Engineering, University of Massachusetts Amherst, Amhest, MA

Surface morphological evolution under the action of external fields has attracted considerable attention over the past two decades. In addition to the interest in a fundamental understanding of field-induced nonlinear response and stability of surface morphology, the problem has been technologically significant in various engineering applications such as microelectronics and nanofabrication. Of particular interest is the driven assembly of confined quantum structures, such as islands grown epitaxially on substrates of different chemical composition than that of the islands. The lattice mismatch between the island and substrate materials induces a misfit strain, which is the source of elastic deformation of the islands. The surface morphological dynamics induced by an externally applied field, such as an electric field, under the simultaneous action of the misfit strain can give rise to morphological patterns, the stability and control of which may have significant impact on fabrication technologies.

In this presentation, we report the results of a theoretical and computational analysis of the current-driven morphological response of coherently strained heteroepitaxial islands on elastic substrates. We have developed a fully nonlinear model for studying the current-driven morphological evolution of one-monolayer-thick coherently strained heteroepitaxial islands on elastically deformable substrates. The model follows a two-dimensional (2D) approximation of island evolution, according to which diffusional mass transport is limited to the island circumference. Within this approximation (i.e., for this mass transport regime), we have carried out a theoretical analysis of the morphological stability of such isolated islands and of the current-driven migration of morphologically stable islands; the theory predicts the dependence of the stable island migration speed on the island size and on the heteroepitaxial system parameters. We have also developed self-consistent dynamical simulators of the driven morphological evolution of such islands, combining front tracking methods with an analytical solution for the corresponding electric-field component tangential to the island's circumference and a numerical computation of the corresponding elastic displacement field. Simulation results are reported together with systematic parametric studies to investigate thermal and size effects on the island's migration and morphological evolution, as well as effects on the island driven dynamical response of elastic properties of the substrate and island materials (such as elastic moduli and misfit strain). We demonstrate a variety of stable asymptotic states in the driven dynamical response of such heteroepitaxial islands and characterize in detail the resulting morphological patterns.

The simulation results validate our theory for stable heteroepitaxial island driven migration. The results also generate experimentally testable hypotheses and motivate experimental measurements that can be compared directly with the theoretical predictions. Finally, the present modeling study sets the stage for designing simulations of the current-driven evolution of populations of heteroepitaxial islands and examination of collective phenomena, such as island coalescence.

See more of this Session: Computational Studies of Electronic and Photonic Materials I
See more of this Group/Topical: Materials Engineering and Sciences Division

(212d) Dielectric Properties of Bismuth Pyrochlores From First Principles Calculations

Beverly Brooks Hinojosa[1], Juan C. Nino[1] and Aravind Asthagiri[2], (1)Department of Materials Science and Engineering, University of Florida, Gainesville, FL, (2)Department of Chemical & Biomolecular Engineering, The Ohio State University, Columbus, OH

There is considerable interest in pyrochlore systems ($A_2B_2O_7$) for use in high-permittivity dielectrics, capacitors, and high-frequency filter applications. The properties of these materials can be tuned through substitutions on the A and B cation sites, resulting in an extensive parameter space. Better understanding of the role of the local atomic structure and dynamics on the macroscopic properties will enable rational design within the vast number of possibilities. For example, the dielectric relaxation in Bi-based pyrochlores were linked to local atomic hopping events and it has been proposed that a highly polarizable cation, chemical substitution, atomic displacement, and the resulting fractional occupancy of equivalent sites, are necessary for dielectric relaxation in pyrochlores. Based on density functional theory (DFT) calculations, we have shown $Bi_2Ti_2O_7$ is expected to have ionic displacements. Being a bismuth based pyrochlore without chemical substitution of multi-valence cations, $Bi_2Ti_2O_7$ is an ideal compound to de-convolute these features and better understand the nature of the dielectric relaxation in pyrochlores. Quantum mechanical calculations were performed on $Bi_2Ti_2O_7$ and other pyrochlores ($Bi_{1.5}ZnNb_{1.5}O_7$, $Ca_{1.5}Ti_{1.5}NbO_7$) with the Fd-3m (No. 227) space group to determine the role of a highly polarizable cation, chemical substitution, and atomic displacement. We will present DFT results on the structural properties of $Bi_2Ti_2O_7$ and other Bi-containing pyrochlores including lattice stability, atomic displacement patterns, atomic hopping events, and the static dielectric permittivity. Based on these results, predictions will be made regarding the likelihood of dielectric relaxation in $Bi_2Ti_2O_7$ which further illustrate the need to synthesize the $Bi_2Ti_2O_7$ pyrochlore without the ferroelectric $Bi_4Ti_3O_{12}$ secondary phase, which is expected to mask the intrinsic dielectric behavior of the pyrochlore. Where possible, these predictions are compared with available experimental results.

See more of this Session: Computational Studies of Electronic and Photonic Materials I
See more of this Group/Topical: Materials Engineering and Sciences Division

(212e) SrTiO3 Based Anode Materials for Solid Oxide Fuel Cells: A Computational Attempt to Understanding and Improving Performance

Andreas Heyden and Suwit Suthirakun, Department of Chemical Engineering, University of South Carolina, Columbia, SC

In recent years, there has been a significant interest in alternative anode materials for solid oxide fuel cells (SOFCs) that are active and stable under hydrocarbon fuel conditions. The requirements for promising SOFC anode materials are good chemical and physical stability under operating conditions, high ionic and electronic conductivity, and good catalytic activity for the anode reaction. Furthermore, the material should not be susceptible to deactivation by both carbon formation and sulfur poisoning. Several alternative anode materials have been proposed that possess fluorite, rutile, tungsten bronze, pyrochlore, perovskite, and spinel structures. Among these anode materials, perovskite and double perovskite oxides are the most promising SOFC anode materials since they exhibit high ionic/electronic conductivity and high catalytic activity for various electrochemical reactions. It is the hypothesis of this research that modern computational tools permit the systematic investigation of the effect of dopants in the perovskite structure on various anode properties including electronic and ionic conductivity, catalytic activity, and resistance to coking and sulfur poisoning. In particular, we are interested here in understanding recently reported conductivity measurements of Ga and Nb doped A-site deficient $SrTiO_3$ [1].

Various configurations of Ga doped $Sr_{0.9}Ti_{0.8}Nb_{0.2}O_{3-x}$ at various concentrations of Ga were investigated to gain a better understanding of the electronic conductivity behavior of the material under SOFC operating conditions. Experimental results suggest that Ga doped $Sr_{0.9}Ti_{0.8}Nb_{0.2}O_3$ displays metallic behavior with a conductivity maximum at 10% Ga doping. Calculated densities of states of these materials suggest that 10% Ga doped $Sr_{0.9}Ti_{0.8}Nb_{0.2}O_3$ exhibits p-type doping behavior which would reduce the electronic conductivity of Nb (n-type) doped $Sr_{0.9}Ti_{0.8}Nb_{0.2}O_3$. On the other hand, in its reduced form $Sr_{0.9}Ti_{0.7}Nb_{0.2}Ga_{0.1}O_{2.9}$ displays increased n-type doping behavior due to the presence of an oxygen vacancy. This effect disappears whenever two Ga atoms are in close proximity.

Ab initio thermodynamic calculations were performed to determine the relative thermodynamic stability of various doped $SrTiO_3$ systems under experimental anode conditions ($P_{O2}= 10^{-20}$ bar). 10% Ga doped systems tend to form oxygen vacancies at elevated temperatures. As a result, the most stable structure of 10% Ga doped $Sr_{0.9}Ti_{0.8}Nb_{0.2}O_3$ is in the reduced form ($Sr_{0.9}Ti_{0.7}Nb_{0.2}Ga_{0.1}O_{2.9}$) and displays n-type doping behavior. To summarize, the experimentally observed conductivity maximum of A-site deficient

10% Ga-doped $Sr_{0.9}Ti_{0.7}Nb_{0.2}Ga_{0.1}O_{3-x}$ can be explained by small amounts of Ga promoting the reduction of the material which increases conductivity. Introducing larger amounts of Ga results in (reduced) Ga-Ga sites that do not increase n-type doping behavior and conductivity.

[1] G. Xiao, et al., *Synthesis and Characterization of A-site Deficient Perovskite* $Sr_{0.9}Ti_{0.8-x}Ga_xNb_{0.2}O_3$, Mater.Res.Bull. (2010), doi:10.1016/j.materresbull.2010.09.044

See more of this Session: Computational Studies of Electronic and Photonic Materials I
See more of this Group/Topical: Materials Engineering and Sciences Division

(212f) The Formation of Polytetrahedral Structures In Elongated Gold Nanowires

Christopher R. Iacovella[1], William R. French[1], Paul R.C. Kent[2] and Peter T. Cummings[3], (1)Chemical and Biomolecular Engineering, Vanderbilt University, Nashville, TN, (2)Oak Ridge National Laboratory, Oak Ridge, (3)Center of Nanophase Materials Sciences, Oak Ridge National Laboratory, Oak Ridge, TN

The observation of quantized conductance in single-atom wide metallic junctions has fueled much interest into understanding the behavior of nanowires undergoing tensile elongation [1]. Quantifying the structural changes that occur as a nanowire elongates is important to elucidating the origin of changes to nanowire properties, such as conductance or the opening of a bandgap, and has additional implications regarding the appropriateness of the assumed geometries used in high level theoretical calculations of nano- and molecular-electronics, e.g., the conductance through single atom wide chains or bridged single molecules.

In this work, we use molecular simulation to study the structural evolution of Au nanowires undergoing tensile elongation. While most of the previous work on gold nanowires has focused on the crystalline nature of the wire, here we focus on identifying the non-crystalline, "amorphous" structures that often form within the necks of the nanowires. To study this under conditions matching experiment, we perform molecular dynamics simulations with two different interatomic Au potentials. We use the tight-binding second-moment approximation (TB-SMA) potential, a common semi-emperical potential used to study Au nanowire elongation [2], as well as the more sophisticated ReaxFF bond-order potential with fitting parameters derived by Keith, et al. [3].

We report the widespread formation of polytetrahedral Z12 (i.e., icosahedron) and Z14 structures with both potentials, quantifying their presence using a shape-matching order parameter based on spherical harmonics [4,5]. We combine these results with density functional theory (DFT) calculations to investigate the energetic description of the polytetrahedral structures observed with TB-SMA and ReaxFF, finding that ReaxFF better captures the energetics predicted by DFT. While both potentials form polytetrahedra, our study suggests that TB-SMA based simulations over-predict their formation, most notably resulting in increased mechanical stability for nanowires with diameters less than 1.5 nm. We further perform calculations of the conductance [6] of the elongated nanowires configurations generated using ReaxFF, which show close agreement with experimental measurements [1], further validating the structural evolution predicted by our simulation results.

[1] J.M. Krans, J.M. van Rutlenbeek, V.V. Fisun, I.K. Yanson, and L.J. de Jongh, *Nature* 375, 767 (1995).
[2] Q. Pu, Y.S. Leng, P.T. Cummings, Rate-Dependent Energy Release Mechanism of Gold Nanowires under Elongation, *JACS* 130, 52, (2008)

[3] J.A. Keith, D. Fantauzzi, T. Jacob, A.C.T. van Duin, Reactive Forcefield for Simulating Gold Surfaces and Nanoparticles, *PRB*, 81, 235404, (2010)
[4] C.R. Iacovella, A.S. Keys, M.A. Horsch, S.C. Glotzer, Icosahedral Packing of Polymer-tethered Nanospheres and Stabilization of the Gyroid Phase, *PRE* 75, 4, (2007)
[5] A.S. Keys, C.R. Iacovella, S.C. Glotzer, Characterizing Structure Through Shape Matching and Applications to Self-Assembly. *Annual Review of Condensed Matter Physics*, 2, (2011)
[6] K. Varga, S.T. Pantelides, Quantum Transport in Molecules and Nanotube Devices, *PRL* 98, 076804 (2007)

See more of this Session: Computational Studies of Electronic and Photonic Materials I
See more of this Group/Topical: Materials Engineering and Sciences Division

(212g) Computational Study of Electronic and Optical Properties of Materials for Energy Applications

Maria Stoica, Energy Environmental and Chemical Engineering, Washington University in St Louis, St Louis, MO and Cynthia S. Lo, Energy, Environmental and Chemical Engineering, Washington University in St. Louis, Saint Louis, MO

We use *ab initio* atomistic DFT modeling as implemented in VASP to determine theoretical values of electronic and optical properties for skutterudites and zinc-based oxides. Among the properties we report are electrical conductivity and Seebeck coefficient. For skutterudites, we study the effect of fractionally filling both CoSb3 and CoAs3 with Ag and Au particles. For zinc oxide, we study the effects of doping ZnO with F and Al, both by direct replacement and interstitial insertion. We compare computationally derived values to experimentally observed values of the unfilled and undoped materials and make recommendations for improvements in estimation methods. We ultimately will use these computational methods to provide predictive insight into experimental materials behavior and guide the design of advanced thermoelectrics and transparent conducting oxides.

See more of this Session: Computational Studies of Electronic and Photonic Materials I
See more of this Group/Topical: Materials Engineering and Sciences Division

(225a) Molecular Simulations of Tryalkoxysilane Functionalized Porous Silicates

Santiago Builes[1], Pedro Lopez-Aranguren[1], Roberta Pacciani[2], Concepcion Domingo[3] and **Lourdes F. Vega**[2], (1)MATGAS Research Center and ICMAB-CSIC, Bellaterra, Barcelona, Spain, (2)MATGAS Research Center and Carburos Metlicos, Air Products Group, Barcelona, Spain, (3)Institut de Ciencia de Materials de Barcelona. Consejo Superior de Investigaciones Cientificas (ICMAB-CSIC), Bellaterra, Barcelona, Spain

Trialcoxysilanes are bifunctional molecules widely used to obtain self-assembled coatings by the chemisorption route in liquid solvents. However, problems arise when the method is used for porous materials due to cross-linking of alkylsilanes in solution, which lead to multilayer deposition, non-uniform surface coverage and pore blocking.

We propose a methodology for the design of models of post-synthesis functionalized silica materials. Using an energy-bias selection scheme for the possible grafting sites, we generate realistic models of the modification of the internal surface of mesoporous silica gels. The amorphous mesoporous supports were chosen for analysis in front of the most studied ordered mesoporous silicas of the type MCM-41 or SBA-15, because amorphous silica materials are considered low-cost adsorbents and could be used in a large amount of bulk applications. A hydrophobic silane was used to impregnate the internal surface to obtain oil adsorbents.

We compare the grafting densities obtained by supercritical silanization and traditional silanization to the maximum theoretical densities obtained by the proposed methodology. We analyze the decrease in grafting density with respect to flat surfaces, in terms of molecules per nm^2, as a result of sterical constrictions for molecules inside narrow irregular pores.

The behavior of N_2 molecules adsorbed on functionalized silicas is analyzed by means of Grand Canonical Monte Carlo (GCMC) simulations. Then, N_2 adsorption isotherms for the support and the functionalized samples between the simulation and the experimental procedures are compared. The comparison of the decrease in the surface area and pore volume serve as an indication of pore blockage in the experimental materials.

The understanding of the grafting of tryalkoxysilanes in silica adsorbents plays a key role on further optimization of the synthesis of these materials. We present our experimental adsorption isotherms and show the potential applications of this kind of materials as oil adsorbents; these results are compared to the simulated isotherms to validate the models and the methodology.

This work was partially financed by the Spanish Government under projects CTQ2008-05370/PPQ, CTQ2011-23255 and CENIT SOST-CO2 (CEN-2008-1027). Additional support from the Catalan Government through 2009SGR-666 and Carburos Metálicos was also provided. S.B. acknowledges a Talent contract from the Catalan Government.

See more of this Session: Industrial Applications of Computational Chemistry and Molecular Simulation I
See more of this Group/Topical: Computational Molecular Science and Engineering Forum

(225b) Monte Carlo Simulation of Fluid Phase Equilibria and of Adsorption In Microporous Materials : Investigation of Binary and Multicomponent Systems of Industrial Interest

Marianna Yiannourakou[1], **Philippe Ungerer**[1], Benoit Leblanc[1] and Paul W. Saxe[2], (1)Materials Design Sarl, Montrouge, France, (2)Materials Design, Inc, Angel Fire, NM

Monte Carlo simulation of fluid phase equilibria and of adsorption in microporous materials : investigation of binary and multicomponent systems of industrial interest.

Marianna Yiannourakou[1], Philippe Ungerer[1], Benoit Leblanc[1] and Paul Saxe[2]

[1]Materials Design Sarl, Montrouge, France

[2]Materials Design Inc., Angel Fire, USA

Several factors make molecular modelling increasingly attractive for industrial applications : the availability of powerful algorithms for complex systems, the development of industrial software allowing efficient use by chemical engineers, the decreasing cost of computing time, and the parametrization of well-tested force fields. The purpose of the talk is to illustrate this trend in the case of Monte Carlo simulation techniques, applied to the equilibrium properties of fluids -including adsorption in microporous solids such as zeolites and model carbons-. For organic fluids we use generic force fields based on united atoms such as TraPPE [1] or AUA [2]. Force field parameters for small molecules (H_2, CO_2, etc.) and for solids are taken from published work, and standard combining rules are selected.

The first type of system investigated is the solubility of gases (H_2, N_2, O_2, CH_4, CO_2, …) in organic liquids (hydrocarbons, ethanol, dimethylether, cresol….). This is an important topic for the conversion of heavy crude oil fractions into high quality fuels and for the co-processing of ligno-cellulosic biomass. We show that biased Widom test insertions allow reliable determination of the Henry constant of gas solubility. Simulation results predict consistently the influence of all factors : nature of the gas, temperature, and nature of the organic liquid. These results provide a quantitative explanation for the solubility behavior : very light gases like H_2 do not display sufficient attraction for the solvent, and they are mostly sensitive to the presence of temporary cavities in the liquid. Their entropy-driven solubility increases with temperature, as a result of the decreasing density of the solvent. Heavier gases like CO_2 feel more strongly the attraction for the liquid solvent, and the effect of temperature is therefore opposite. Gibbs ensemble simulations allow a very consistent extrapolation of these results to high pressures, whenever data are available (especially considering alcohols, ethers..). They allow also to model the vapour phase composition at high pressure.

The second type of systems considered in this work is the adsorption of multicomponent fluids in zeolites, which are often used in the industry for purifications and delicate separations . Grand Canonical ensemble Monte Carlo techniques that are used for this purpose. The assumptions made while building the sodium faujasite models NaY and NaX (framework structure, location of sodium cations in zeolites, force field parameters…) follow from published work [3]. Simulation results are compared with experimental adsorption data on pure compounds (water, alkanes, aromatics, ethanethiol), and binary systems, with very good results with some exceptions. Once these validations are made, the behavior of complex multicomponent systems in condensed phases (high pressure natural gas, liquid mixture) can be considered. For this purpose, compressibility factor at high pressure may be used as an indirect test of the gas model, and biased test insertions are used to determine the chemical potential of each molecular type. The adsorption behavior of the multicomponent gas may then be investigated. This is illustrated by the example of natural gases comprising up to 12 components. The sensitivity of adsorption to minute amounts of heavy or polar compounds (such as water, methanethiol, liquid hydrocarbons) is discussed.

The examples treated in this work are systems for which analytical models (equations of state, activity coefficient models,…) need abundant experimental data to be parametrized. Our results support the idea that molecular modeling is now a cost-effective tool to perform difficult extrapolations. Other comparable applications (supercritical fluids, sequestration of CO_2 for coal bed methane recovery) are cited in conclusion.

1. Maerzke, K.A., et al., *TraPPE-UA Force Field for Acrylates and Monte Carlo Simulations for Their Mixtures with Alkanes and Alcohols.* Journal of Physical Chemistry B, 2009. **113**(18): p. 6415-6425. and references cited.

2. Ferrando, N., et al., *Transferable Force Field for Alcohols and Polyalcohols.* Journal of Physical Chemistry B, 2009. **113**(17): p. 5985-5995, and references cited.

3. diLella, A., et al., *Molecular simulation studies of water physisorption in zeolites.* Phys. Chem. Chem. Phys., 2006. **8**: p. 5396-5406.

See more of this Session: Industrial Applications of Computational Chemistry and Molecular Simulation I
See more of this Group/Topical: Computational Molecular Science and Engineering Forum

(225c) Modeling Molecular Recognition In Polymer Self-Assembly

Jed W. Pitera, Science & Technology, IBM Almaden Research Center, San Jose, CA

Molecular recognition elements (hydrogen bonds, ion pairs, metal-ligand complexes) provide strong, selective, anisotropic interactions useful for tuning the structure, properties, and functionality of advanced polymeric materials. However, the strong interactions that make molecular recognition elements useful present a challenge for traditional simplified polymer models. We have developed a mixed molecular mechanics-coarse grained model to treat sparse molecular recognition elements in polymer melts and solutions. The bulk of the polymer is treated with a coarse grained model of bead-spring chains interacting with a soft Flory-type potential. I will detail the model as well as its application to explain how a 24-Dalton change in a 10000 kD antibacterial polymer can radically alter its morphology.

See more of this Session: Industrial Applications of Computational Chemistry and Molecular Simulation I
See more of this Group/Topical: Computational Molecular Science and Engineering Forum

(225d) Computer-Aided MOLECULAR Design: Methodology & APPLICATIONS

K. M. Yerramsetty, B. J. Neely and K. A. M. Gasem, School of Chemical Engineering, Oklahoma State University, Stillwater, OK

The demand for newly-designed molecules that enhance existing processes and satisfy more stringent operating requirements in technology has been increasing. However, the design of molecules with desired properties challenges engineers attempting to meet the needs of various industries, including pharmaceuticals, polymers, petrochemicals and construction. The traditional approach of identifying molecules with desired properties involves testing thousands of molecules for their chemical and physical properties, which is an expensive and laborious undertaking. Hence, design techniques, such as computer-aided molecular design (CAMD), have found wide application in recent years.

In this work, an automated CAMD technique that combines genetic algorithms (GAs) as the search tool and three-dimensional non-linear quantitative structure-property relationship (QSPR) models as the prediction platform was developed. Specifically, evolutionary algorithms (EAs) in combination with artificial neural networks (ANNs) were used to select the most significant molecular descriptor inputs to the QSPR models, while identifying simultaneously the optimal model architecture. The developed CAMD methodology was validated using different applications. Herein, we report our findings on the design of novel chemical penetration enhancers (CPEs) for enhancing insulin permeation.

See more of this Session: Industrial Applications of Computational Chemistry and Molecular Simulation I
See more of this Group/Topical: Computational Molecular Science and Engineering Forum

(225e) Molecular Simulation Strategies for Large Scale Thermodynamic Data Generation

Gbor Rutkai[1], Monika Thol[2], Roland Span[2] and Jadran Vrabec[1], (1)Institute of Thermodynamics and Energy Technology, University of Paderborn, Paderborn, Germany, (2)Institute of Thermodynamics, Ruhr-University Bochum, Bochum, Germany

Today almost every effort that aims at technological process design and optimization requires a sufficient amount of reliable thermodynamic data for the involved chemical compounds. Traditionally, thermodynamic properties are obtained from experiments and are subsequently expressed in terms of equations of state (EOS). The development of a highly accurate EOS requires a broad basis of reliable thermodynamic data containing as many different properties as possible. Despite the extensive effort that was invested in measurements over more than a century, the data availability today is still surprisingly low. Among the about 1000 chemical pure compounds that are in technological use [1], a complete thermodynamic knowledge is available for about 10, and an advanced or satisfactory knowledge is available for less than 100 [2]. Considering mixtures, where the number of relevant systems is orders of magnitude larger, the data availability is much worse.

Recent progress in molecular simulation has shown that molecular models (force fields) have powerful predictive capabilities with respect to thermodynamic data [3]. Such information can also be straightforwardly accessed for fluids and states which are experimentally difficult to investigate. Moreover, molecular simulation allows for the generation of large data sets containing consistent information on arbitrary thermodynamic properties at low cost. As there is a dire need for raw thermodynamic data for EOS optimization, the idea to develop a new generation of EOS which are partly based on molecular simulation data comes naturally. In practice, however, the generation of extensive data sets that contain as many different thermodynamic properties as possible may be cumbersome. Standard textbook approaches in the molecular simulation literature indicate specific statistical mechanical ensembles for particular thermodynamic properties [4]. It is true that certain properties have simpler statistical analogs in certain ensembles and may be difficult to derive in other ensembles. However, any thermodynamic property that can be measured in a given statistical mechanical ensemble can also be measured in any other statistical mechanical ensemble [5]. This is a direct consequence of the physical equivalence of various forms of the thermodynamic fundamental equation, e.g. entropy $S = S(N, V, E)$, Helmholtz energy $A = A(N, V, T)$ or Gibbs energy $G = G(N, p, T)$. As a consequence, any arbitrary thermodynamic property can be expressed as a combination of higher-order partial or mixed derivatives of any of the chosen fundamental equation representation from: $S(N, V, E)$, $A(N, V, T)$ or $G(N, p, T)$, etc. [6, 7]. Since the *NVT* ensemble is a convenient option with respect to parameter settings, the representation $A = A(N, V, T)$ was chosen here, considering the appropriate derivatives of the Helmholtz energy. This formalism was implemented in the molecular simulation tool *ms*2 [8] for electro-neutral Lennard-Jones based force fields with superimposed electrostatic sites. It can be shown that basically every static thermodynamic property can be determined using only *NVT* ensemble molecular simulations offering a new route to develop empirical EOS on the basis of hybrid data sets, composed of experimental data and molecular simulation data.

References

[1] K.E. Gubbins, N. Quirke, “Molecular Simulation and Industrial Application”, *Gordon and Breach*, Amsterdam (1996)

[2] R. Span, “Multiparameter Equations of State”, *Springer*, Berlin (2000)

[3] http://www.fluidproperties.org, Industrial fluid properties simulation collective (2011)

[4] M. Lagache, P. Ungerer, *Phys. Chem. Chem. Phys.*, 3, 4333-4339 (2001)

[5] H.W. Graben, J.R. Ray, *Mol. Phys.*, 80, 1183-1193 (1993)

[6] R. Lustig, *J. Chem. Phys.*, 100, 3048-3059 (1994)

[7] K. Meier, S. Kabelac, *J. Chem. Phys.*, 124, 064104-064114 (2006)

[8] S. Deublein et al., *Comp. Phys. Comm.*, in press (2011)

See more of this Session: Industrial Applications of Computational Chemistry and Molecular Simulation I
See more of this Group/Topical: Computational Molecular Science and Engineering Forum

(225f) Characterization of Thermal Stability In Endoglucanases with Bioinformatics, Molecular Simulation, and Rosetta Design

Christina M. Payne[1], Deanne W. Sammond[1], Yannick J. Bomble[1], Michael E. Himmel[1], Michael F. Crowley[1] and Gregg T. Beckham[2], (1)Biosciences Center, National Renewable Energy Laboratory, Golden, CO, (2)National Bioenergy Center, National Renewable Energy Laboratory, Golden, CO

The desire for increased protein thermal stability in commercial applications of cellulases motivates continued research focusing on the molecular aspects of protein stability. Starting with a well-defined set of published data for mesophilic, thermophilic, and hyperthermophilic cellulase enzymes belonging to the Families 5 and 12 glycoside hydrolases, we have applied multiple techniques to gain insight into the structural features that confer thermal stability. Bioinformatics, along with sequence and structural alignments, have been used as a preliminary means of identifying features relevant to thermal stability. Rosetta simulations were then used to identify characteristic protein folding metrics in order to narrow down the critical structural features, which were subsequently evaluated by molecular dynamics. Here, we present insights into the engineered molecular basis for increased thermal stability from molecular dynamics simulations of selected mutant structures compared to known thermal stable endoglucanases.

See more of this Session: Industrial Applications of Computational Chemistry and Molecular Simulation I
See more of this Group/Topical: Computational Molecular Science and Engineering Forum

(225g) Molecular Dynamics Studies of the Structure and Dynamics of An Endoglucanase Cel9A In Ionic Liquid Solutions

Hanbin Liu, Supratim Datta, Blake A. Simmons and Kenneth Sale, Joint BioEnergy Institute, Emeryville, CA 94608 and Sandia National laboratories, Livermore, CA, Emeryville, CA

Molecular Dynamics Studies of the Structureand Dynamics of an Endoglucanase Cel9A in Ionic Liquid Solutions

Hanbin Liu, Supratim Datta, Blake A Simmons and Ken Sale

Joint BioEnergy Institute, Emeryville, CA 94608 and SandiaNational laboratories, Livermore, CA

Ionic liquids (ILs) are currently under intense study as solventsand as additives to macromolecules and have recently show great promise as a pretreatmentof lignocellulosic biomass for the production of biofuels. Pretreatment ofbiomass is followed by saccharification using a cocktail of cellulase enzymes (glycosidehydrolases) that catalyze the breaking of the b1-4glycosidic bonds of the cellulose, which was liberated from the biomass by theionic liquid, to glucose. This process means that these cellulases must betolerant of the ionic liquid, making the study of the structures and dynamicsof cellulases in the present of ionic liquids of great current interests inbiofuels research, especially in industrially relevant processes in whichbiomass pretreatment is consolidated with the saccharification step(simultaneous pretreatment and saccharification). In this work, we investigatedthe origins of the tolerance of a thermophilic family 9 cellulase (Cel9A) toionic liquids. Cel9A is an endoglucanase that also displays exoglucanaseactivity and has been experimentally shown to be functional in bioprocessing relevantconcentrations of the ionic liquid 1-ethyl-2methyl imidazolium acetate [C2mim][OAc].Using all atom molecular dynamics simulations of Cel9A over a wide range of [C2mim][OAc]concentrations (weight range from 0% IL to 50% IL), we show that the cel9Astructure is stable in [C2mim][OAc] over the entire 80ns simulation at these ILconcentrations. We also show that, due to the slow diffusion of [C2mim][OAc] andinteractions between [C2mim][OAc] and Cel9A, [C2mim][OAc] stabilizes thedynamics of Cel9A. During our simulations, [C2mim][OAc] diffused into theactive site of the enzyme and interacts directly with the two catalyticresidues, indicating that [C2mim][OAc] may also promote enzymatic activity byaltering either the dynamics of the overall active site, the dynamics of thecatalytic residues or the electrostatic environment of the active site; we arecurrently investigating these possibilities.

See more of this Session: Industrial Applications of Computational Chemistry and Molecular Simulation I
See more of this Group/Topical: Computational Molecular Science and Engineering Forum

(258a) Mechanistic Modeling of Simultaneous Enzymatic Saccharification of Cellulose and Xylan Using Continuous Distribution Kinetics

Ambarish Nag[1], Andrew J. Griggs[2], Jonathan J. Stickel[2], Michael A. Sprague[1] and James J. Lischeske[2], (1)Computational Science Center, National Renewable Energy Laboratory, Golden, CO, (2) National Bioenergy Center, National Renewable Energy Laboratory, Golden, CO

Cost effective production of biofuels and biochemicals often involves enzymatic saccharification, which has been the subject of intense research and development. In this work, we develop a mechanistically based kinetic model for the enzymatic hydrolysis of biomass composed of cellulose, xylan, and lignin. The prototypical biomass particle in our model has a cellulosic core that is partially covered by xylan and lignin that impede the access of glucanase enzymes to the cellulosic core. Our model hence incorporates both glucanase and xylanase enzymatic saccharification of surface xylan and the cellulosic core. For degradation of insoluble cellulose chains we consider the actions of endoglucanase (endo-β-1,4-glucanase,) and exoglucanase (exo-β-1,4-cellobiohydrolase). Likewise for insoluble xylan we consider the actions of endoxylanase (endo-β-1,4-xylanase) and exoxylanase (xylan β-1,4-xylosidase), the latter enzyme hydrolyzing soluble xylan oligomers as well. The enzyme system also includes beta-glucosidase for the hydrolysis of soluble glucan oligosaccharides. The kinetics for the depolymerization of cellulose and xylan chains are described using population-balance equations, which are capable of describing the evolution of the distribution of chain sizes during hydrolysis, without solving equations for all chemical species present in the reacting mixture. This approach affords computationally efficient simulations, enabling the use of the model for process design and optimization. In combination with experiments, the kinetic model is used to study the transformations of the cellulose and xylan chain-size distributions during hydrolysis by an enzyme cocktail of all the enzyme species.

See more of this Session: Reaction Kinetics and Transport Fundamentals for Biomass Conversion
See more of this Group/Topical: Sustainable Engineering Forum

(258b) Kinetic Modeling of Cellulose Hydrolysis with First Order Inactivation of Adsorbed Cellulases

Zhuoliang Ye and R. Eric Berson, Chemical Engineering, University of Louisville, Louisville, KY

The rate of glucose release during enzymatic hydrolysis of cellulose slows down significantly as the reaction proceeds, and the mechanism is not yet clear. It is, therefore, desirable to develop a kinetic model that accurately describes the enzymatic hydrolysis reaction. In this work, a model based on Langmuir binding kinetics that also incorporates first order inactivation of adsorbed cellulase was developed that accurately predicts rates and yields during the hydrolysis reaction for two different cellulose substrates. Substrate concentrations of 20g/L and less were tested in order to remove any potential effects due to product inhibition or mass transfer limitations. A key result of the modeling suggests that Vmax decreases over time due to deactivation of the adsorbed enzyme. Rate constants associated with product formation and enzyme inactivation from our model yield are comparable to reported values elsewhere. The model predicted results within 10% of experimental results for both Solka Floc and Sigmacell substrates. Key model parameters were similar for the two substrates: the apparent hydrolysis rate (k_2) and inactivation rate (k_f) were 33 min^{-1} and 0.38 h^{-1}, respectively, comparable to values reported elsewhere. The rate constants were within 5% for the two different substrates, further strengthening the value of the model and implying that a common but yet undefined rate-limiting step associated with loss of enzyme activity likely exists in the pathway of cellulose hydrolysis. It is proposed that this rate-limiting step may be either dissociation or isomerization of the enzyme.

See more of this Session: Reaction Kinetics and Transport Fundamentals for Biomass Conversion
See more of this Group/Topical: Sustainable Engineering Forum

(258c) Observing and Modeling Cellulosic Substrate Depolymerization by Commercial Enzyme Cocktails Using Confocal Fluorescence Microscopy

Jeremy S. Luterbacher[1], Jose M. Moran-Mirabal[2] and Larry P. Walker[2], (1)Department of Chemical and Biomolecular Engineering, Cornell University, Ithaca, NY, (2)Department of Biological and Environmental Engineering, Cornell University, Ithaca, NY

Understanding and modeling the depolymerization mechanisms of cellulosic and lignocellulosic substrates by cellulase cocktails is a critical step towards further optimizing the production of monosaccharides from biomass. The Spezyme CP cellulase cocktail combined with the Novo 188 b-glucosidase blend was used to depolymerize bacterial microcrystalline cellulose (BMCC), filter paper cellulose and untreated and H_2O-CO_2 pretreated lignocellulosic substrates (hardwood and switchgrass), which were immobilized on a glass surface. The enzyme mixture was supplemented with a small fraction of fluorescently labeled *T. Reseii* CBH I, which served as a reporter to track cellulase diffusion into and binding onto the internal physical structure of the cellulosic substrate. All reported experiments were conducted at 50°C, the optimal temperature for maximum hydrolytic activity of the enzyme cocktail. BMCC was observed throughout degradation by labeling it with a fluorescent dye. Native autofluorescence was used in a similar way to track filter paper and lignocellulosic substrates. This method allowed us to measure the binding of cellulases *in situ* and follow the temporal morphological changes of cellulose and lignocellulosic substrates during their depolymerization by a commercial cellulase mixture.

In the case of BMCC, a kinetic model was developed and fitted to fluorescence intensity data obtained through image processing. This model was successfully used to predict the bulk sugar concentrations that were liberated over time from BMCC in a separate experiment and measured by liquid chromatography. A more complete mass transfer and kinetic model is currently under development for the more complex substrates (filter paper and lignocellulosic biomass).

See more of this Session: Reaction Kinetics and Transport Fundamentals for Biomass Conversion
See more of this Group/Topical: Sustainable Engineering Forum

(258d) Effects of Mixing Quality On the Kinetics of Biomass Liquefaction

David M. Lavenson[1], Emilio J. Tozzi[1], Tina Jeoh[2], Michael J. McCarthy[3] and Robert L. Powell[1], (1) Chemical Engineering and Materials Science, University of Calfiornia Davis, Davis, CA, (2)Biological and Agricultural Engineering, University of Califonrnia, Davis, Davis, CA, (3)Biological and Agricultural Engineering, University of Calfiornia Davis, Davis, CA

Abstract: Biochemical conversion is a promising route to produce fuels and chemicals from cellulosic biomass. A key step in the process is biomass liquefaction, where enzymes are used to break down pretreated cellulosic particles. During this process, drastic changes in the mechanical properties of the cellulosic slurry cause a transition from a solid-like material to a pumpable liquid. A fast liquefaction is desirable to achieve reductions in equipment size and cost. To assess the effect of mixing quality on the speed of liquefaction, cellulases were added to static fiber beds in vials. Different initial mixing conditions were used. One of the conditions was a well-mixed slurry where the enzyme is uniformly distributed in the slurry. The opposite extreme condition was addition of a thin layer of enzymes on one end of the vial, representing a completely unmixed enzyme. Samples were incubated for 96h at 50C. The state of liquefaction of the fiber beds was assessed using a mechanical penetrometer test. Large differences in the force-displacement curves were observed between untreated control slurries and enzymatically treated slurries. More importantly among the treated ones, the liquefaction times were much shorter for well-mixed slurries than for poorly mixed ones. Liquefaction data was also compared with reducing sugars measured throughout the incubation period to correlate with product yields.

See more of this Session: Reaction Kinetics and Transport Fundamentals for Biomass Conversion
See more of this Group/Topical: Sustainable Engineering Forum

(258e) Coupled Fluid-Dynamics and Population-Balance Kinetic Models for Enzymatic Hydrolysis of Biomass

Michael A. Sprague[1], Jonathan Stickel[2] and Andrew Griggs[2], (1)Computational Science Center, National Renewable Energy Laboratory, Golden, CO, (2)National Bioenergy Center, National Renewable Energy Laboratory, Golden, CO

A population-balance kinetic model developed previously for the enzymatic hydrolysis of lignocellulosic biomass is extended to include fluid mixing and transport. The kinetic model is focused on the depolymerization of insoluble cellulose chains by an endoglucanase and an exoglucanase. Population-balance models combined with a continuous-distribution representation provide computational efficiency, which is critical when considering the addition of fluid-mixing effects. The resulting coupled equations are solved with a high-order open-source computational fluid dynamics (CFD) package. The CFD model is validated against experimental and analytical data for a simple vane mixer. Preliminary results examining the effects of fluid mixing on dilute enzymatic hydrolysis are presented and compared to existing experimental data. This work provides a foundation for future efforts to model pilot- and commercial-scale bioconversion-reactors.

See more of this Session: Reaction Kinetics and Transport Fundamentals for Biomass Conversion
See more of this Group/Topical: Sustainable Engineering Forum

(258f) Flow and Transport Properties In High-Solids Biomass Slurries

James J. Lischeske and Jonathan J. Stickel, National Bioenergy Center, National Renewable Energy Laboratory, Golden, CO

Economic biochemical conversion of cellulose to sugars necessitates minimal addition of water to biomass slurries during conversion, resulting in high-solids systems. The resulting slurries are effectively wet-granular materials. These materials presents difficulties for reactor engineering, as their flow and transport properties, along with the mechanical properties that govern them, are poorly understood. This work examines the nature of high-solids flow, along with the heat and mass transfer properties of biomass and related model materials. Methods were developed to measure the volume fraction of solid, liquid and gaseous phases of high-solids slurries. The effect of phase composition on flow was systematically examined in a rotating vessel. Qualitative observations, as well as accessible quantitative information about size distribution of aggregates, were made. Further, a model for measuring bed diffusion in biomass was developed, and experiments were performed to measure the bed diffusion coefficient in model materials and biomass. Finally, heat transfer in a packed bed of biomass was examined, and conductivity coefficients were calculated and related to phase composition.

See more of this Session: Reaction Kinetics and Transport Fundamentals for Biomass Conversion
See more of this Group/Topical: Sustainable Engineering Forum

(289a) Efficient Determination of Force Field Parameters Using a Physically Based Equation of State

Thijs van Westen, Thijs Vlugt and Joachim Gross, Delft University of Technology, Delft, Netherlands

One of the most prominent problems when applying molecular simulations to represent real substances is the choice of (or the optimization of) the force field parametrization [1]. With ``parametrization'', we here refer to both the choice of a functional form (such as Lennard-Jones or Buckingham potentials for van der Waals interactions) and the values of its parameters. Usually, interactions can be divided into bonded and non-bonded interactions. Although the parametrization that describes the bonded interactions is accurately obtained from quantum chemical calculations or spectroscopic data [2, 3], it is difficult and computationally expensive to obtain accurate parameters describing the non-bonded van der Waals interactions from quantum mechanics; in practice, the non-bonded parameters are usually obtained by fitting molecular simulation results to experimental data like vapor-liquid equilibrium (VLE) data. Since the calculation of the vapor-liquid equilibria depends on the van der Waals interactions of all atom types in the system, the fitting of molecular simulation results to VLE data requires many time-consuming molecular simulations; therefore this is currently a cumbersome and computationally expensive task.

In this work, we present a novel method to significantly reduce the computational burden required for fitting force field parameters to experimental VLE data. In sharp contrast to conventional optimization approaches, we use additional physical insight on how computed VLE data depends on the force field parameters in molecular simulations. We use an analytical equation of state, the perturbed chain statistical associating fluid theory (PC-SAFT) [4], to predict the results of Gibbs Ensemble molecular simulations without actually performing these simulations. We show that all parameters in the PC-SAFT approach are directly related to the force field parameters, thereby making it possible to optimize force field parameters in a different, much faster framework (*i.e.* PC-SAFT). Since PC-SAFT calculates VLE data at least a factor 10^6 faster than molecular simulation, the presented optimization procedure results in a drastic decrease of the computational burden.

As a proof of concept, the method is illustrated by optimizing transferable Lennard-Jones parameters for the n-alkane series. The target is to obtain united atom force field parameters that describe the vapor-liquid part of the phase diagrams of n-butane and n-hexane. Optimizing a vector **p** of 4 force field parameters $\mathbf{p} = (\varepsilon_{CH2CH2}, \varepsilon_{CH3CH3}, \sigma_{CH2CH2}, \sigma_{CH3CH3})$ we obtain excellent agreement of coexisting densities, vapor pressure and caloric properties using only two Gibbs ensemble simulations. It is important to note that as we use additional physical insight in the optimization procedure, the number of molecular simulations needed hardly depends on the initial guess for the vector **p**, *i.e.* choosing initial force field parameters that deviate 40% from the optimal values leads to an almost equally efficient convergence.

References

[1] Frenkel, D.; Smit, B., *Understanding Molecular Simulation: from Algorithms to Applications*, 2nd ed. Academic Press; San Diego, 2002.

[2] Allinger, N.L.; Yuh, Y.H.; Lii, J.H. *J. Am. Chem. Soc.* **1989**, *111*, 8551-8566.

[3] Martin, M.G.; Siepmann, J.I. *J. Phys. Chem. B* **1998**, *102*, 2569-2577.

[4] Gross, J.; Sadowski, G. *Ind. Eng. Chem. Res.* **2001**, *40*, 1244-1260.

See more of this Session: Development of Intermolecular Potential Models
See more of this Group/Topical: Engineering Sciences and Fundamentals

(289b) Force Field Development for Uranyl Systems

Neeraj Rai, Surya P. Tiwari and Edward J. Maginn, Chemical & Biomolecular Engineering, University of Notre Dame, Notre Dame, IN

Our understanding of actinide materials chemistry is rather limited compared to most other elements of the periodic table largely due to the complex electronic structure of actinides which is dominated by the interplay of 5f, 6p, 6d, and 7s orbitals and relativistic effects. This complexity is also responsible for rich actinide chemistry, and presents an opportunity to exploit this diversity to design novel materials tailored to perform well in extreme environments often present in nuclear energy production cycle. Molecular simulations of the structure, energetics, and reactivity of actinide-containing materials are instrumental in developing an atomistic understanding of the role of actinide ions in complex crystalline, nanoscale materials, and aqueous solutions. Simulations provide structural and thermodynamic data that are difficult to obtain experimentally.

The accuracy of structural and thermodynamic data obtained via molecular simulations depends solely on the ability of the model interaction potential, more commonly referred to as the force field, to mimic the true interaction potential of the system. To this end, we will present a method to determine intermolecular potential for actinides which combines strengths of two highly accurate quantum mechanical methods: coupled--cluster of single and double excitations with perturbative treatment of triples (CCSD(T)) and symmetry adapted perturbation theory (SAPT). Results for selected urnayl systems will be presented. The force field obtained in this manner will lay the foundation for studying actinide system in the nanosecond time scales for very large system sizes enabling us to calculate thermodynamic, structural and transport properties of these systems.

See more of this Session: Development of Intermolecular Potential Models
See more of this Group/Topical: Engineering Sciences and Fundamentals

(289c) Ensuring Reliability, Reproducibility and Transferability In Atomistic Simulations: The Knowledgebase of Interatomic Models (openKIM.org)

Ellad B. Tadmor[1], Ryan S. Elliott[1], James P. Sethna[2], Ronald E. Miller[3], Chandler A. Becker[4], Valeriu Smirichinski[1] and Trevor J. Wennblom[5], (1)Department of Aerospace Engineering and Mechanics, University of Minnesota, Minneapolis, MN, (2)Department of Physics, Cornell University, Ithaca, NY, (3)Department of Mechanical and Aerospace Engineering, Carleton University, Ottawa, ON, Canada, (4) Metallurgy Division, Material Measurement Laboratory, National Institute of Standards and Technology, Gaithersburg, MD, (5)Silicon Life Sciences, LLC, Minneapolis, MN

Atomistic simulations in materials science play a key role in realistic scientific and industrial applications. However, the predictive capability of these approaches hinges on the accuracy of the model (potential or force field) used to describe the atomic interactions. Modern models are optimized to reproduce electronic structure results for a dataset of representative atomic configurations. However, no standardized approach currently exists for quantifying the range of applicability of an interatomic model or estimating the accuracy of its predictions. This makes it difficult or even impossible to select an appropriate model for a given application. A second difficulty is that since computer implementations of interatomic models do not subscribe to coding standards, it is hard to connect a given model to a desired simulation tool. This talk will describe a current NSF-CDI funded effort to create an open source online tool that addresses these limitations: the Knowledgebase of Interatomic Models (http://openKIM.org). KIM will allow users to compare model predictions with reference data, to generate new predictions by uploading simulation test codes, and to download models conforming to application programming interface (API) standards which are being developed in collaboration with atomistic simulation community. Inthis talk I will give an overview of the KIM project and describe recent developments.

See more of this Session: Development of Intermolecular Potential Models
See more of this Group/Topical: Engineering Sciences and Fundamentals

(289d) KIM Application Programmming Interface As a Standard for Molecular Simulations

Valeriu Smirichinski[1], **Ryan S. Elliott**[2], Ellad B. Tadmor[2], James P. Sethna[3], Ronald E. Miller[4] and Chandler A. Becker[5], (1)Department of Aerospace Engineering and Mechanics, University of Minnesota, Minneapolis, MN, (2)Aerospace Engineering and Mechanics, University of Minnesota, Minneapolis, MN, (3)Physics, Cornell University, Ithaca, NY, (4)Department of Mechanical and Aerospace Engineering, Carleton University, Ottawa, ON, Canada, (5)Metallurgy Division, Material Measurement Laboratory, National Institute of Standards and Technology, Gaithersburg, MD

Nanoscale modeling of materials often involves the use of molecular simulations or multiscale methods. These approaches frequently use empirical (fitted) interatomic potentials to represent the response of the material. Currently, no standardized approach exists for estimating the accuracy of interatomic potentials. In addition, a lack of standardization in programming interfaces for potentials and the lack of a systematic infrastructure for archiving them makes it difficult to use potentials for new applications and to reproduce published results.

The *Knowledgebase of Interatomic Models* (KIM) (http://openKIM.org) is a 4-year NSF CDI program which attempts to address these limitations. One difficulty in the use of interatomic models is that simulation codes and interatomic potentials are written in a variety of programming languages. In order to allow the programs and model subroutines to be mutually compatible, it is necessary to define a robust a*pplication programming interface* (API) designed to accommodate the requirements of molecular simulations. The KIM project has developed such an API in collaboration with key members of the molecular simulation community. The API is based on the concept of *descriptor files*, which define all variables and methods needed for input and output for a given application or model. The API addresses issues such as naming conventions for models (unique names), neighbor-list objects and their interface, system of units, unit conversion, and a complete description of variables and methods.

In this talk we will present a brief overview of the KIM framework and then provide details on the KIM API protocol. We invite the molecular simulation community to participate in the KIM project and contribute to the standardization efforts under way.

See more of this Session: Development of Intermolecular Potential Models
See more of this Group/Topical: Engineering Sciences and Fundamentals

(289e) Monte Carlo Calculations of the Virial Coefficients for Development of Potential Models

Katherine R. S. Shaul, Andrew J. Schultz and David A. Kofke, Chemical and Biological Engineering, University at Buffalo, The State University of New York, Buffalo, NY

The second and third virial coefficients, B_2 and B_3, have long played a role in the development of simple pair potentials because these quantities can be computed rapidly as integrals over configuration space and extracted reliably from experimental data. Consideration of the virial coefficients during model development is expanding to larger numbers of molecules and more complex model forms as the advancement of targeted Monte Carlo methods facilitates their computation. Path-integral Monte Carlo has enabled computation of the uniquely quantum contributions to B_2 and B_3 for monatomic gases [1], which are required for meaningful comparisons to experimental data at low temperature. Mayer-sampling Monte Carlo has made possible computation of classical virial coefficients for a wide range of models, including B_4 for the TraPPE-UA model of icosane [2], B_5 for the Gaussian charge polarizable model of water [3], and B_8 for the Lennard-Jones model [>4].

Here we present two extensions of the Mayer-sampling Monte Carlo approach designed to facilitate development of potential models. The first provides an estimate of the error in the classical virial coefficient for an *ab initio* potential model relative to that of the quantum-mechanical model it approximates and identifies configurations appropriate for inclusion within the fitting procedure. The second is designed to facilitate path-integral Monte Carlo calculations of quantum contributions to the virial coefficients beyond third order.

[1] G. Garberoglio and A. H. Harvey, "Path-integral calculation of the third virial coefficient of quantum gases at low temperatures," *Journal of Chemical Physics,* vol. 134, 2011.

[2] A. J. Schultz and D. A. Kofke, "Virial coefficients of model alkanes," *Journal of Chemical Physics,* vol. 133, 2010.

[3] K. M. Benjamin, *et al.*, "Fourth and Fifth Virial Coefficients of Polarizable Water," *Journal of Physical Chemistry B,* vol. 113, pp. 7810-7815, 2009.

[4] A. J. Schultz and D. A. Kofke, "Sixth, seventh and eighth virial coefficients of the Lennard-Jones model," *Molecular Physics,* vol. 107, pp. 2309-2318, 2009.

(289f) Reliability of Embedded-Atom Potentials for Boiling Points of Metals

Lev Gelb and Somendra Nath Chakraborty, University of Texas at Dallas, Richardson, TX

We determine the normal boiling points of a wide range of metals as predicted by available embedded-atom potentials. Embedded-atom potentials are empirical many-body potentials widely used in studies of solid and liquid metals and metal nanoparticles. These potentials are almost exclusively parameterized again solid-state properties, with occasional inclusion of small-cluster data and/or melting points. Vapor pressures and related coexistence data are not normally used in the parameterizations, and liquid-vapor phase diagrams have only been determined for a few such models [1,2].

While complete experimental LVE data are available only for the alkali metals, mercury, and a few other systems, reasonably accurate boiling points are available for most pure metals, and vapor pressures are available for most elements up to at least 2500 K. We present and discuss boiling points and associated properties determined for parameterizations of the Sutton-Chen [3], "quantum-corrected" Sutton-Chen [4] and original Finnis-Sinclair [5] type potentials for a range of metals. Data were obtained from both standard Gibbs-Ensemble calculations as well as using the adiabatic (NPH) extension of the Gibbs Ensemble [6], which is shown to be quite effective in accurately determining boiling temperatures. Shifts in LVE data due to potential truncation and possible methods for including electronic excitations are also discussed.

[1] D. Bhatt, et al., JPC-B 110 (2006) 26135.
[2] T. Aleksandrov, C. Desgranges, J. Delhommelle, FPE 287 (2010) 79.
[3] A. P. Sutton and J. Chen, Phil. Mag. Lett. 61 (1990) 139.
[4] S.-N. Luo et al., Phys Rev. B 68 (2003) 134206.
[5] G. J. Ackland et al., Phil. Mag. A 56 (1987) 735.
[6] T. Kristof, J. Liszi, Chem. Phys. Lett. 261 (1996) 620.

See more of this Session: Development of Intermolecular Potential Models
See more of this Group/Topical: Engineering Sciences and Fundamentals

(289g) Development of the TraPPE Force Field for Substituted Arenes, Heteroarenes, and Cyclic Alkanes and Ethers

J. Ilja Siepmann[1], Neeraj Rai[2], Katie A. Maerzke[3], Rebecca K. Lindsey[2] and Samuel J. Keasler[2], (1) Depts. of Chemistry and of Chemical Engineering & Materials Science, University of Minnesota, Minneapolis, MN, (2)Department of Chemistry, University of Minnesota, Minneapolis, MN, (3) Department of Chemical and Biomolecular Engineering, Vanderbilt University, Nashville, TN

The transferable potentials for phase equilibria (TraPPE) force field is extended to substituted arenes and heteroarenes by combining an explicit-hydrogen representation for the arene rings and a united-atom representation for various substituent groups. Partial charges are assigned to each specific compound, whereas Lennard-Jones parameters are transferred. New united-atom parameters are presented to extend the TraPPE-UA force field to 5- and 6-membered cyclic alkanes and ethers.

See more of this Session: Development of Intermolecular Potential Models
See more of this Group/Topical: Engineering Sciences and Fundamentals

(289h) Development and Testing of a CO_2H_2o Potential for Gas Hydrate and Liquid Phases

Srinath Velaga, Chemical Engineering, National Energy Technology Laboratory/West Virginia University, Morgantown, WV and **Brian Anderson**, Chemical Engineering, West Virginia University, Morgantown, WV

Gas hydrate reservoirs are receiving increased attention as potential locations for CO_2 sequestration, with CO_2 replacing the methane that is recovered as an energy source. In this scenario it is very important to correctly characterize the cage occupancies of CO_2 to correctly assess the sequestration potential as well as the methane recoverability. In order to predict accurate cage occupancies, the guest-host interaction potential must be represented properly. In previous efforts, the potential parameters used for CO_2 hydrate equilibrium calculations were obtained by fitting to experimental hydrate equilibrium data. These fitted parameters do not match with those obtained by second virial coefficient or gas viscosity data. *Ab initio* quantum mechanical calculations provide an independent means to directly obtain accurate intermolecular potentials.

A potential energy surface (PES) between H_2O and CO_2 has been computed at the MP2/aug-cc-pVTZ level and corrected for basis set superposition error (BSSE), an error caused due to the lower basis set, by using approximately half of the counterpoise correction. Intermolecular potentials were obtained by fitting the Exponential-6 and Lennard-Jones 6-12 models to the *ab initio* PES, correcting for many-body interactions. Using this potential, the pure CO_2 hydrate equilibrium pressure was predicted with an average absolute deviation of less than 2% from the experimental data. Predictions of the small cage occupancy ranged from 22-38% and the hydration number for the CO_2 hydrate was calculated to be above 7.0, whereas the large cage is fully occupied. Through this work we show that intermolecular interactions occurring on the nanometer scale influence phase equilibria and macroscopic properties. This *ab initio* intermolecular potential for the CO-$_2$–H_2O interaction has been tested using molecular dynamics simulations against other intermolecular potentials from the literature by calculating gas hydrate reference chemical potentials, the densities of pure CO_2 liquid and vapor, and the solubility of CO_2 in water.

See more of this Session: Development of Intermolecular Potential Models
See more of this Group/Topical: Engineering Sciences and Fundamentals

(289i) Development of An Optimized Intermolecular Potential for Sulfur Dioxide

Jeffrey J. Potoff, Chemcial Engineering and Materials Science, Wayne State University, Detroit, MI, MaryBeth H. Ketko, Chemical Engineering and Materials Science, Wayne State University, Detroit, MI and Ganesh Kamath, Computer Science and Mathematics Division, Oak Ridge National Laboratory, Oak Ridge, TN

Sulfur dioxide is an important industrial pollutant, formed primarily from the combustion of sulfur containing coal. SO_2 reacts with water to form sulfuric acid, and with increasingly strict environmental regulations, there is significant interest in the sequestration of SO_2 from flue gas in addition to CO_2. One promising technology for the sequestration of SO_2 is the use of ionic liquids. Recent publications have shown significant amounts of SO_2 may be physically adsorbed in ionic liquids[1]. However, given the multitude of potential combinations of anion and cation, the design of an optimal solvent of SO_2 capture by trial and error is likely to be an arduous process. Computational methods have recently been used to investigate the interactions of SO_2 with ionic liquids[2-4], however, the force fields used for SO_2 have not undergone rigorous validation.

In this work, a new force field for sulfur dioxide, capable of predicting accurately the vapor_liquid equilibria, critical properties, vapor pressure, and heats of vaporization is presented. The new force field reproduces the saturated liquid densities, vapor pressures and heats of vaporization to within 0.5, 2, and 2% of experiment, respectively. The predicted critical properties and the normal boiling point are in excellent agreement with experimental results. Pair distribution functions are calculated for the S_S, S_O, and O_O interactions are in close agreement with neutron and X-ray scattering experiments. In addition to the new force field, similar calculations are performed for four $SO\text{-}_2$ intermolecular potentials proposed by Sokolic et al. (Sokolic, F.; Guissani, Y.; Guillot, B. J. Phys. Chem. 1985, 89, 3023], which show that these models work reasonably well near the state point where they were originally parametrized, but large errors in the predicted coexistence properties are displayed at higher and lower temperatures. Errors in the vapor-pressure suggest that the Sokolic force fields underestimate electrostatic contributions to the intermolecular potential. Comparison of the radial distribution functions show the local structure is only weakly affected by the different force field parameters.

1. Anderson, J.L., et al., *Measurement of SO_2 solubility in ionic liquids.* Journal of Physical Chemistry B, 2006. **110**(31): p. 15059-15062.

2. Baer, M., et al., *Interpreting Vibrational Sum-Frequency Spectra of Sulfur Dioxide at the Air/ Water Interface: A Comprehensive Molecular Dynamics Study.* Journal of Physical Chemistry B, 2010.

114(21): p. 7245-7249.

3. Wick, C.D., T.M. Chang, and L.X. Dang, *Molecular mechanism of CO_2 and SO_2 molecules binding to the air/liquid interface of 1-butyl-3-methylimidazolium tetrafluoroborate ionic liquid: a molecular dynamics study with polarizable potential models.* J Phys Chem B, 2010. **114**(46): p. 14965-71.

4. Siqueira, L.J., et al., *Shielding of ionic interactions by sulfur dioxide in an ionic liquid.* J Phys Chem B, 2008. **112**(20): p. 6430-5.

See more of this Session: Development of Intermolecular Potential Models
See more of this Group/Topical: Engineering Sciences and Fundamentals

(300a) The Factory of the Future: Integrating Multiscale Modeling and Experiments to Produce New, Better Nanocomposite Materials

Maurizio Fermeglia, Paola Posocco, Radovan Toth, Daniel R. Nieto, Francesca Santese and Sabrina Pricl, Mose-DI3, University of Trieste, Trieste, Italy

Nanocomposite materials hold the potential to redefine traditional thinking of the performances and fields of application of polymers. Indeed, nanocomposites can offer improved properties of thermoplastics including flame retardancy, tensile strength, mechanical moduli, barrier and heat distortion temperature, just to name a few. However, developing state-of-the-art manufacturing technologies required to commercialize these opportunities still remains one of biggest challenge for the plastics industry.

The main goal of the 6th Framework Program EU Integrated Project (IP) MULTIHYBRIDS was to develop an innovative factory of the future with integrated technologies for the preparation of advanced specialty materials based on industrially important new polymer hybrids and nanocomposites. The objective-driven approach was the in-process tailoring of materials with successive validation of the approach through on-line characterisation of the process, properties, and targeted performance of the newly produced nanomaterials.

Special innovations developed during MULTIHYBRIDS included on-line extrusion monitoring, new melt processing technology ("sol/gel extrusion"), and new approaches to exploit functional fillers and to prepare easy-to-disperse masterbatches. The in-situ compatibilization was the key to allow selective nanoreinforcement of polymer blends such as rubber-toughened thermoplastics with improved toughness/stiffness balance and increased heat distortion temperature. The process management for in-line monitoring was totally different from those currently employed off-line tests.

Moreover, during the entire project, multiscale molecular modelling was used for interpreting and predicting the properties of the materials of interest. Information derived from atomistc level simulations were used as input for the subsequent level of computation, namely mesocale simulations. At each step of the progress of the work, computer simulations at all scales were integrated in the project, and the calculated results were compared to the available experimental evidences.

The performances of a nanocomposite depends on many factors: (i) the chemistry of the components, (ii) the size and shapes of the nanofiller, (iii) the degree of dispersion and (iv) the nanofiller loading. All these issues are relevant and it is of paramount importance to be able to predict the effect of such independent variables on the final properties of the material. This contribution considers all these aspects and applies different molecular modeling methods, such as atomistic Molecular Mechanics and

Molecular Dynamics, the mesoscale Dissipative Particles Dynamics and the macroscale Finite Element Method for the prediction of mechanical, thermal and barrier properties (taken as representative of the performances of a nanostructured material) for systems based on maleated polypropylene. The nanofillers considered are: montmorillonite, hydrotalcite, bohemite, sepiolite, carbon nanotubes and titania nanoparticles.

Finally, the process management object of this IP brought unambiguous advantages in terms of time and energy consumption and waste: process adjustments can be made in real-time minimising the need of time-consuming off-line tests (total time typically in the order of hours) and the production of unsuitable (waste) materials. Furthermore, the maximisation of simultaneous nanocharge funcionalisation/ modification and nanocomposite formation can allow solvent-free processes favouring environmentally friendly productions.

See more of this Session: Industrial Applications of Computational Chemistry and Molecular Simulation II
See more of this Group/Topical: Computational Molecular Science and Engineering Forum

(300b) Modeling Industrial Ethane Steam Cracking From First Principles

Wenjie Sun, Chemical and Biomolecular Engineering, National University of Singapore, Singapore 117576, Singapore and **Mark Saeys**, Chemical and Biomolecular Engineering, National University of Singapore, Singapore, Singapore

Many important processes such as combustion, atmospheric chemistry, polymerization, and steam cracking proceed via a complex gas phase radical reaction mechanism, often involving hundreds of important reactions [1]. Building accurate models and determining the required kinetic and thermodynamic parameters for such processes is challenging. Using state-of-the-art quantum chemical calculations and methods from statistical mechanics, it has become possible to accurately predict kinetic and thermodynamic parameters for well-defined gas phase reactions [e.g., 2]. These techniques then make it possible to quantitatively model complex radical processes fully from first principle, i.e., without the input of experimental data. To illustrate that the accuracy that can be obtained with *ab initio* computational chemistry methods has become sufficient to describe a complex, high temperature gas phase radical process, the industrial steam cracking of ethane was modeled [4]. Steam cracking is the most important industrial process for the production of light olefins, which are the building blocks for the chemical, plastic, and pharmaceutical industry. A relatively small reaction network of 150 reversible elementary reactions and 20 species was constructed and all the thermodynamic and kinetic parameters were obtained from *ab initio* CBS-QB3 and W1U calculations. After accounting for hindered internal rotations, quantum tunneling, variational corrections to transition state theory, and the pressure dependence of the rate coefficients, the predicted C_2H_6, C_2H_4, H_2 and CH_4 yields and reactor outlet temperature are within 10 % of experimental values. This study hence demonstrates the feasibility of constructing a predictive kinetic model for gas phase chemistry fully from first principles.

References

1. Broadbelt LJ, Pfaendtner J., AIChE J., 51, 2112 (2005)
2. Vandeputte AG, Sabbe MK, Reyniers MF, Van Speybroeck V, Waroquier M, Marin GB, J. Phys. Chem. A 111, 11771 (2007); Sun W, Saeys M, J. Phys. Chem. A, 112, 6918 (2008)
3. Sun W, Saeys M, AIChE J, doi: 10.1002/aic.12446; Sabbe MK, Van Geem KM, Reyniers M-F, Marin GB, AIChE J, 57, 482 (2011)

See more of this Session: Industrial Applications of Computational Chemistry and Molecular Simulation II
See more of this Group/Topical: Computational Molecular Science and Engineering Forum

(300c) Molecular Modeling of a Key Step In Metal-Mediated Controlled Free Radical Polymerization of Acid Monomers

Susan Fitzwater, Engineering and Process Sciences, The Dow Chemical Company, Spring House, PA

Metal-mediated controlled free radical polymerization processes such as ATRP and SET-LRP offer a route to a variety of controlled structure polymers. The practical utility of these processes has been somewhat limited by their limited ability to incorporate acid monomers. It has been presumed that acid monomers interfere with the control mechanism by protonating key ligands on the metal. However, it has been demonstrated recently that SET-LRP of methyl methacrylate (MMA) can proceed in an acid environment, with kinetics similar to those observed in the absence of acid. Copolymerization of moderate amounts of methacrylic acid (MAA) with MMA could be achieved via SET-LRP, but the reaction kinetics were significantly slower than those for MMA homopolymerization in the presence of acid. This suggests that it is the incorporation of acid monomer, not the presence of acid, which inhibits metal-mediated controlled free radical polymerization.

The key activation step in metal-mediated controlled polymerization processes is assumed to be cleavage of a terminal C—X (X = halogen) bond to form a C radical. The radical-tipped chain can add monomer and so is active. The C—X –tipped chain is dormant. Molecular modeling suggests that under some conditions, the dormant chain can undergo heterolytic cleavage of the C—X bond, forming a halide anion and a dormant/inactive polymeric cation. For (meth)acrylate polymers, the polymeric cation is stabilized by a lactone-like ring. If the penultimate monomer unit is a (meth)acrylic acid anion, the ring structure is particularly stable. This finding has implications for the choice of polymerization conditions for metal-mediated controlled free radical polymerization of monomeric acids.

See more of this Session: Industrial Applications of Computational Chemistry and Molecular Simulation II
See more of this Group/Topical: Computational Molecular Science and Engineering Forum

(300d) Computational Study of the Cyclohexanone-Based Initiation Mechanism In Thermal Polymerization of Methyl Acrylate and Methyl Methacrylate

Shi Liu[1], Sriraj Srinivasan[1], Michael C. Grady[2], Masoud Soroush[3] and Andrew M. Rappe[1], (1) Makineni Theoretical Laboratories, Department of Chemistry, University of Pennsylvania, Philadelphia, PA, (2)DuPont Marshall Lab, Philadelphia, PA, (3)Department of Chemical and Biological Engineering, Drexel University, Philadelphia, PA

The free radical polymerization of acrylates and methacrylates has been the subject of great interest to both fundamental and industrial research due to their various applications such as automobile coatings and adhesives[1-5]. The spontaneous thermal polymerization of acrylates and methacrylates is a promising approach to produce novel resins economically[6-10]. The high-temperature prompted secondary reactions such as back-biting reactions reduce the molecular weight of produced resins. Resins with low average molecular weights require less solvent, typically volatile organic compound, in order to be brushable and sparable. The absence of an external initiator also saves the cost for expensive initiator and after-reaction purification process as no-additional moiety is introduced[10]. Junker et al. [9] also reported that highly uniform macromonomers with unsaturated end-groups were obtained in the high temperature auto-initiated polymerization of acrylates. These highly functional macromonomers can further participate in polymerization leading to the synthesis of polymers with complex architectures. However, the absence of traditional controlling agent in spontaneous thermal polymerizations to some extent limits our ability to control the reactions. Here, we are suggesting to take advantage of solvent effect to achieve the controlled spontaneous thermal polymerization as the reaction medium can influence the reaction via various ways to different extents.

Cyclohexanone (CH), a widely used solvent, has been reported to be capable of initiating the polymerization of methyl methacrylate (MMA) at 75°C[11,12]. A recent experimental study also found that spontaneous thermal polymerization of methyl acrylate (MA) at 120°C has the highest reaction rate and final conversion in cyclohexanone compared to polymerization conducted in dimethyl sulfoxide and xylene[2]. All these experimental studies point out that cyclohexanone might provide additional initiation mechanisms generating free radicals.

To the best of our knowledge, no experimental or computational study has convincingly characterized the CH-mediated initiation mechanism in details. In the present work, four different types of mechanisms, the Kaim mechanism[11], the Flory mechanism[13], the α-position hydrogen addition reaction and the Mayo mechanism[14] are studied with density functional theory(DFT) and second-order Møller–Plesset (MP2) calculations to elucidate the true nature of cyclohexanone-based initiation reaction in MA and MMA. Transition state for each mechanism is located with B3LYP/6-31G* and assessed with MP2/6-31G*. Activation energies and rate constants are calculated and compared to determine which mechanism is more likely to be responsible for the initiation. Our results show that Diels-Alder adduct formed from a cyclohexanone and a monomer can transfer a hydrogen atom to a second monomer, generating free radicals for initiation. We propose that cyclohexanone via the Mayo mechanism, together with the monomer self-initiation reactions[3,5], is likely to explain the cyclohexanone-facilitated polymerization in thermal polymerization of MA and MMA.

References

[1] Adamsons, K., Blackman, G., Gregorovich, B., Lin, L., Matheson, R. Progress in organic coatings, 1998, 34, p. 64.

[2] Srinivasan, S., Kalfas, G., Petkovska, V. I., Bruni, C., Grady, M. C., Soroush, M., Experimental Study of the Spontaneous Thermal Homopolymerization of Methyl and n-Butyl Acrylate, J. Appl. Polym. Sci., 2010, 118, p. 1898.

[3] Srinivasan, S., Lee, M. W., Grady, M. C., Soroush, M., Rappe, A. M., Computational Study of the Self-Initiation Mechanism in Thermal Polymerization of Methyl Acrylate, J. Phys. Chem. A, 2009, 113, p. 10787.

[4] Srinivasan, S., Lee, M. W., Grady, M. C., Soroush, M., Rappe, A. M., Self-Initiation Mechanism in Spontaneous Thermal Polymerization of Ethyl and n-Butyl Acrylate: A Theoretical Study, J. Phys. Chem. A, 2010, 114, p. 7975.

[5] Srinivasan, S., Lee, M. W.; Grady, M. C., Soroush, M., Rappe, A. M., Computational Evidence for Self-Initiation in Spontaneous High-Temperature Polymerization of Methyl Methacrylate, J. Phys. Chem. A, 2011, 115, p. 1125.

[6] Walling, C., Briggs, E. R., Mayo, F. R., The kinetics of the thermal polymerization of styrene, J. Am. Chem. Soc., 1946. 68(7), p.1145.

[7] Grady, M.C., Quan, C., Soroush, M. Thermally initiated polymerization process. US Patent Application Number 60/484,393, filed on July 2, 2003

[8] Quan, C., Soroush, M., Grady, M.C., Hansen, J.E., Simonsick, W.J. Characterization of thermally polymerized n-butyl acrylate and ethyl acrylate, Macromolecules, 2005, 38(18), p. 7619.

[9] Zorn, A. M.; Junkers, T. , Barner-Kowollik, C. , Synthesis of a Macromonomer Library from High-Temperature Acrylate Polymerization, Macromol. Rapid Commun., 2009, 30, p. 2028.

[10] Grady, M.C., Simonsick, W.J., Hutchinson, R.A., Studies of higher temperature polymerization of n-butyl methacrylate and n- butyl acrylate, Macromol. Symp., 2002, 182, p. 149.

[11] Kaim, A., Polymerization of Vinyl Monomers Initiated with Cyclohexanone, Journal of Polymer Science Part C-Polymer Letters, 1984, 22, p. 203.

[12] Kaim, A., Kinetics of Polymerization of Methyl Methacrylate Initiated with Cyclohexanone. Part I, J. Polym. Sci. Pol. Chem., 1984, 22, p.1891.

[13] Flory, P.J., The mechanism of vinyl polymerizations, J. Am. Chem. Soc., 1937, 59, p.241.

[14] Mayo, F.R., The dimerization of styrene, J. Am. Chem. Soc., 1968, 90(5), p. 1289.

See more of this Session: Industrial Applications of Computational Chemistry and Molecular Simulation II

See more of this Group/Topical: Computational Molecular Science and Engineering Forum

(300e) Chain Transfer to Polymer Reactions In Thermal Polymerization of Methyl Acrylate: Computational Study

Nazanin Moghadam, Chemical and Biological Engineering, Drexel University, Philadelphia, PA, Masoud Soroush, Department of Chemical and Biological Engineering, Drexel University, Philadelphia, PA, Sriraj Srinivasan, Arkema, Philadelphia, PA, Andrew M. Rappe, Makineni Theoretical Laboratories, Department of Chemistry, University of Pennsylvania, Philadelphia, PA and Michael C. Grady, DuPont Marshall Lab, Philadelphia, PA

In the past sixty years, North American regulations on volatile organic content (VOC) of coatings have been driving changes in basic nature of resins in the paint and coatings industries [1, 2], to replace high molecular weight, non-functional polymer solutions with low molecular weight, high functionalized polymer solutions [3]. High-temperature free-radical polymerization involves a large set of complex competing reactions.

Chain transfer to polymer during free-radical polymerization can lead to the formation of mid-chain radicals and highly branched polymers [4]. The formation of mid-chain radicals, and short and long chain branches at different initial monomer concentration was studied using nuclear magnetic resonance spectroscopy in free-radical solution polymerization of n-butyl acrylate [5]. Recently, a macroscopic model with new rate expressions for inter- and intra-molecular transfer to polymer was used to determine the overall polymerization rate, average chain length, and level of branching in thermal polymerization of n-butyl acrylate [6]. However, the mechanism of chain transfer to polymer and the location of the radical center on a polymer chain is not yet well understood.

Previous studies [6-11] have shown the successful use of computational quantum chemistry to better understand self-initiation and propagation reactions in polymerization systems. Before this study there was no computational study of the mechanism of chain transfer to polymer reactions in self-initiated high-temperature free-radical polymerization of alkyl acrylates.

This paper presents a computational study of chain transfer to polymer reactions in self-initiated high temperature polymerization of methyl acrylate. Energy barriers and rate constants of chain transfer to polymer reactions are predicted using density functional theory calculations. Pathways for the reactions of interest are determined through intrinsic reaction coordinate calculations. The effects of polymer chain length on energy barriers and rate constants of the reactions are also determined.

References:

[1] Superintendent of Documents, Title 1, US Government Printing Office, Washington, DC, 1990, p.1.

[2] Superintendent of Documents, Clean Air Act Amendments of 1990, Title 111, US Government Printing Office, Washington, DC, 1990, p.236.

[3] Adamsons, K.; Blackman, G.; Gregorovich, B.; Lin, L., Matheson, R.; Oligomers in the Evolution of Automotive Clearcoats : Mechanical Performance Testing as a Function of Exposure, *Progress in Organic Coatings* 1998, 34, 64-74.

[4] Lovell, P.A.; Shah, T.H.; Heatley, F.; Chain Transfer to Polymer in Emulsion Polymerization of n-Butyl Acrylate Studied by Carbon -13 NMR Spectroscopy and Gel Permeation Chromatography, Polym. Commun. 1991, 32, 98-103.

[5] Ahmad, N.M.; Heatley, F.; Lovell, P.A.; Chain Transfer to Polymer in Free-Radical Solution Polymerization of n-Butyl Acrylate Studied by NMR Spectroscopy, Macromolecules, 1998, 31, 2822-2827.

[6] Nikitin, A.N.; Hutchinson, R.A.; Kalfas, G.A.; Richards, J.R.; Bruni, C.; The effect of Intramolecular Transfer to Polymer on Stationary Free-Radical Polymerization of Alkyl Acrylates, Macromolecular, 2009, 18, 247-258.

[7] Heuts, J.P.A.; Gilbert, R.G.; Radom, L.; A Prior Prediction of Propagation Rate Coefficients in Free Radical Polymerizations: Propagation of Ethylene, Macromolecules, 1995, 28, p. 8771-81.

[8] Khuong, K.S.; H. Jones, W.H.; Pryor, W.A.; Houk, K.N.; The Mechanism of Self-Initiated Thermal Solution of a Classic Problem, J. Am. Chem. Soc. 2005. 127, 1265-77.

[9] Yu, X.; Pfaendtner, J.; Broadbelt, L.J.; Ab Initio Study of Acrylate Polymerization Reactions: Methyl Methacrylate and Methyl Acrylate Propagation, J. Phys. Chem. A. 2008, 112, 6772-82.

[10] Srinivasan, S.; Lee, M.W.; Grady, M.C.; Soroush, M.; Rappe, A.M.; Computational Study of the Self-Initiation Mechanism in Thermal Polymerization of Methyl Acrylate, J. Phys. Chem. A 2009, 113, 10787-94.

[11] Srinivasan, S.; Lee, M.W.; Grady, M.C.; Soroush, M.; Rappe, A.M.; Self-Initiation Mechanism in Spontaneous Thermal Polymerization of Ethyl and n-Butyl Acrylate: A Theoretical Study, J. Phys. Chem. A 2010, 114, 7975-83.

See more of this Session: Industrial Applications of Computational Chemistry and Molecular Simulation II

See more of this Group/Topical: Computational Molecular Science and Engineering Forum

Analysis of Warm Membrane- and Adsorbent-Based Carbon Dioxide Capture

David Couling[1]*, Kshitij Prakash*, Robert Brasington†, and William H. Green*

* Department of Chemical Engineering, Massachusetts Institute of Technology

† MIT Energy Initiative, Massachusetts Institute of Technology

Coal will continue to be a significant part of the world's energy portfolio in the near future, and its use will see continued growth—especially in the developing world [1, 2]. Unfortunately, coal power technology has the significant disadvantage of the high level of contaminants in the feedstock, including nitrogen and sulfur compounds, trace amounts of heavy metals, and a large amount of carbon, which is released as carbon dioxide. Integrated Gasification Combined Cycle (IGCC) with Carbon Capture and Sequestration (CCS) presents a method to achieve the tight emission limits on criteria pollutants (NO_x, SO_x, CO, PM_{10}) and mitigate greenhouse gas emissions by capturing CO_2.

All commercially available syngas cleanup technologies for IGCC with CCS involve cooling the stream to low temperatures and employing scrubbing with physical or chemical solvents. This not only increases the capital costs of the IGCC plant through the addition of large heat exchangers, it also potentially decreases the overall efficiency of the plant due to the loss of availability of the hot syngas stream [3, 4]. Therefore, a strong desire exists to develop syngas cleanup technologies that are able to perform at elevated temperatures. A basic schematic of hypothetical warm syngas cleanup within an IGCC plant is shown below.

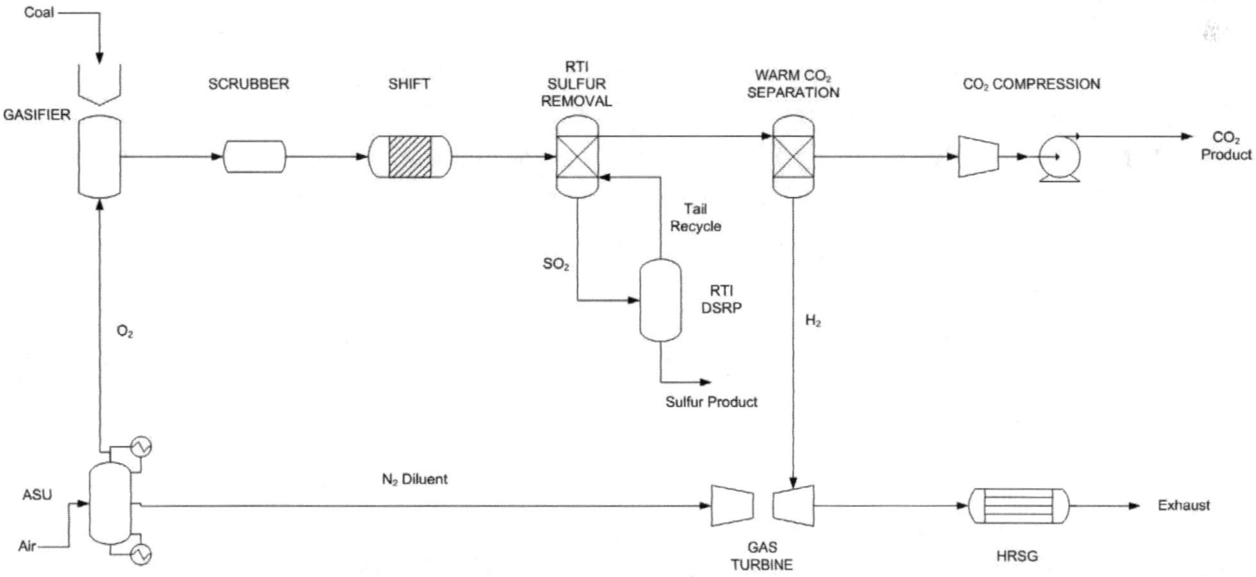

Figure 1 Generic IGCC flowsheet with warm syngas cleanup

---

[1] dcouling@mit.edu

The sulfur removal in the above schematic is performed using a zinc oxide-based sorbent developed by the Research Triangle Institute and Eastman Chemical [4].

Because of the large volumes of CO_2 generated in an IGCC plant and the growing concerns over greenhouse gas emissions, much of the effort directed at developing syngas cleanup technologies at elevated temperatures focuses on CO_2 capture. The most commonly proposed methods for the capture of carbon dioxide at elevated temperatures involve the use of membranes or adsorbent materials. Unfortunately, because these proposed technologies do not exist commercially, the potential gains in efficiency that could arise through the use of these proposed technologies are not well-known.

In this work we develop models for several proposed warm CO_2 separation technologies: Pd alloy-based H_2-permeable membranes, polymeric CO_2-permeable membranes, and solid CO_2 adsorbents. We then use these models to evaluate how the different parameters, including membrane permeability, selectivity, and adsorbent heat of adsorption, affect the performance of these different technologies. Finally, we integrate the CO_2 separation models with promising parameter values into an overall IGCC model developed within Aspen Plus in order to evaluate the parasitic energy losses present with each novel technology. Using this Aspen Plus model we are able to effectively compare the different warm CO_2 separation technologies with each other and with a base case physical solvent absorption system (Selexol®) and propose avenues of future research.

H_2 Membranes

H_2 transport through Pd alloy membranes may be divided into a series of different stages. Initially the H_2 adsorbs onto the metal surface and dissociates into two protons. These protons then diffuse into the lattice and permeate across the membrane. They then recombine into a H_2 molecule on the surface of the permeate side of the membrane, and this resultant H_2 molecule desorbs from the membrane surface into the bulk. Following the findings of Ward and Dao that the diffusion of the H atom through the Pd alloy metal is likely to be the rate-dominating process [5], the flux through the membrane can then be written as shown in Equation (1).

$$J_{H_2} = \frac{P}{l}\left(p_{H_2,r}^{0.5} - p_{H_2,p}^{0.5}\right) \tag{1}$$

In the above equation, P refers to the permeability of the H_2 through the membrane, l the membrane thickness, and p_{H_2} the partial pressure of H_2 on the retentate or permeate side. The permeability is often determined experimentally; in the case of H_2 permeating through a Pd alloy membrane, the permeability taken from the literature for our model was 2.2×10^{-9} mol m^{-1}

s^{-1} $Pa^{-0.5}$ [6]. This particular permeability value was chosen because it was measured in the presence of CO, CO_2, and H_2O and accounts for some of the competitive adsorption effects of these species on the H_2 permeability. N_2 gas, taken from the Air Separation Unit in the IGCC plant (see Figure 1), functions as the sweep gas for this membrane system, increasing the driving force for the H_2 permeation. We assumed the permeabilities of the other components of the syngas stream were negligible.

An important operational parameter to investigate with any membrane technology is the pressure ratio, or the ratio of the total retentate pressure to the total permeate pressure.

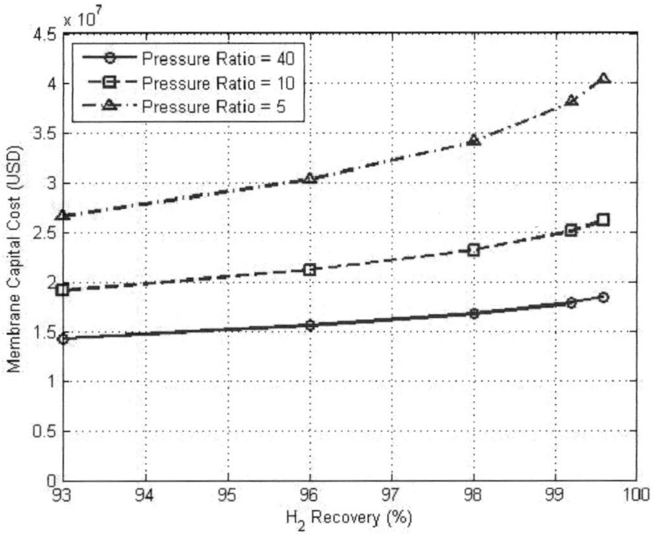

Figure 2 Membrane capital cost, assuming a price of \$2800/m^2, for different pressure ratios as a function of H_2 recovery

Determining a favorable pressure ratio for a given process is dependent on two competing factors: efficiency and membrane cost. At lower pressure ratio, the H_2 permeate stream is produced at a higher pressure, meaning less energy is needed to recompress the permeate to the gas turbine delivery pressure. Unfortunately, producing the permeate at higher pressure dramatically increases the required membrane area, in turn increasing the overall capital cost of the membrane unit. This latter effect is demonstrated in Figure 2 above.

The sweep gas flow rate is another important parameter to consider.

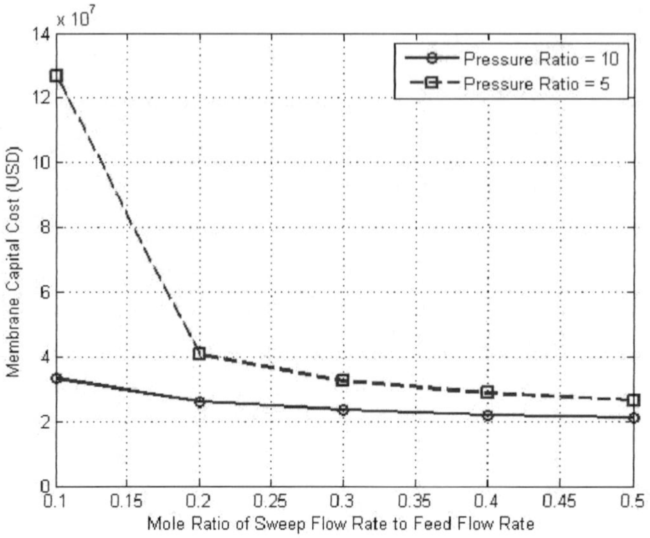

Figure 3 Membrane capital cost as a function of the sweep gas to syngas ratio

A higher sweep gas flow rate may lower the required membrane area, but additional heat energy must be supplied to heat the sweep gas flow rate to the membrane temperature. As a result, it may be beneficial to keep the sweep gas flow rate low. As Figure 3 shows, having at least a 5:1 syngas to sweep gas ratio is sufficient to maintain a relatively low capital cost of the membrane unit.

CO_2 Membranes

In contrast to the Pd alloy membranes, most polymeric CO_2 membranes operate with a solution-diffusion mechanism. In this case the flux is proportional to the difference in the partial pressures in the permeate and retentate streams.

$$J_{CO_2} = \frac{P}{l}\left(p_{CO_2,r} - p_{CO_2,p}\right) \tag{2}$$

The permeability in Equation (2) was taken to be 3650 Barrer (1 Barrer = 10^{-11} (cm^3 STP) cm cm^2 s^{-1} $mmHg^{-1}$) with the CO_2/H_2 selectivity equaling 100 (estimated for a high partial pressure of CO_2 case) [7]. Pan determined that when modeling pressure-driven solution-diffusion transport across a membrane, the flux of permeating species is influenced by the total permeate pressure and feed composition but is independent of the bulk permeate concentration [8]. The effect of back-diffusion on the permeate side is assumed to be negligible and therefore using a sweep gas to lower the CO_2 partial pressure in the bulk does not increase the driving force across the membrane. Some part of the captured high purity CO_2 is usually recirculated as the sweep gas. This avoids the need for any further separations in the permeate stream and usage of pressurized steam as a sweep gas. A model simulating this behavior was developed using the solution method of Murad Chowdhury et al. [9].

The effect of pressure ratio was also modeled for this CO_2 membrane case. Unlike the H_2 membrane case, because components other than CO_2 are able to permeate the membrane, the pressure ratio influences more than just the membrane area.

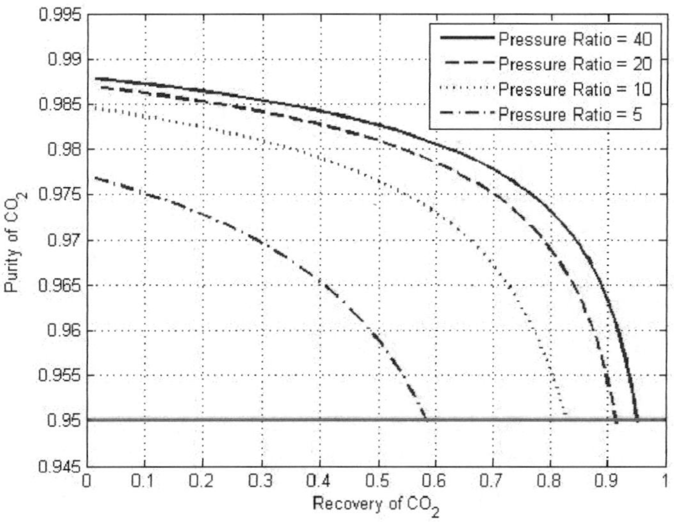

Figure 4 Dry purity of the CO_2 product stream as a function of the recovery of CO_2 within the membrane unit for different pressure ratios. The permeability of CO_2 is 3650 Barrer, and the CO_2/H_2 selectivity is 100

It is important to note that the "Recovery of CO_2" axis in Figure 4 refers to the fraction of CO_2 recovered within the membrane unit itself and does not account for unconverted CO or other sources of CO_2 within the overall IGCC plant. The red line in Figure 4 depicts the target purity of the CO_2 product stream for sequestration, and any CO_2 stream with a purity below this target is unable to be sequestered without further processing. Therefore, in order to meet the same 90% overall CO_2 capture specification that is met by a Selexol unit (corresponding to roughly 92.6% on the figure above), the pressure ratio needs to be slightly more than 20.

CO_2 Adsorbents

For the purposes of our model, the CO_2 adsorption reaction was assumed to be a simple chemisorption reaction, where CO_2 reversibly binds with an unknown sorbent X.

$$CO_2 + X \rightleftharpoons CO_2X \tag{3}$$

For example, if "X" were calcium oxide (CaO), the resulting product would be calcium carbonate ($CaCO_3$). Rather than applying the model to known sorbent materials, however, the model was instead used to evaluate what material and chemical properties of a hypothetical adsorbent would result in a favorable process when integrated into the IGCC system. For example, one might initially think that a large heat of adsorption would be desirable in order to help ensure that the CO_2 reacts with the adsorbent; however, if the heat of adsorption is too large then the adsorbent will not be able to be efficiently regenerated. The mass balance equations are shown below in Equations (4) and (5), with C referring to the gas phase concentration (in mol m^{-3}) and q referring to the solid phase loading (in mol kg^{-1}).

$$\varepsilon \frac{\partial C}{\partial t} = -\varepsilon \frac{\partial (uC)}{\partial z} - (1-\varepsilon)\rho_{sorbent} \frac{dq}{dt} \tag{4}$$

$$\frac{dq}{dt} = k_{LDF}\left(q^* - q\right) \tag{5}$$

The velocity u was related to the pressure in the adsorption unit using the Ergun equation, and the equilibrium loading q^* was determined using the Langmuir isotherm.

The temperature of the adsorption unit added an interesting design variable. Because adsorption reactions are exothermic to overcome the decrease in entropy, heat is released upon adsorption. If the adsorption unit is not operated isothermally, this released heat will raise the temperature of the unit, making future adsorption reactions less thermodynamically favorable due to this higher temperature. Conversely, the regeneration process is endothermic, causing the temperature of the bed to decrease. This decrease in temperature causes future desorption reactions to be less thermodynamically favorable. As a result, a nonisothermal adsorption unit has a much lower working capacity than an isothermal unit. This can be further seen in Figure 5.

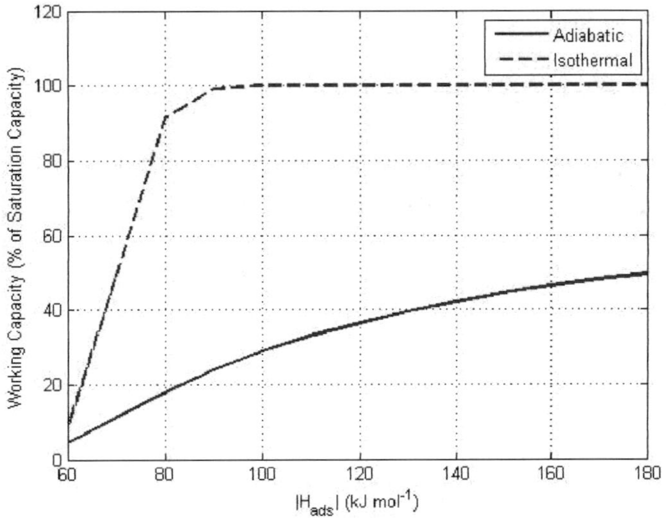

Figure 5 Working capacity as a function of the heat of adsorption for both adiabatic and isothermal adsorption units

Figure 5 depicts an estimate of the working capacity of the adsorption unit based on the partial pressure of CO_2 present in the gas phase and, in the case of the adiabatic unit, the temperature rise in the unit. In both cases the entropy of adsorption was assumed to be a constant value of -160 J mol^{-1} K^{-1}, which is a typical value for adsorption on Group I and Group II metals [10]. There clearly exists a sizeable difference between the working capacities of an isothermal unit versus an adiabatic unit; as a result, both types of unit were modeled in our process.

The models developed for H_2 membranes, CO_2 membranes, and CO_2 adsorbents were all integrated into an Aspen Plus IGCC model in order to evaluate their parasitic energy requirements. The results of these simulations will be discussed.

References

1. *International Energy Outlook*. 2010, Energy Information Administration.
2. Katzer, J., et al., *The Future of Coal*. 2007, Massachusetts Institute of Technology.
3. *Current and Future Technologies for Gasification-Based Power Generation, Volume 2: A Pathway Study Focused on Carbon Capture Advanced Power Systems R&D Using Biuminous Coal*. 2009, U.S. Department of Energy, Office of Fossil Energy, NETL.
4. Schlather, J., D. Denton, and B. Turk. *Comparison of a New Warm-Gas Desulfurization Process versus Traditional Scrubbers for a Commercial IGCC Power Plant*. in *Gasification Technologies Conference*. 2007. San Francisco, CA.
5. Ward, T. and T. Dao, *Model of Hydrogen Permeation Behavior in Palladium Membranes*. Journal of Membrane Science, 1999. **153**: p. 211-231.
6. Peters, T., et al., *High Pressure Performance of thin Pd-23%Ag/Stainless Steel Composite Membranes in Water Gas Shift Gas Mixtures; Influence of Dilution, Mass Transfer, and Surface Effects on the Hydrogen Flux*. Journal of Membrane Science, 2008. **316**: p. 119-127.

7. Taniguchi, I., et al., *Facile Fabrication of a Novel High Performance CO_2 Separation Membrane: Immobilization of Poly (Amidoamine) Dendrimers in Poly (Ethylene Glycol) Networks.* Journal of Membrane Science, 2008. **322**: p. 277-280.
8. Pan, C., *Gas Separation by High-Flux, Asymmetric Hollow-Fiber Membrane.* AIChE Journal, 1986. **32**: p. 2020-2027.
9. Murad Chowdhury, M., et al., *A New Numerical Approach for a Detailed Multicomponent Gas Separation Membrane Model and Aspen Plus Simuation.* Chemical Engineering and Technology, 2005. **28**: p. 773-782.
10. *CRC Handbook of Chemistry and Physics*, ed. D.R. Lide. 2008: Taylor and Francis [distributor].

(300g) A Theoretical Study of SO2 Capture by Hydroxyl Ammonium Ionic Liquids

Zhang Suojiang Sr.[1], Hongyan He[2] and Xiaoqian Yao[2], (1)State Key Laboratory of Multiphase Complex Systems, Institute of Process Engineering, Chinese Academy of Sciences, Beijing, China, (2) Institute of Process Engineering, Chinese Academy of Sciences, Beijing, China

Room-temperature ionic liquids, simply known as ionic liquids (ILs), are defined as a class of organic salts with melting points below 100°C. Owing to some unique physical and chemical properties, ILs have been initially proposed as replacements for conventional organic solvents in a variety of fields such as separation, organic synthesis, catalysis, electrochemistry, fuel and solar cells, lubricants, and functional materials, etc.

Recently, it was found that hydroxyl ammonium ionic liquids show good performances in SO2 separation science. In this study, the interaction between sulfur dioxide (SO_2) and room temperature ionic liquid tri-(2-hydroxy ethyl)-ammonium (THEA) lactate (LAC) (THEAL) have been investigated theoretically by using quantum chemical calculation (QM) at B3LYP/6-31++G* level. The relevant geometrical characteristics, energy properties as well as the characters of the intermolecular hydrogen bonds (HB) have been systematically discussed. The calculation results give us a deeper understanding of the factors which determine the high solubility of SO_2 in THEAL. The most stable geometries and binding energies indicate a strong association of the SO_2 and the THEAL molecule, especially the LAC anion. The results demonstrate that the LAC anion is the main factor for the absorption of SO_2.

See more of this Session: Industrial Applications of Computational Chemistry and Molecular Simulation II
See more of this Group/Topical: Computational Molecular Science and Engineering Forum

(312a) Optimization Studies Based On CFD Modeling of a Multiple Tube Reactor for Solar-Thermal Processes

Janna Martinek and Alan W. Weimer, Department of Chemical and Biological Engineering, University of Colorado, Boulder, CO

A numerical model of an indirectly-irradiated solar receiver with five tubes enclosed within a reflective cavity has been developed and experimentally validated for the gasification of carbon black. The radiation model consists of a combination of Monte Carlo and finite volume methods and is coupled with a three-dimensional CFD model treating heat, mass, momentum and species transport for the two-phase fluid/particle mixture. This model is used to investigate the impact of factors describing receiver configuration and operating conditions on overall receiver performance. Thirteen individual factors are considered with four factors describing operating conditions (gas flow rate, carbon feed rate, solar power input, and steam to carbon ratio) and nine two-level factors describing receiver geometry. The geometric factors are defined relative to a base design in which all tubes are arranged in a semicircle around the back cavity wall and allow for variations in cavity size, and number, radius, and arrangement of tubes with both staggered and non-staggered arrangements considered along with factors allowing for offset of all or part of the tube semicircle toward the window. Designs are evaluated in the framework of a 2^{13-8} fractional factorial design on the basis of receiver solar to chemical efficiency and reaction conversion for both absorbing and reflective cavity walls.

See more of this Session: Modeling and Analysis of Chemical Reactors
See more of this Group/Topical: Catalysis and Reaction Engineering Division

(312b) CFD Based Simulation of Heat Transfer In a Fixed Bed Reactor Tube

Mohsen Behnam[1], Anthony G. Dixon[1], Michiel Nijemeisland[2] and E. Hugh Stitt[2], (1)Chemical Engineering Department, Worcester Polytechnic Institute, WORCESTER, MA, (2)Johnson Matthey Catalysts, Billingham, United Kingdom

Low *N* (tube to particle diameter ratio) fixed bed reactors are commonly used in the chemical industry for extremely endothermic and exothermic reactions due to the better heat transfer. In the past, lots of researches have been carried out to simulate the heat transfer in fixed bed reactors by using k_r/k_f and h_w. Most models for k_r/k_f and h_w are based on a direct correlation with *Re* and *N*. Although most models seem to predict k_r/k_f very well, but there is a lot of problem still exists to predict h_w [1-2]. The main objective of this study was to develop a new heat transfer model with emphasis of industrial application for conduction and convection heat transfer in fixed bed tubes without using k_r/k_f and h_w.

Fluid flow in a fixed bed of tube to particle ratio 3.96, 5.96 and 7.99 were obtained for packed bed column by solving the 3D Navier–Stokes and using commercial software FLUENT 6.3.26. The heat transfer was modeled by using pseudo-homogenous energy equation. COMSOL 3.5 was used to simulate heat transfer based on pseudo-homogenous equation in 2D. Effective radial thermal conductivity was calculated from Zehner-Schlünder model as function of bed porosity position, fluid thermal conductivity and solid thermal conductivity [3]. The convection heat transfer was calculated in the 2D flow fluid from the CFD run.

Results were obtained for Reynolds numbers in the range 240–1900. The accuracy of the new model has been validated by analytical solution. The temperature calculated by the pseudo-homogenous energy equation showed reasonable quantitative agreement with values predicted by CFD model.

[1] S. A. Logtenberg, A. G. Dixon, "Computational fluid dynamics studies of fixed bed heat transfer", Chemical Engineering and Processing 37 (1998) 7–21.

[2] A.G. Dixon, "An improved equation for the overall heat transfer coefficient in packed beds", Chem. Eng. Process. 35 (1996) 323.

[3] P. Zehner, E.-U. Schlunder, "Die Effective Warmeleitfahigkeit Durchstromter ugelschuttungen bei Massigen und Hohen Temperaturen", Chem. Ing. Tech. 45 (1973) 272.

See more of this Session: Modeling and Analysis of Chemical Reactors
See more of this Group/Topical: Catalysis and Reaction Engineering Division

(312c) Effect of Spatial Segregation On Commensalistic Cultures - Series Reactors

Satish J. Parulekar, Chemical and Biological Engineering, Illinois Institute of Technology, Chicago, IL

A comprehensive analysis of static and dynamic behavior of a mixed culture in two bioreactors in series is presented considering anaerobic digestion involving acidogens (X) and methanogens (Y) as the example bioprocess. A single continuous culture may operate at up to seven steady states, including up to four coexistence steady states, with only one coexistence steady state being locally stable. The bioreactors in the two-reactor system are identical in terms of composition of and dilution rate with respect to fresh feed. A two-reactor system may admit up to forty nine steady states, which are comprised of up to forty coexistence steady states, at least at very low interaction rates (R). When physically realizable, up to five coexistence steady states are locally stable. The static and dynamic analysis of the two-reactor system is facilitated by appropriate grouping of large number of steady states arising for very low R into nine clusters and grouping of the clusters into three cluster groups. Numerical illustrations reveal the rich steady state structure of the bioprocess for bioreactors in series. The two bioreactors can operate at up to five locally stable coexistence steady states over certain ranges of R. The two-reactor system is operationally more flexible and more robust vis-a-vis single reactor as concerns maintenance of mixed culture. Emergence of additional steady states at intermediate R reveals that series bioreactors are an example of a complex system.

See more of this Session: Modeling and Analysis of Chemical Reactors
See more of this Group/Topical: Catalysis and Reaction Engineering Division

(312d) Modeling the Start-up Phase of Fischer Tropsch Synthesis In a Fixed Bed Reactor: Strategizing the Optimum Operation

Syed Ali Gardezi, Chemical Engineering, University of South Florida, Tampa, FL, Babu Joseph, Chemical and Biomedical Engineering, Clean Energy Research Center, University of South Florida, Tampa, FL and John T. Wolan, University of South Florida, Tampa, FL

To date, very little work has been done to model Fischer Tropsch Synthesis startup. In this study, finite difference approximation is used to model the intra-pellet reaction and heat transport. Finite element analysis was used to model the catalyst inter-pellet heat and mass transport. Vapor-liquid equilibrium was assumed to be attained at each pore, fugacity coefficients were used to develop a quantitative correlation. The goal was to estimate the time required to fill the catalyst pores. This time represents the unsteady transient phase which requires fine temperature control to avoid reaction run away in an actual working packed bed reactor.

In the resulting kinetic model, coupled material and energy balance are used to assess the optimum reactor startup procedures. Another goal is to identify the impact of different types of inert packing used to control heat transfer and conversion in the fixed bed reactor with a view towards avoiding the hot spots and steep temperature gradients. The combined effect of heat transfer and kinetics present a challenging problem for both the bed design and for optimal startup procedure.

See more of this Session: Modeling and Analysis of Chemical Reactors
See more of this Group/Topical: Catalysis and Reaction Engineering Division

(312e) Modeling and Optimization Studies of Combined LNT-SCR Catalyst Systems

Arun S. Kota, Dan Luss and Balakotaiah Vemuri, Department of Chemical and Biomolecular Engineering, University of Houston, Houston, TX

Abstract:

Aone-dimensional two-phase model with global rate expressions for variousreactions is used to study the performance of combined LNT-SCR catalystsystems. The LNT system modeled is that of Pt/BaO/ Al_2O_3catalyst with two different BaO storage sites,designated as fast and slow sites. LNT model includes reactions that describeNO oxidation, storage of NO_2 in the form of nitrates during leanstorage cycle and reduction to N_2 in the subsequent rich cycle withH_2 as a reductant. The SCR systemconsidered is that of Cu-ZSM5 and the model includes NO oxidation, ammoniastorage and oxidation, selective catalytic reduction of NO, NO_2 andformation of N_2O. Model predicts the spatio-temporalconcentration profiles of reactants and products inside the channel, which givesinsight into the performance of each LNT and SCR catalyst during operation. Ithas been found that series arrangement of alternate LNT and SCR catalyst, bydividing them into equal halves, substantially improves NO_xconversion. For fixed total catalyst length as the number of divisionsincrease, NO_x conversion increases andreaches a limit which is theoretically equivalent to that of mixed catalyst. Itis also observed that for fixed length of LNT catalyst better DeNO_x performance is obtained by increasing thelength of SCR catalyst in the alternate arrangement. Spatio-temporalprofiles indicate complete storage of NH_3 on SCR catalyst which canbe efficiently used to reduce NO_x slippingout of LNT. Various catalyst arrangements, the effect of different catalystloading and rich cycle time are investigated in the present work.

See more of this Session: Modeling and Analysis of Chemical Reactors
See more of this Group/Topical: Catalysis and Reaction Engineering Division

(312f) A Parameterized Iron-Zeolite SCR Model Calibrated to Reactor Data

Seth DeLand[1], Gordon Parker[1], **Jason Keith**[2] and John Johnson[1], (1)Department of Mechanical Engineering - Engineering Mechanics, Michigan Technological University, Houghton, MI, (2) Department of Chemical Engineering, Michigan Technological University, Houghton, MI

A numerical model was developed to capture the behavior of the SCR catalyst for varying inlet conditions. The model form was developed based on analysis of reactor data, and simulated over a temperature range reflective of typical exhaust gas temperatures. The experimental protocol was designed to capture the effects of different NH3/NOx ratios and different NOx compositions, as well as ammonia storage and oxidation. Feedback from the modeling effort was used to modify portions of the protocol in order to further investigate key areas. Two different types of storage sites were used to represent the adsorption and desorption of ammonia. An adsorption equation that includes ammonia inhibition was used to capture the slow approach of the catalyst to steady state storage. Two different states were used to represent the gas distribution in the catalyst: a bulk-gas concentration and a surface concentration. Species move between the bulk-gas and surface states via mass transport. Initially, the model used only the Standard (NH3-NO), Fast (NH3-NO-NO2), and Slow (NH3-NO2) reactions to capture the reduction of NOx by stored ammonia. Based on experimental data, reactions were added to extend the model beyond the typical SCR reactions. The single-channel, isothermal model was discretized axially to form a system of ordinary differential equations that could be numerically integrated. An optimization routine was used to determine the reaction rate parameters that provided the best fit to the reactor data, and the resulting model is presented. We discuss changes that were made to the original model form, which allowed the model to capture the dynamics seen in the experimental data.

See more of this Session: Modeling and Analysis of Chemical Reactors
See more of this Group/Topical: Catalysis and Reaction Engineering Division

(318a) Accurate Treatment of Electrostatics During Molecular Simulations In Nanoporous Crystals without Assigning Point Charges to Framework Atoms: Application In Material Selection for Flue Gas Separation

Taku Watanabe and David Sholl, School of Chemical and Biomolecular Engineering, Georgia Institute of Technology, Atlanta, GA

Molecular simulations are now routinely performed to study gas adsorption and diffusion in metal-organic frameworks (MOFs). Traditionally, the electrostatic potential energy surface (EPES) in the pore of a MOF is reproduced by assigning the electrostatic charges to the framework atoms. Despite some recent advance in the charge assignment methods, point charge models are inherently an imperfect representation of the material's true EPES. We have shown how the EPES determined directly from DFT calculations without assigning point charges can be used to accurately describe molecular adsorption in MOFs. Examples will be given including adsorption isotherms and molecular diffusivities of CO_2 and N_2 in selected MOFs including ZIF-8 and ZIF-90. We will also discuss recent advances in using information from ab initio Molecular Dynamics to find useful information regarding the internal flexibility of MOFs without requiring development of classical force fields. We will show how models based on these methods are being combined with experimental efforts by our collaborators to develop robust, scalable MOF-based membranes for carbon capture.

See more of this Session: Molecular and Multiscale Modeling for Renewable Energy Systems
See more of this Group/Topical: Computational Molecular Science and Engineering Forum

(318b) Molecular Dynamics Studies of Diffusion and Solvation of Rubidium Bromide Solutions In Nanoconfinement

Katherine A. Phillips, Jeremy C. Palmer and Keith E. Gubbins, Chemical and Biomolecular Engineering, North Carolina State University, Raleigh, NC

The need for improved energy storage devices has increased due to demands for renewable energy sources. Supercapacitors, or electrochemical double-layer capacitors (EDLCs), have become a viable option as a leading energy storage device due to their high charge efficiency, charge/discharge times, and long cycle life. While EDLCs have many attractive properties, they cannot store large amounts of energy. Carbide derived carbons (CDCs) have shown an improvement of energy storage in EDLC applications due to a dramatic increase in capacitance in pores ranging from 0.7 to 1.1nm[1]. The increase is thought to occur due to the stripping or deformation of the hydration shell as the ions enter the nanopores[1]. In the present work we analyze the hydration of ions in simple nanopore geometries to gain an improved understanding of the solvation structures inside the EDLC.

Using molecular dynamics (MD) simulations, we studied the impact of confinement on the solvation of rubidium and bromide ions in aqueous solutions confined in slit-shaped pores ranging from 0.65 to 1.6 nm in size, which encompasses the range typically found in CDCs and other nanoporous EDLC materials [2]. In the MD simulations, the rubidium and bromide ions were modeled as charged Lennard-Jones particles as prescribed by Koneshan, *et al.*[3]. We investigated 0.1 M, 0.5 M and 1.0 M solutions of these ions using the SPC/E[4] model for water. We found the average number of water molecules hydrating each of the ions and the structure of the solvation shell, which is characterized by the ion-water radial distribution function. The effect on confinement on the dynamics was also investigated by calculating the self-diffusion coefficients of the ions. Finally, the solvation of ions in an amorphous carbon structure was studied in order to determine the propensity of ions to reside in pores of specific sizes. Our results indicate that although materials with smaller pores have significantly increased capacitances, the diffusion, of ions is hindered by the confinement, which is potentially detrimental to enhancing the power density of EDLCs.

[1] Chmiola, J.; Yushin, G.; Gogotsi, Y.; Portet, C.; Simon, P.; Taberna, P. L. *Science*, **2006**, *313*, 1760.

[2] Simon, P.; Gogotsi, Y.; Nature Materials, **2008**, *7*, 845.

[3] Koneshan,S.; Rasaiah, J. C.; Lynden-Bell, R. M.; Lee, S. H. *J. Phys. Chem. B*, **1998**, *102*, 4193.

[4] Berendsen, H. J. C.; Grigera, J. R.; Straatsma, T. P. *J. Phys. Chem.*, **1987**, *91*, 6269.

(318c) An Electronic Structure Based Understanding of Amine-Carbon Dioxide Interactions for CO2 Capture

Anita S. Lee and John R. Kitchin, Department of Chemical Engineering, Carnegie Mellon University, Pittsburgh, PA

Gas scrubbing using amine solvents is one of the most promising first generation CO_2 capture technologies, due to the chemical and process knowledge gained through their long term use in natural gas. However amine scrubbing has high parasitic energy costs due to equipment heating, amine regeneration, and captured CO_2 compression costs. Thus, minimizing this energy requirement is an obvious place to start when developing enhanced amines for CO_2 capture. The energy associated with amine regeneration is a function of the CO_2 capacity of the amine as well as the reaction energy. Early studies of amine-CO_2 interaction explained capacity and bond strength through two acid/base reactions to form carbamates (2:1 amine:CO_2 interaction) and bicarbonates(1:1 amine:CO_2 interaction), with bicarbonates favored for sterically hindered secondary amines and tertiary amines. The picture is not that simple, however, as significant variations in the equilibrium CO_2 capacity among structurally similar amines and amines of similar pH have been reported. Understanding the origin of these variations in the experimental equilibrium capacity is the motivation behind this work. At equilibrium, the capture capacity can be characterized by the Gibbs free energy of reaction. Using Density Functional Theory (DFT) we studied the reactions of CO_2 with structurally similar amines and amines of similar pH to form carbamic acid, carbamate and bicarbonate ions, determined the reaction energies for each amine with CO_2 relate these energies to electronic structure properties of the amine. We show that the reaction energy associated with amine-bicarbonate formation correlates well with the amine electronegativity. Additionally, we identify a means to manipulate amine electronegativity through functionalizing the amine with different functional groups. These results provide us with a fundamental understanding of one of the reactions critical to CO_2 capture and a means to control the reaction energy through modifying the amine itself.

See more of this Session: Molecular and Multiscale Modeling for Renewable Energy Systems
See more of this Group/Topical: Computational Molecular Science and Engineering Forum

(318d) Engineering Protein-Carbohydrate Binding Affinity Via Glycosylation: A General Strategy to Improve Cellulase Performance

Courtney B. Taylor[1], Clare McCabe[1], Lintao Bu[2], Michael E. Himmel[3], Michael F. Crowley[4] and Gregg T. Beckham[5], (1)Chemical and Biomolecular Engineering, Vanderbilt University, Nashville, TN, (2)National Bioenergy Center, National Renewable Energy Lab, Golden, CO, (3)Biosciences Center, National Renewable Energy Lab, Golden, CO, (4)Biosciences Center, National Renewable Energy Laboratory, Golden, CO, (5)National Bioenergy Center, National Renewable Energy Laboratory, Golden, CO

Carbohydrate-binding modules (CBMs) display specificity to mono- and polysaccharides for docking of enzyme complexes to carbohydrates. CBMs are found ubiquitously as components of cellulase and hemicellulase enzymes that degrade cell wall polysaccharides in nature. For this purpose, CBMs have evolved to recognize specific cell wall polysaccharides via hydrophobic interactions and hydrogen bonding. Additionally, CBM binding affinity to carbohydrates has been shown to be proportional to enzyme activity, and thus increasing binding affinity is an important component of engineering improved cellulases. To that end, here we apply molecular simulation and free energy calculations to examine the binding affinity of a well-characterized Family 1 CBM to cellulose. To validate our approach, we quantify the impact of aromatic-carbohydrate interactions on the CBM binding affinity. Our predictions are in excellent quantitative agreement with previous experimental results showing that single-point mutations can alter the CBM binding affinity by approximately 2-fold. Following this validation, we demonstrate that a single, native *O*-linked mannose increases the CBM binding affinity to cellulose by 40-fold, which has not been considered explicitly in previous work. Furthermore, we show that the addition of a single artificial glycan on the CBM can have the same order of magnitude impact on binding affinity. These results have significant implications for protein engineering, in that modifying glycosylation patterns, for example by using heterologous host expression, manipulation of culture conditions, or introduction of artificial glycosylation sites, can drastically alter protein binding affinity to carbohydrate substrates, and may thus be a general strategy to improve cellulase performance.

See more of this Session: Molecular and Multiscale Modeling for Renewable Energy Systems
See more of this Group/Topical: Computational Molecular Science and Engineering Forum

(318e) Multi-Scale Models for Poly (1, 3-cyclohexadiene) (PCHD) Polymer

Qifei Wang[1], David J. Keffer[1], Suxiang Deng[2] and Jimmy W. Mays[2], (1)Chemical and Biomolecular Engineering, University of Tennessee Knoxville, Knoxville, TN, (2)Chemistry, University of Tennessee Knoxville, Knoxville, TN

Abstract

Atomistic and coarse-grained (CG) models of Poly (1, 3-cyclohexadiene) (PCHD) were developed and implemented in Molecular Dynamics (MD) simulations of PCHD chains with different architectures. In the atomistic model, two PCHD chains are cross linked by a sulfur-sulfur bond. Sulfonic acid groups are evenly distributed along the chain. The architecture is specifically aimed for application as a proton exchange membrane used in fuel cells. An atomistic force field for this architecture was tested and applied in the atomistic MD simulation of PCHD for the first time. The atomistic simulations generate density, chain radius of gyration and chain end-to-end distance distribution. To further study the structural properties of longer chain systems, a CG model was proposed. The bonded structural probability distribution functions (PDFs) and nonbonded pair correlation function (PCF) of the CG beads were obtained from the atomistic simulation results. The bonded CG potentials are obtained by simple inversion of the corresponding PDFs. The CG nonbonded potential is parameterized to the PCF using the Iterative Boltzmann inversion (IBI) method. The CGMD simulations of PCHD chains using potentials from above method satisfactorily reproduce the structural properties from atomistic MD simulation of the same system. A comparison of the CG potentials and structural distribution functions from the PCHD melt and a hydrated PCHD membrane is presented.

See more of this Session: Molecular and Multiscale Modeling for Renewable Energy Systems
See more of this Group/Topical: Computational Molecular Science and Engineering Forum

(318f) Band Gap Engineering In Donor-Acceptor Conjugated Copolymers

Ying-Chieh Hung, Department of Chemical Engineering, National Taiwan University, Taipei, Taiwan and Shiang-Tai Lin, Chemical Engineering, National Taiwan University, Taipei, Taiwan

Here we report the principles of the change in the electronic properties (especially, HOMO/LUMO, band gap, and band width) by introducing different amounts of electron donating (D) or electron accepting (A) units to the backbone of a conducting polymer. We performed systematic studies on thiophene (T) based copolymers containing thienopyrazine (TP) or methoxythiophene (OT) at varying compostions using *ab initio* density functional theory (DFT). We found that

(1) The increase of acceptor (TP or OT) content in the mainchain leads to a smaller band gap of conjugated copolymers.

(2) For moderate D-A contrast system (e.g., T-OT), band gap varies linearly with composition at similar D-A ratio region; however, there is little composition effect at high D and A concentration regions.

(3) For strong D-A contrast system (e.g., T-TP), LUMO and band gap drop rapidly when a small amount of acceptor (TP) is added. The band gap become composition insensitive at high acceptor concentration regions.

(4) An extremely narrow bandwidth of conduction band would be observed for strong D-A contrast system especially at low concentration of π-electron acceptor.

Based on these results, we demonstrate an effective molecular design strategy by adjusting D-A composition toward desired electronic properties for donor-acceptor conjugated copolymers.

See more of this Session: Molecular and Multiscale Modeling for Renewable Energy Systems
See more of this Group/Topical: Computational Molecular Science and Engineering Forum

(318g) Structure and Diffusion of Furans and Other Cellulose-Derived Compounds In Solvents Via MD Simulation

Brooks D. Rabideau, Laboratoire de Matriaux et des Structures du Gnie Civil, Champs sur Marne, France and Ahmed E. Ismail, Mechanical Engineering, RWTH Aachen University, Aachen, Germany

There now exists a large push towards the development of sustainable, environmentally friendly energy sources. The conversion of cellulose derived from biomass into biofuels is one such promising route. As part of this process, a variety of intermediary compounds that appear critical to the successful expansion of the process to an industrial scale have been identified. These intermediary compounds consist of either different furans or acids derived from cellulose, as well as the cellulosic strands themselves. We examine the structure and diffusion of these platform chemicals in 1-butyl-3-methylimidazolium chloride ([BMIM]Cl), an ionic liquid, and water via molecular dynamics (MD) simulations. Specifically, we study the solvation properties of these chemicals by examining their pair correlation functions in each solvent. We determine the diffusion constants of these compounds, and explore the dispersibility of these chemicals within these different solvents as well as the hydrogen bonding networks that develop. Finally we explore the effect of the chain length of cellulose on its behavior in these solvents.

See more of this Session: Molecular and Multiscale Modeling for Renewable Energy Systems
See more of this Group/Topical: Computational Molecular Science and Engineering Forum

DESIGN CHALLENGES AND MODELING FOR AN EFFICIENT LIQUID HYDROGEN STORAGE TANK FOR AUTONOMOUS SYSTEMS

*Richard O. Stroman and Michael W. Schuette**
Naval Research Laboratory, Washington DC
**Global Strategies Group (North America) Inc., Washington DC*

Introduction

Small autonomous systems are rapidly adding capability and taking on new roles due to advances in navigation, controls, communications and sensing. Despite these improvements, however, the mission duration remains limited by the amount of useful energy available. There are two ways to provide more energy: raise the conversion efficiency or store more energy at the outset.

In recent years some vehicles have used fuel cells to convert fuel to electrical power for the propulsion plant and payload, because they tend to be more efficient than combustion engines in the 0.5 – 5 kW power range. One such vehicle is the Naval Research Laboratory's Ion Tiger, a 15.9 kg unmanned air vehicle (UAV) with a proton exchange membrane fuel cell (PEMFC) power system. PEMFCs convert H_2 to electrical power with about 50% efficiency, which in combination with the high specific energy of H_2 (33,400 Wh/kg) makes them ideal for long endurance systems.

The Ion Tiger demonstrated a 26 hr flight using 0.5 kg of gaseous H_2 (GH_2) stored at 5000 psi in a 22 L, 3.8 kg pressure vessel. [1] The pressure vessel and regulator together had to weigh less than 4 kg to satisfy the aircraft weight budget, which dictated a minimum storage efficiency of 12 wt% H_2 when combined with the goal of 24+ hr flight. Higher storage efficiency is necessary to increase the Ion Tiger flight endurance beyond 1 day, and our experience suggests that GH_2 is not a solution. Increasing the H_2 pressure would provide more H_2 storage in the same volume, but would require a disproportionately heavier pressure vessel and decrease storage efficiency. Furthermore, pressures above 5000 psi can require heavier materials not susceptible to H_2 embrittlement. Therefore, we opted to explore liquid H_2 (LH_2) as an alternative to GH_2.

The mass density of LH_2 is 62.6 g/L at 80 psi and 25 K, which is nearly three times larger than that of GH_2 at 5000 psi and 298 K (23.0 g/L). [2] The greater mass density of LH_2 could in principle provide nearly three times the storage efficiency of the present GH_2 system. LH_2 has been largely abandoned for most fuel cell applications because boil off cannot be eliminated entirely, so long term storage is problematic. For UAVs, however, LH_2 presents an interesting opportunity because they can be launched immediately after fueling and then continuously burn fuel in flight. The high mass efficiency of an LH_2 tank may enable unique small, long endurance electric UAVs. The major challenges when designing an LH_2 tank for an aircraft are to build a lightweight container with very low heat leak from ambient, and to accommodate transients in the fuel withdrawal rate.

Three aspects of the LH_2 storage tank design problem are described here: design of the insulation, design of the vent tube assembly (VTA), and the selection of an operating scheme. Validation of the insulation and VTA design models is also addressed.

Design Considerations

The tank design process was dominated by two (non-rigorous) optimization goals: maximize fuel storage and minimize venting. Fuel storage is directly related to the system weight – if the system weighs less, more of the weight budget can be allocated to fuel. Venting is related to the rate of heat leak from ambient – a lower rate of heat leak is less likely to raise tank pressure to the point at which it must be vented. Therefore the practical design goals were to minimize weight and minimize the heat leak rate.

There are additional reasons to minimize weight and heat leak rate. Lighter weight aircraft tend to require less propulsion power, be more maneuverable, and carry heavier payloads. With the rate of heat leak from ambient minimized, that the rate of boil off can be controlled over the widest possible range to match the withdrawal rate. Reducing the heat leak rate provides more control flexibility, "turn down" capability and endurance.

A dewar consisting of two nested vessels separated by a vacuum space is used in most high performance applications where exceptionally low heat leak is required. The vacuum eliminates conduction and convection heat transfer between the inner and outer vessels, and multi-layer insulation (MLI) is often added to reduce the rate of radiation heat transfer. Two heat conduction paths between the inner and outer vessels are largely unavoidable: the inner vessel supports and the structure(s) for filling and venting the inner vessel. The first design task is to determine the amount of MLI and the vacuum quality required by estimating the rate of radiation heat leak as a function of those parameters. The second is to design the vent tube by estimating the rates of conduction and radiation heat leak it introduces.

The fuel consumption rates for a typical Ion Tiger flight are shown in Table 1. Ideally the rate of LH_2 boil off would vary to match the rate of withdrawal, but because it is determined by the temperature difference between the LH_2 and ambient, it is nearly constant. To accommodate transients in the withdrawal rate one can allow the tank pressure to fluctuate, use a heater to "trim" the rate of boil off, or both. The third design task is to determine how large a heater is required, how often it should turn on, and how large an open space (filled with GH_2) above the LH_2 is required to damp transients in the withdrawal rate by allowing the tank pressure to vary.

Table 1. Fuel flow and heating rates required in a typical Ion Tiger flight.

Phase	H_2 Flow Rate [g/s]	Heat Rate, Vaporize LH_2 [W]	Heat Rate, Warm GH_2 [W]
Warm Up	8.33×10^{-4}	0.34	3.0
Climb Out	1.17×10^{-2}	4.78	41.4
Cruise	5.67×10^{-3}	2.32	20.1
Landing	8.33×10^{-4}	0.34	3.0

The total heat leak design goal was 90% of the heat rate required to boil off LH_2 at cruise power, shown as 2.32 W in Table 1. This limit kept the design goal feasible while providing some turn down capability. Early estimates of the relative rates of heat leak through each path led to the detailed heat leak budget in Table 2.

Table 2. Heat leak budget.

Path	Transfer Mode	Rate [W]
Supports	conduction	0.025
Insulation	radiation	1.889
VTA	conduction and radiation	0.175
Total	all	2.089

Design of the Insulation

A 1-D steady state heat transfer model was developed to determine how many radiation shields would be necessary to satisfy the heat leak budget. The model accounted for radiation heat transfer among layers of double aluminized Mylar insulation, using the Stefan-Boltzmann law [3-5] and absorbtivity data provided by the manufacturer. Conduction through the rarefied gas (assuming He at a pressure of 1×10^{-5} torr) [6] and through the Dacron net spacer were also included. Boundary conditions were the temperatures of the inner and outer vessels, assumed to be 25 K and 298 K, respectively. A nonlinear equation solver (the *fsolve* function in MATLAB) was used to determine which temperature profile through the stack of radiation shields would conserve energy, and the radiation flux through the stack. The number of MLI layers was increased until the desired radiation flux was reached.

Comparison to experimental data showed that the model consistently underpredicted the rate of heat transfer, and hence the number of MLI layers required. The discrepancy was attributed to insufficient fidelity in the spacer heat conduction model – the spacer reduces the insulation effectiveness by providing a conduction path, and the rate of conduction was underpredicted. The rate of heat conduction through the spacer layers depends on the contact area between the woven spacer and the radiation shields, and on the flux area, neither of which were well characterized. Attempts to fit these parameters were unsuccessful, so a published empirical model [7] was used instead. The empirical model predictions compared well with a test rig built for model validation (see Validation section) and confirmed that considerable effort is necessary to produce an accurate physics based MLI model. The empirical model should be sufficient for most designs.

Design of the Vent Tube Assembly

A 1-D steady state center differencing model was developed to estimate the rate of heat transfer through the VTA as a function of tube material, tube length, wall thickness and gas flow rate. The VTA heat leak model accounted for conduction through the tube wall, convection to the H_2 gas and radiation exchange between the tube and outer vessel (see Figure 1).

The tube and gas channel were broken into discrete volumes with the rates of heat transfer among them determined by Fourier's Law for conduction, Newton's Law of Cooling for convection and the Stefan-Boltzmann law for radiation. [3-5] Constant temperature boundary conditions were imposed at each end of the tube and at the gas entrance to the tube. A nonlinear equation solver (the *fsolve* function in MATLAB) was used to determine which tube and gas temperature profiles $T(x)$ conserved energy in each discretization. The resulting tube temperature profile was used to determine the heat flow rate into the inner vessel.

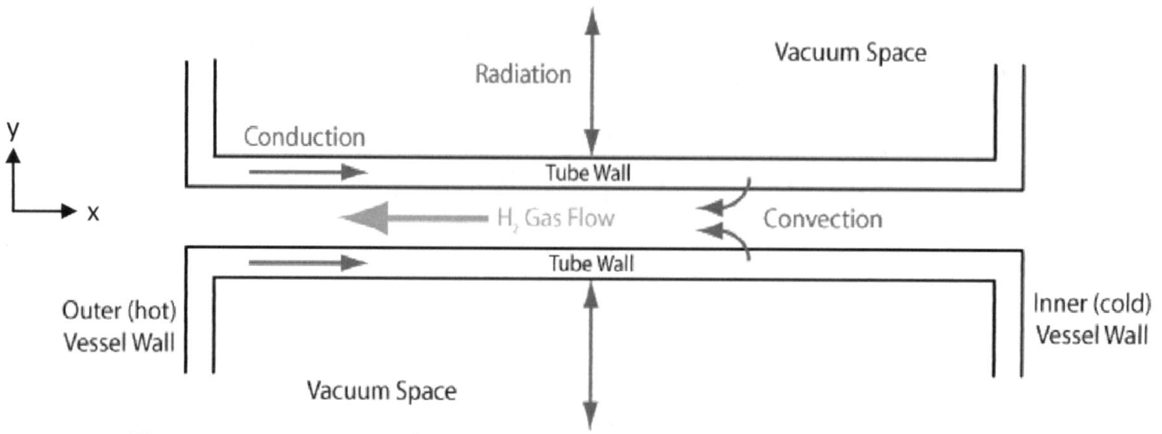

Figure 1. Heat transfer modes included in the VTA heat leak model.

The gas and tube thermal conductivities *k* varied considerably through the 25 – 298 K temperature range experienced by the tube, so the model interpolated among data from an online NIST database [8] to determine $k_{tube}(T)$ *and* $k_{H2}(T)$. The gas velocity (and hence the rate of convective heat transfer) was calculated in each discretization by accounting for the variation in gas density. The gas density was determined as a function of temperature and pressure by interpolating among data which functioned as an empirical equation of state. [2] It was assumed that the VTA was positioned in the vacuum space between the MLI and outer vessel, so it absorbed radiation from the outer vessel directly and via reflection from the MLI. This geometry was approximated by treating the VTA as if it were enclosed by a black body, i.e. assuming that the MLI was a perfect reflector. A typical radiation absorbtivity for each tube material was chosen from the literature, but the calculated temperature profiles showed only a small dependence on the value selected because the net rate of radiation heat transfer was small. The numbers of tube and gas discretizations were increased until the result ceased to vary with the number of discretizations.

An example result from the VTA heat leak model is shown in Figure 2, where the heat leak rate due to various tube lengths and gas flow rates have been plotted. Each curve in Figure 2 assumes the same tube material, inside diameter and outside diameter.

The importance of heat leak being intercepted by the gas through convection is clear – for each gas flow rate, there is a length beyond which the entire conduction and radiation heat leaks are intercepted. Given a maximum allowable rate of vent tube heat leak, expected gas flow rate and tube geometry, results such as in Figure 2 can be used to determine the minimum tube length necessary. It can also be used to explore different tube materials and combinations of inside diameter and gas flow rate.

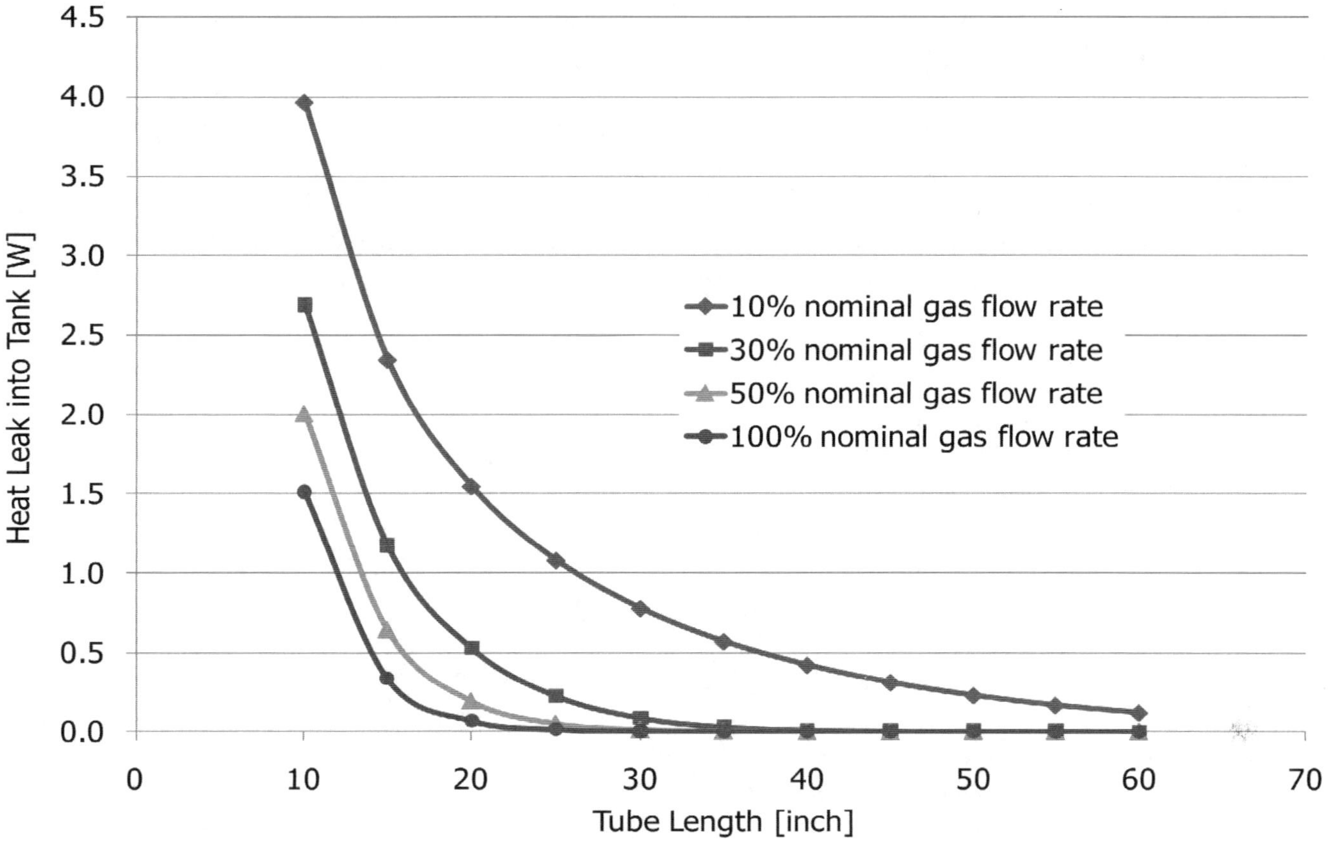

Figure 2. Predicted VTA net heat leak rate as a function of tube length and GH_2 flow rate.

A safety concern related to the VTA is the possibility of a catastrophic loss of vacuum which would cause rapid LH_2 vaporization, and if the gas cannot escape, rising tank pressure. In order to avoid a boiling liquid expanding vapor explosion (BLEVE) the VTA must vent GH_2 at the failure boil-off rate without exceeding the burst pressure of the inner vessel. The failure heat flux and boil off rate were estimated by assuming that gaseous N_2 from the air would condense on the outside of the inner vessel. This established temperature boundary conditions at the inside (25 K) and outside (77 K) surfaces of the inner vessel wall, which allowed an estimate of the heat flux to be calculated. Note that neglecting the thermal resistance associated with the LH_2 boiling made the heat flux estimate conservative (high).

The method outlined by [9] was used to estimate the maximum flow rate through tubes of various diameter and length. In each case the gas pressure in the inner vessel was assumed to be 75% of the inner vessel burst pressure. For the chosen tube length and inside diameter, the predicted gas flow rate was sufficient to avoid a BLEVE when compared to the predicted failure boil off rate.

Validation of Heat Transfer Models

A test rig was built to validate the model; the rig consisted of two concentric aluminum cylinders separated by a vacuum space. The inner cylinder was wrapped with MLI and separated from the outer cylinder by low thermal conductivity supports. The inner cylinder was filled with liquid N_2 (LN_2) and the total rate of heat transfer to the inner cylinder was estimated from the rate of LN_2 boil off.

The total rate of heat leak was measured for three different boundary temperatures: 207±1 K, 298±4 K and 387±3 K. The relative amounts of conduction and radiation heat transfer were estimated by assuming a dependence on ($T_h - T_c$) for conduction and ($T_h^4 - T_c^4$) for radiation. The rate of radiation heat transfer agreed with the empirical model to within 10% using a degradation factor of 7.4, which describes the deviation from ideal behavior introduced by seams and penetrations. In all of the measurements, the flow rate of LN_2 was sufficient to ensure that the heat leak rate was dominated by radiation heat transfer, confirming the expectation that MLI plays a major role in the design.

Load Based Transient Analysis

Fuel consumption rate data from GH_2 powered Ion Tiger flights were analyzed to estimate the transient performance required of the LH_2 tank. All aircraft have varying propulsion power requirements which depend on vehicle mass, drag and the way in which it maneuvers, but because the Ion Tiger is relatively small, the propulsion power also varies substantially with weather and thermal phenomena. Due to these variations, it is important to accurately estimate the range and typical rate of change in the fuel withdrawal rate.

To accommodate periods of low fuel consumption rate, a pressure relief valve (PRV) must be included in the tank design. Tank pressure increases when the boil off rate exceeds the consumption rate; when the pressure reaches the upper limit for the inner vessel, the PRV opens to release GH_2. Having a larger open space above the liquid allows more LH_2 to boil off before the PRV must open, thereby decreasing the amount of lost fuel, but this comes at the cost of storing less LH_2 at the outset. Having a lower overall heat leak rate and using a heater in periods of high fuel consumption is preferred, because then a smaller gas volume can be used to damp transients. The smaller gas volume leaves more space for LH_2 storage.

The design goal was to maintain tank pressure between 40 and 80 psig; to do this a transient performance model was developed. The model used propulsion load data and heat leak rate as inputs and the tank fill fraction and pressure as initial conditions. It integrated the difference between the LH_2 boil-off rate and the rate of fuel consumption, and then used an empirical equation of state [2] to estimate the tank pressure at each time step. When the pressure reached the upper limit, the tank vented, and when the pressure reached the lower limit, the heater turned on. Deadband of ½ the pressure range was assumed for the heater controller – the heater turned off at the nominal tank pressure. Model outputs were the predicted tank pressure, GH_2 blow-off rate and heater duty cycle as functions of time.

Figure 3 shows the predicted tank pressure as a function of time given an Ion Tiger fuel consumption curve, total heat leak rates of 2.0, 2.1 and 2.2 W, no PRV or heater, and a gas volume of 4 L. Note that changes in the heat leak rate as small as ±100 mW integrated over 26 hr meant the difference between exceeding the pressure limits of the tank and starving the fuel cell. The same is true to a greater extent when integrated over a 72 hr flight. Figure 3 shows the clear need for a low heat leak rate from ambient, a PRV, and a heater for adequate control of the LH_2 storage system. The high heat leak case exceeded the 80 psia inner vessel pressure limit and the low heat leak case could not support the fuel consumption rate over the entire flight.

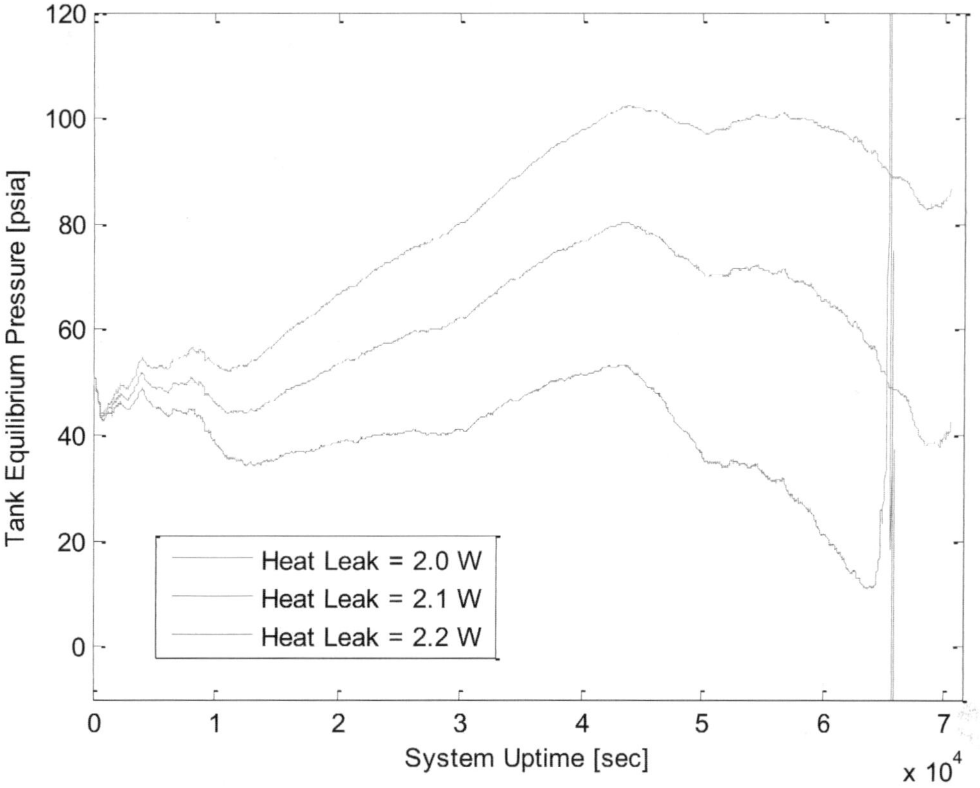

Figure 3. Predicted tank pressure assuming heat leak rates of 2.0, 2.1 or 2.2 W and a gas volume of 4 L. The large transient in the 2.0 W curve is an integration error due to the tank pressure falling below ambient pressure, which cause the equation of state to fail.

Assuming a total heat leak rate of 2.09 W, the transient performance model was used to compare two design scenarios: a large gas space and no heater vs. a small gas space with a heater. In the first case, the gas volume and PRV damp are the only means to damp transients, so the gas volume must be larger. In the second case, the heater supports the pressure during periods of high withdrawal rate, so the gas space can be smaller. The model predicted that adding a heater would improve tank pressure control and ensure it remained within the desired 60±20 psi range, as shown in Figure 4. The smaller gas space led to an increase in venting, but it was more than offset by larger volume of fuel stored at the outset. The tank with the heater was predicted to hold 18% more LH_2 at the end of the 26 hr simulated flight, as shown in Table 3.

Table 3. Tank performance in a simulated 26 hr flight, with and without a heater. * indicates predicted values.

Case	Initial fuel volume [L]	Total Heat Leak [W]	Fuel vented* [g]	Total Capacity* [g]
No heater, 4 L gas volume	18.0	2.09	0.0	1161
Heater, 0.5 L gas volume	21.5	2.09	19.4	1367

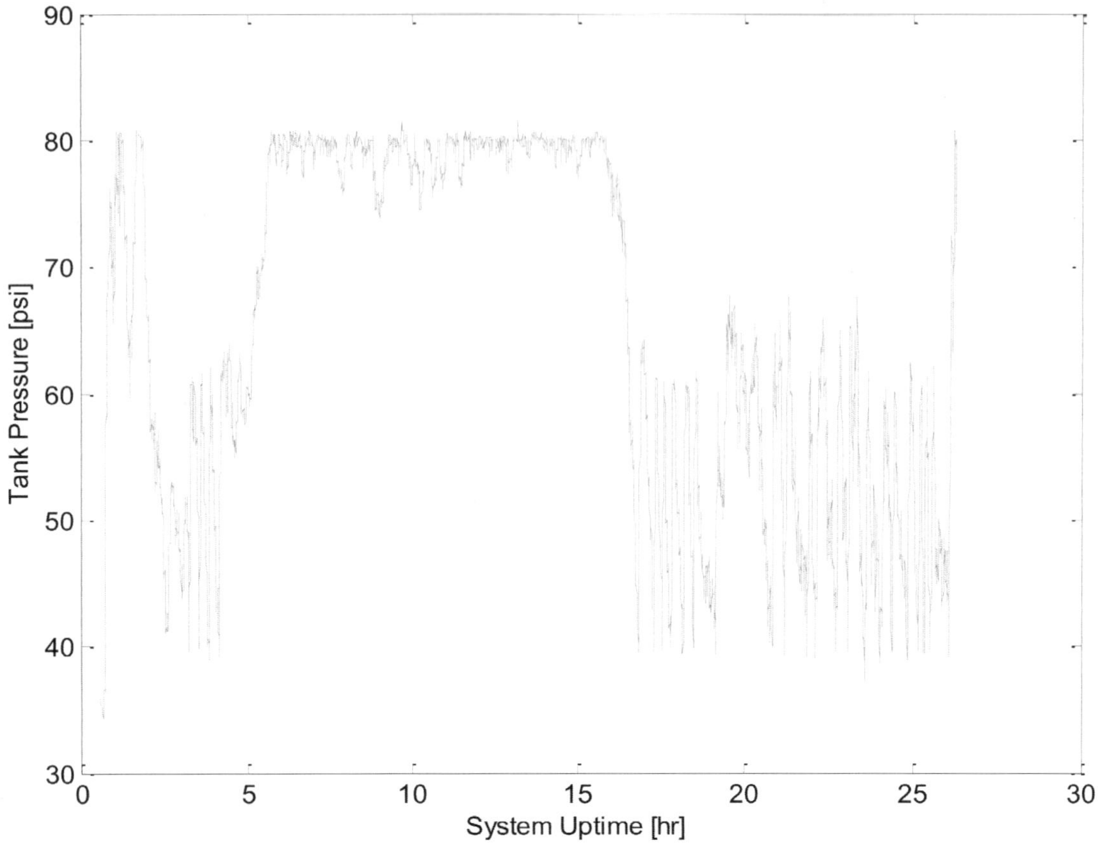

Figure 4. Predicted tank pressure with a heat leak rate of 2.09 W, 0.5 L gas volume, PRV set to 80 psia and heater activated at 40 psia.

Conclusions

The design of a LH_2 storage tank for small aircraft applications is challenging in large part because of the restrictions on weight and heat leak rate. For the Ion Tiger LH_2 storage tank design, these two restrictions drove most of the material selection, mechanical design and thermal design decisions. Steady state heat transfer and transient flow models can provide valuable insights into the performance of proposed designs, and guide the design process, once the basic design concepts have been established. As shown by the insulation modeling, however, validation of the models is essential and empirical models are sometimes preferred to physics based approaches.

Acknowledgements

The authors gratefully thank the Office of Naval Research for funding this work.

References

1. K. Swider-Lyons, R.O.Stroman, G. S. Page, J. F. Mackrell, J. A. Rodgers, M. W. Schuette. *The Ion Tiger Fuel Cell Unmanned Air Vehicle*. in *Proceedings of the 44th Power Sources Conference*. 2010.

2. *NIST Thermophysical Properties Database*. [cited 2011; Available from: http://webbook.nist.gov/chemistry/fluid/.
3. Incropera, F.P. and D.P. DeWitt, *Fundamentals of heat and mass transfer*. 5th ed. 2002, New York: J. Wiley. xix, 981 p.
4. Kays, W.M., M.E. Crawford, and B. Weigand, *Convective heat and mass transfer*. 4th ed. McGraw-Hill series in mechanical engineering. 2005, Boston: McGraw-Hill Higher Education. xxx, 546 p.
5. Brewster, M.Q., *Thermal radiative transfer and properties*. 1992, New York: Wiley. xxii, 543 p.
6. Barron, R.F., *Cryogenic systems*. 2nd ed. Monographs on cryogenics. 1985, New York; Clarendon Press. xvii, 507 p.
7. Gilmore, D.G., *Satellite thermal control handbook*. 1994, El Segundo, Calif.: Aerospace Corporation Press.
8. Marquardt, E.D., J.P. Le, and R. Radebaugh, *Cryogenic material properties database.* Cryocoolers 11, 2001: p. 681-687.
9. C. K. Liu, L.G.N., A. F. Manikowski, Jr. *Sizing an emergency venting system for a cryogenic dewar.* in *Advances in Cryogenic Engineering*. 1986. Cambridge, MA: Plenum Press.

(378b) Out-Gassing Study and Species Measurement for Cryogenic Carbon Fiber Hydrogen Storage Vessels

Joseph W. Reiter, Rajeshuni Ramesham and Alexander W. Raymond, Jet Propulsion Laboratory, Pasadena, CA

Out-gassingStudy and Species Measurement for Cryogenic Carbon Fiber Hydrogen Storage Vessels[1]

Joseph Reiter, Rajeshuni Ramesham, and AlexanderRaymond

JetPropulsion Laboratory, California Institute of Technology, Pasadena, CA 91109,USA

Cryo-adsorption andcryo-compressed hydrogen storage are two promising technologies for vehicularhydrogen storage. These low-temperature storagemethods have the benefits of increasing hydrogen density and materialadsorption capacity, which enhance both the gravimetric and volumetricperformance of these systems. In bothcases, carbon-fiber wrapped tanks appear to be preferable to metal pressurevessels for vehicular applications because of their higher gravimetriccapacity; however, maintaining adequate thermal isolation for sustaineddormancy at cryogenic temperatures is challenging with carbon compositevessels. Cryogenic pressure vessels aretypically surrounded by a vacuum jacket and multi-layer insulation (MLI.) The effectiveness of MLI depends on theextent to which high-vacuum is maintained in the vacuum jacket. Even moderate vacuum degradation can resultin significant gas conduction across the separated layers and will render theMLI ineffective. Carbon compositeout-gassing is studied in this preliminary investigation with relation to itseffects on vacuum insulation degradation.Species and total out-gassing rates will be presented on both agravimetric and on an areal basis using the sampleporous surface area provided by the BET (**Brunauer-Emmett-Teller**) method. Temperature dependence of the out-gassingrate will be deduced by conducting experiments at cryogenic operatingtemperature (77 K), room temperature (298 K), and elevated temperature (358 K). A discussion of the allowable out-gassingrates will be presented along with potential mitigation strategies.

Acknowledgments:The researchwork described in this paper has been carried out at the Jet Propulsion Laboratory,California Institute of Technology, Pasadena, CA. We acknowledge theDepartment of Energy (DOE) for supporting this effort.

See more of this Session: Hydrogen Storage Systems: Designs, Analysis, and Safety Aspects

(378d) Removal of Heat From a Hydrogen Storage Media by a Novel Microscale-Based Heat Exchanger

Christopher Loeb[1], **Agnieszka Truszkowska**[1], Bruce Hardy[2], Richard Chahine[3] and Goran N. Jovanovic[1], (1)Chemical, Biological and Environmental Engineering, Oregon State University, Corvallis, OR, (2)Computational Sciences, Savannah River National Laboratory, Aiken, SC, (3) Hydrogen Research Institute, Trois-Rivieres, QC, Canada

High volumetric storage density of hydrogen gas is a major challenge in the progress towards hydrogen-fueled automobiles. Of the various storage methods, pressurized gas tanks, liquid hydrogen storage, liquid chemical systems, and solid-state storage, activated carbon adsorbents show great potential. A reliable thermal management system within the storage tank is vital to provide quick, safe, and efficient charging and a controlled release of hydrogen gas on demand.

The current work investigates room temperature adsorption of nitrogen gas (model gas for hydrogen) in activated carbon bed and compares the temperature distribution within the storage media without thermal management, and with the addition of a microscale-based heat exchanger. Experimentally obtained temperature distribution data is used to assess the effect of charging time, and removal of generated heat within the system. A mathematical model is used to numerically simulate the experimental process.

The maximum height of activated carbon bed positioned between two heat exchange devices, while providing adequate thermal management, is investigated by varying the height of the adsorption bed and monitoring nitrogen-charging time and temperature distribution.

A novel microscale-based heat exchanger design takes advantage of 250 μ m high features, which facilitate the flow of cooling fluid. The heat exchanger is fabricated by diffusion bonding of patterned stainless steel shims, resulting in an overall 250 μ m thick resistance to thermal transport. The heat exchanger design also provides uniform gas distribution into the bed of storage media. The device's ability to provide a uniform temperature at the contacting surface is verified using infrared thermal imaging.

Experimental results indicate a dramatic improvement in charging time due to the fast and efficient removal of generated heat.

See more of this Session: Hydrogen Storage Systems: Designs, Analysis, and Safety Aspects
See more of this Group/Topical: International Congress on Energy 2011

(378e) Mitigating Strategies for Reactive Solids-Based Hydrogen Storage Systems

Joseph W. Pratt[1], Joseph G. Cordaro[2], Daniel E. Dedrick[3], Michael P. Kanouff[1] and Yehia (John) F. Khalil[4], (1)Thermal/Fluid Science and Engineering, Sandia National Laboratories, Livermore, CA, (2) Materials Chemistry, Sandia National Laboratories, Livermore, CA, (3)Hydrogen and Combustion Technology, Sandia National Laboratories, Livermore, CA, (4)Chemical Engineering Group, Physical Sciences Department, United Technologies Research Center (UTRC), East Hartford, CT

Solid-state materials such as metal hydrides and complex hydrides have been shown to be viable candidates for on-board hydrogen storage systems. However, some complex metal hydrides such as sodium alanate ($NaAlH_4$) in the powder form, exhibit unfavorable exothermic reactions when exposed to air or water. Mitigating this reactivity and its associated risks in the event of inadvertent exposure (e. g., tank rupture during a vehicular collision) is an important step for achieving widespread use and public acceptance of these technologies.

Several types of accident scenarios, exposures, and mitigating methods have been investigated in the past. This paper focuses on a specific scenario and mitigating strategy: exposure of the reactive solid to oxygen, and the effect of an active material-polymer composite mixture on its exothermic behavior.

Sodium alanate is selected as the representative of reactive solid hydrogen storage materials that exhibit unfavorable exothermicity, and is the "active material," although the mitigation methods developed are applicable to other similar materials. Previously, we reported on the use and effectiveness of polystyrene polymers synthesized (polymerized) along with the active material to create a composite. This paper presents our work on other polymers using the same methodology, comparing their effectiveness to the active material alone and to the previous work with polystyrene. Additive effectiveness is measured by the amount of heat released during controlled exposure to oxygen. Measurements are done both before and after subjecting the composite material to representative operating conditions of adsorption/desorption cycling, to understand the impacts (if any) of operating conditions on the stability and effectiveness of the composite material. SEM imagery is also used as a tool to better understand the method of mitigation and reasons for degradation (if any) of the composite. Additional tests were performed to investigate the impact of exposing the hydride (both in powder and powder compact forms) to vehicle fluids and road sprays and also to assess the sensitivity of the mitigated material to mechanical impact.

Daniel E. Dedrick, Joseph G. Cordaro, Michael P. Kanouff, Craig L. Reeder, Joseph, W. Pratt, and Y. F. Khalil, "Mitigation Technologies for Hydrogen Storage Systems based on Reactive Solids," presented at AIChE Annual Meeting, Salt Lake City, UT, Nov. 8-12, 2010.

(378f) Can a Liquid Hydrogen Storage Tank BLEVE?

Subramanya Nayak, Sergio Garcia, Victor Carreto and Sam Mannan, Chemical Engineering, Texas A&M University, College Station, TX

Abstract

A BLEVE (boiling-liquid expanding-vapor explosion) occurs when a tank containing liquefied gas ruptures, resulting in the instantaneous release of the vapor, boiling liquid, missiles (tank fragments), and blast wave[1]. The most common type of the BLEVE is caused when the external wall of the tank is exposed to a fire[1]. The continuous heat transfer from the fire to the tank wall increases the temperature and pressure inside the tank, until the tank fails. This sudden failure of the tank results in an explosive vaporization of its contents.

In passenger cars the LH_2 is stored as a cryogenic liquid at -240° C and at moderate pressures (20 to 30 bars)[2]. It is believed that the LH_2 will not BLEVE, since it is stored at such a low temperature[3]. However, LNG, which is also stored as a cryogenic liquid presents a potential for BLEVE[4]. Consequently, in this study we have theoretically investigated if the LH_2 storage tanks will BLEVE.

A theoretical model considering transient mass and energy transfer is developed and solved numerically[5]. It is assumed that the LH_2 tank is engulfed in a fire and the fire is caused due to spilled gasoline or hydrogen. Different tank materials and insulations are tested. The transient model has a capability to predict the time required for the LH_2 tank to BLEVE. This prediction will be useful for the fire fighters and emergency responders when dealing with LH_2 tank engulfed in a fire. Finally the consequences due to the BLEVE of LH_2 tank are also studied.

References

1. Abbasi, T. and S. A. Abbasi (2007). "The boiling liquid expanding vapour explosion (BLEVE): Mechanism, consequence assessment, management." Journal of Hazardous Materials 141(3): 489-519.
2. Rigas, F. and S. Sklavounos (2005). "Evaluation of hazards associated with hydrogen storage facilities." International Journal of Hydrogen Energy **30**(13-14): 1501-1510.
3. C. J. H. van den Bosch, R. A. P. M. W. (2005). Methods for the calculation of physical effects - due to the releases of hazardous materials (liquids and gases) - 'Yellow Book'. The Hague,

Committee for the Prevention of Disasters.
4. Pitblado, R. (2007). "Potential for BLEVE associated with marine LNG vessel fires." Journal of Hazardous Materials 140(3): 527-534.
5. Runnels, D. S. R. (2008). AFFTAC: Analysis of Fire Effects on Tank Cars.

See more of this Session: Hydrogen Storage Systems: Designs, Analysis, and Safety Aspects
See more of this Group/Topical: International Congress on Energy 2011

(379a) Development and Validation of ReaxFF Forcefields for Mo Ni O S H Interactions Under High Pressure Conditions

David Newsome[1], Alex Vasenkov[1], Debasis Sengupta[1] and Adri van Duin[2], (1)CFD Research Corporation, Huntsville, AL, (2)Department of Mechanical and Nuclear Engineering, Penn State University, State College, PA

Currently, the development of novel materials for advanced fossil energy systems remains slow because it is driven by a trial-and-error experimental approach. Atomistic Molecular Dynamic (MD) simulations have the potential to accelerate this development through the prediction of mechanical properties and corrosion resistance of new materials, reducing the time-to-market for new economically-viable materials to be used in fossil fuel systems. The success of MD simulations depends critically on the fidelity of interatomic potentials. Existing potentials typically are not able to account for reactions, or are not applicable for high-temperature simulations, or can be used only for modeling tiny nano-scale clusters.

These deficiencies were successfully addressed in this study by (1) developing QM-based Mo, Mo-O, Mo-Ni-O, Mo-H_2S and Mo-O-H_2S ReaxFF reactive potentials, (2) validating developed potentials against the Mo equation of state and temperature-dependent density and Young's modulus in the temperature range of 300 K to 1500 K, (3) reproducing experimentally established trends of Ni-Mo alloy resistance to oxidation, and (4) sulfur agglomeration in molybdenum and molybdenum trioxide. These newly developed ReaxFF potentials offer a compromise between high-level QM description and computational speed.

Mechanisms of initial oxidation in Mo, Ni, and Mo-Ni were compared. Specifically, it was found that Mo oxidation involves oxygen transport inside the metal slab and large heat release melting the surface. Oxidation of Mo slab proceeded at a significantly higher rate than the oxidation of Ni slab. Interactions of H_2S with Mo and MoO slabs were also modeled to test Mo-H_2S and Mo-O-H_2S ReaxFF force fields. Some key trends were identified based on the test results: (1) H_2S consistently diffused into Mo and MoO slabs and decomposed there to form S and H, (2) Mo-Mo metal bonds were replaced by S-Mo bonds leading to S agglomeration in the slab, and (3) H preferably desorbed from Mo slab at temperatures above 1400 K and formed gaseous H_2.

Major applications of the ReaxFF database include (1) Design of refractory alloys for new turbine blades, (2) Design of 'smart' materials with unprecedented levels of aerodynamic efficiencies for aviation and astronautics, (3) Design of materials for new generation of electronic devices where the development time is of crucial importance, and (4) Chemical process design where computer-aided development could potentially revolutionize the current technology by, for example, making possible catalytic "dream" reactions. Currently, the materials used in the above-listed applications constitute a market worth of tens of billions dollars.

See more of this Session: Industrial Applications of Computational Chemistry and Molecular Simulation III
See more of this Group/Topical: Computational Molecular Science and Engineering Forum

(379b) Direct Calculation of Octanol-Water Partition Coefficients From Adaptive Force Bias Molecular Dynamics Simulations

Navendu Bhatnagar, Chemical Engineering, Wayne State University, Detroit, MI and Jeffrey J. Potoff, Chemcial Engineering and Materials Science, Wayne State University, Detroit, MI

Partition coefficient (log K_{ow}) of a solute between 1-octanol and water [1] is an important parameter to predict a spectrum of solute properties such as pharmacokinetic characteristics in biological systems, drug partitioning in biophases, bioaccumalation [2,3], and is also used as a key input in quantitative structure-activity relationships (QSAR) [4]. Several techniques have been devised over the years for prediction of log K_{ow} which include experimental [5], semi-empirical [6], and QSAR based procedures. However, experimental methods are expensive and time consuming, QSAR is relatively fast but produces inaccurate results for molecules which are dissimilar to the ones in training set [4]. Molecular simulations are an effective tool for the determination of free energy differences that may be used to determine log K_{ow} [7]. Free energy perturbation (FEP) and thermodynamic integration (TI) [8] in conjunction with Monte Carlo (MC) [9,10] or molecular dynamics (MD) [11] simulations have been used extensively to determine free energies of solvation. Both free energy perturbation and thermodynamic integration are limited, though, by the need for a reference solute, which may create both systematic and statistical errors in the determination of octanol-water partition coefficients.

In this work, free energy change and log K_{ow} is predicted for transfer of n-alkanes across the 1-octanol| water and the air|water interface using a recently developed method known as adaptive biasing force (ABF) [12], which is based on unconstrained molecular dynamics simulations. Calculations are performed by using two separate approaches, where in the first one the solute is transferred directly from water to 1-octanol resulting providing the absolute free energy difference directly whereas in the second approach, a thermodynamic cycle is constructed and solute is transferred from water and 1-octanol phases to vapor phase in separate calculations resulting in relative free energy difference. Molecular dynamics simulations are performed with NAMD version 2.7b3 using united atom TraPPE [13] and the all atom CHARMM27 force field at a constant temperature of 298.0 K. Predictions from the ABF method are in close agreement with previous calculations that used thermodynamic integration [11].

References

1. Hansch, C.; Fujita, T. *J. Am. Chem. Soc.* (1964), 86, 1616.
2. Sangster, J. *Octanol-Water Partition Coefficients: Fundamentals and Physical Chemistry*, John Wiley & Sons, 1997

3. Connell, D. W. *Bioaccumalation of Xenobiotic Compounds,* CRC Press, 1990.
4. Hansch, C.; Leo, A. *Exploring QSAR: Fundamentals and Applications in Chemistry and Biology*, ACS, 1995.
5. Bergstrom, C. A. S.; Norinder, U.; Luthmann, K.; Artursson, P. *Pharmaceut. Res.*,(2002), 19, 182.
6. Leo, A.; Hansch, C.; Elkins, D. *Chem. Rev*. (1971), 71, 525.
7. Jorgensen, W.; Briggs, J.; Contreras, M. *J. Phys. Chem.* (1990), 94, 1683.
8. Kirkwood, J. G. *J. Chem. Phys.* (1935), 3, 300.
9. Chipot, C.; Pohorille, A. *Free Energy Calculations: Theory and Applications in Chemistry and Biology*, Springer, 2007.
10. Chen, B.; Siepmann, J.I. *J. Am. Chem. Soc.* (2000), 122, 6464.
11. Garrido, N.M; Queimada, A.J.; Jorge, M.;Macedo, E.A.; Economou, I.G. *J. Chem. Theory Comput.* (2009), 5, 2436.
12. Darve, E.; Pohorille, A. *Mol. Simul.* (2002), 28, 113.
13. Martin, M.G.; Siepmann, J.I. *J. Phys. Chem. B* (1998), 102, 2569.

See more of this Session: Industrial Applications of Computational Chemistry and Molecular Simulation III
See more of this Group/Topical: Computational Molecular Science and Engineering Forum

(379c) Calculating the Permeability Coefficients of Mixed Matrix Membranes of Polydimethylsiloxane and Silicalite Crystal to Various Ethanol-Water Solutions Using Molecular Simulations

Paul F. Harten, Sustainable Technology Division, U.S. Environmental Protection Agency, Cincinnati, OH

The permeability coefficients of mixed matrix membranes of polydimethylsiloxane (PDMS) and silicalite crystal are taken as the sum of the permeability coefficients of membrane components each weighted by their associated mass fraction. The permeability coefficient of a membrane component can be expressed as the product of an absorption coefficient and a diffusion coefficient. Absorption coefficients are calculated using Grand Canonical Monte Carlo (GCMC) simulations, but are done with respect to the initial alcohol-water mixtures to which the membrane is exposed. In contrast, the diffusion coefficients are calculated using molecular dynamics, but are done with respect to the alcohol-water mixtures absorbed by the membrane. The results of these calculations are compared with available experimental data.

See more of this Session: Industrial Applications of Computational Chemistry and Molecular Simulation III
See more of this Group/Topical: Computational Molecular Science and Engineering Forum

(379d) Combined Use of Molecular Dynamics and Monte Carlo for the Prediction of Thermodynamic Properties

Marianna Yiannourakou[1], David Rigby[2], Philippe Ungerer[1], Benoit Leblanc[1] and Paul W. Saxe[3], (1) Materials Design, Sarl, Montrouge, France, (2)Materials Design, Inc, San Diego, CA, (3)Materials Design, Inc, Angel Fire, NM

The industrial importance of a wide variety of compounds (from simple molecules to more complex structures) and the constant industrial need for synthesizing new and improved materials to be used in a range of different pressure and temperature conditions raises the need for accurate, efficient, systematic and rapid property prediction. Current advances in computational power, molecular simulation methodologies and forcefields (all atom and united atom) have rendered the use of molecular simulation as an increasingly powerful tool for the prediction of the thermodynamic properties of materials.

Since *combined* application of molecular dynamics (MD) and Monte Carlo (MC) techniques to compute a range of thermodynamic properties remains relatively unexplored (especially using the same force field), we have used these methods for the calculation of thermodynamic properties of pure compounds ranging from hydrocarbons (alkanes, olefins, aromatic hydrocarbons) to oxygenated molecules (alcohols, aldehydes, ketones, ethers), amines, amides and halogens. We accordingly demonstrate the use of the LAMMPS MD [1] and Gibbs MC [2] simulation programs in a complementary manner to probe a wide variety of properties; some can be uniquely calculated from MD (transport properties like viscosity, diffusion coefficient, thermal conductivity), while others can be uniquely calculated from MC (boiling point and critical point temperatures, chemical potential) and certain properties can be calculated using both techniques (pressure, density, heat of vaporization, specific heat, joule-thomson coefficient). Comparison of the calculated properties with experimental data demonstrates the efficiency of the combined use of MD and MC to characterize pure compounds. Our study illustrates the level of accuracy that is achievable with all-atom (extended PCFF [3], OPLS-AA [4]) and united-atom (TraPPE [5], AUA [6]) forcefields, for each property of interest. For certain functional groups, such as brominated aromatic compounds, it was necessary to re-paramaterize existing force fields; accordingly we will present a detailed description of improved new parameters and property predictions for these materials.

The ability of property prediction for pure compounds is also a necessary step towards simulation and property prediction for binary and multi-component mixtures. Test cases demonstrate that treatment of mixtures via molecular simulation is an effective means of study of the properties of the mixtures and also provides insight to the microscopic details that lead to the macroscopic behavior of such systems.

[1] S. Plimpton, J .Comp. Phys., 117, 1 (1995).

[2] MedeA-Gibbs v.2.6, Materials Design (Gibbs License IFP Energies Nouvelles-CNRS-Universite Paris-Sud)
[3] H. Sun, S.J. Mumby, J.R. Maple and A.T. Hagler, J. Am. Chem. Soc.. 116, 2978 (1994)
[4] W.L. Jorgensen, D.S. Maxwell and J. Tirado-Rives, JACS, 118, 11225-11236 (1996)
[5] J.M. Stubbs, J.J. Potoff and J.I. Siepmann, J. Phys. Chem. B, 108, 17596-17605 (2004) and references therein
[6] N. Ferrando, V. Lachet, J-M. Teuler and A. Boutin, J. Phys. Chem. B, 113, 5985-5995 (2009); J. Perez-Pellitero, PhD Thesis, ISBN: 978-84-691-0377-7/DL: T.2195-2007 and references therein

See more of this Session: Industrial Applications of Computational Chemistry and Molecular Simulation III
See more of this Group/Topical: Computational Molecular Science and Engineering Forum

(379e) Nano-Particle Adhesion In PEM Fuel Cell Electrodes

Qianping He, Department of Chemical and Biomolecular Engineering, University of Tennessee, Knoxville, Knoxville, TN, David C. Joy, Materials Science and Engineering, University of Tennessee, Knoxville, Knoxville, TN and David J. Keffer, Chemical and Biomolecular Engineering, University of Tennessee, Knoxville, Knoxville, TN

In the search for sustainable fuels, hydrogen remains of interest for long-range vehicular applications. There are numerous technical challenges that must be overcome in order to achieve significant market penetration. The durability of PEM fuel cells remains one of the most significant barriers for the commercialization of PEM fuel cells. Among various failure modes that shorten the lifetime of the PEM fuel cells, the degradation of carbon support Pt catalyst (Pt/C) due to carbon support corrosion is one of the most important factors [1]. Carbon is known to undergo electrochemical oxidation to surface oxides during fuel cell operation time. The presence of oxygen-containing groups can weaken the interaction between the carbon support and the catalytic nanoparticles, causing sintering of platinum nanoparticles and thus influence the conductivity of electrons and protons in the fuel cell and affect the lifetime of the fuel cell [2].

In the work described here, molecular dynamic simulations are performed to calculate the adhesion energy and force of platinum nanoparticles on carbon surfaces. To investigate how the carbon corrosion will affect the adhesion, we use three types of carbons surfaces: epoxy oxidized carbon surface, hydroxyl oxidized carbon surface and perfect carbon surface. Different shapes and sizes of nanoparticles (Pt) are applied to study their shape and size effects on the adhesion energy. Oxidized carbon surface of different oxidation rate are used to study the oxidation effect on adhesion energy. To mimic the real catalyst layer environment, water and Nafion are added as essential components. Simulation results are then compared with experimental results measured by nano-scale force sensitive manipulators, which are placed inside a SEM and operated while imaging the area of interest. Nano manipulators can be used to measure the adhesion of the catalyst particles to the substrate by determining the force required to dislodge them. Adhesion force between different types and sizes of platinum and three different carbon surfaces with varying oxidation ratios are explored to better understand the mechanism of how will carbon corrosion affect the fuel cell durability.

Acknowledgments

This work was supported by the STAIR program at the University of Tennessee, funded by NSF under agreement number: DGE 0801470.

[1]. Roen, L. M.; Paik, C. H.; Jarvic, T. D. Electrochem. Solid-State

Lett. 2004, 7, A19.

[2]. Coloma, F.; Sepulveda-Escribano, A.; Fierro, J. L. G.; Rodriguez-

Reinoso, F. Langmuir 1994, 10, 750.

See more of this Session: Industrial Applications of Computational Chemistry and Molecular Simulation III
See more of this Group/Topical: Computational Molecular Science and Engineering Forum

(379f) Multicomponent Maxwell-Stefan Diffusivities At Infinite Dilution

Thijs JH Vlugt[1], Xin Liu[2] and Andre Bardow[2], (1)Delft University of Technology, Delft, Netherlands, (2)RWTH Aachen University, Aachen, Germany

Diffusion plays a crucial role in (bio)chemical processes. It is usually difficult to obtain Maxwell-Stefan diffusivities from experiments as well as molecular simulation. Therefore, predictive models based on easily measurable quantities are highly desired. The Vignes equation is commonly used to describe the concentration dependence of Maxwell-Stefan diffusivities. In mixtures containing at least three components, the generalized Vignes equation requires the value of the quantity $D_{ij}^{x_k\text{->}1}$, which describes the friction between components i and j when both are infinitely diluted in component k. Over the past decades, several empirical models were proposed for estimating this quantity from binary diffusion data, and all of these are lacking a sound theoretical basis [1]. Here, we show that $D_{ij}^{x_k\text{->}1}$ actually exists, i.e. its value does not depend on the molar ratio x_i/x_j, and we derive an analytical expression for $D_{ij}^{x_k\text{->}1}$ based on the linear response theory and the Onsager relations [2]. We find that $D_{ij}^{x_k\text{->}1}$ can be expressed in terms of self-diffusivities and integrals over velocity cross-correlation functions. By neglecting the latter terms, we obtain a convenient predictive model for $D_{ij}^{x_k\text{->}1}$ in terms of self-diffusivities.

Molecular Dynamics simulations are used to validate the assumptions made in this model. The following test systems are considered: a ternary system consisting of particles interacting using Weeks-Chandler-Andersen (WCA) interactions and the ternary systems n-hexane-cyclohexane-toluene and ethanol-methanol-water. Our results show that: (1) for the WCA system as well as the system n-hexane-cyclohexane-toluene, neglecting the integrals over velocity cross-correlation functions results in accurate predictions for $D_{ij}^{x_k\text{->}1}$; (2) for the WCA system, our model prediction is superior compared to the existing models for $D_{ij}^{x_k\text{->}1}$; (3) in the ethanol-methanol-water system, the integrals over velocity cross-correlation functions cannot be neglected due to the presence of hydrogen bonds. Models for predicting $D_{ij}^{x_k \to 1}$ in this system will require detailed information on the collective motion of molecules; (4) our model provides an explanation why the Maxwell-Stefan diffusivity between adsorbed components in a porous material is usually very large.

References

[1] X. Liu, T.J.H. Vlugt, A. Bardow, Fluid Phase Equilibria, (2011), 301, 110-117.

[2] X. Liu, A. Bardow, T.J.H. Vlugt, Ind. Eng. Chem. Res., (2011), in press.

See more of this Session: Industrial Applications of Computational Chemistry and Molecular Simulation III
See more of this Group/Topical: Computational Molecular Science and Engineering Forum

(379g) Molecular Dynamics Study of Drug Diffusion In Hydrogels : Effect of Cross-Linking

Ravi C. Dutta and Beena Rai, Process Engneering Department, Tata Research Development and Design centre, Pune, India

Hydrogels, unique class of macromolecular networks that can hold a large fraction of an aqueous solvent within their structures are presently under investigation as a carrier in controlled delivery system for bioactive molecules, because of their similar physical properties as that of living tissues. Anionic hydrogels are used in the design of controlled release devices for site specific drug delivery of therapeutic proteins to large intestine since most of these proteins cannot sustain the acidic environment in stomach.

The functionality of the hydrogel system depends on both its structural and its dynamic properties. Structural aspects of the polymeric gel concern chain conformations, physical and/or chemical junctions, mesh size and the “structuring”/“destructuring” of water molecules due to the interaction of water with the hydrophyllic/hydrophobic polymeric components. Molecular structure of the hydrogel, can thus be characterized by cross-linking density which is a function of several parameters such as type and amount of cross-linking agent, temperature, method of cross-linking etc. The cross-linking density affects gel properties such as swelling, diffusion/release of molecules, stability and density. Experimental estimation of cross-linking density is expensive and time consuming.

In order to design gels for a given application such as drug-release, estimation of cross-linking density is important. Quite often one has to adapt a trial and error methodology to experimentally deduce the optimum cross-linking density for a given polymer – drug system, which is definitely time consuming and expensive. Computer simulations provide an excellent route to study polymer networks. While on one hand, they can reduce trial and error experimentation; on the other hand, they provide insightful information at molecular as well as micro length scales which are often not extracted from experiments. Molecular dynamics (MD) simulations which consider atomic details have been very successful in exploring various phenomena occurring at pico to nanosecond time scales and relating these phenomena to the structure and molecular chemistry. MD simulations also possess an inherent advantage over continuum modeling in addressing nanoscale interactions between water/drug and polymer networks which are crucial for designing gels for certain specialized applications such as control drug release.

In order to develop a robust and scalable MD methodology to study the effects of method of cross-linking and cross-linking density on diffusion of drug molecules in a polymeric network, we have chosen a model hydrogel system consisting of cross-linked *Diglycidyl ether of bisphenol A(DGEBA) / Isophorone diamine*(IPD). We present the effect of method of cross-linking and system size on cross-linking density. The diffusion of drug and water molecules in cross-linked polymer network have been studied for different cross-linking densities and system sizes. The effect of cross-linking density on gel properties such as swelling, diffusion/release of molecules, stability and density has been compared with experimental results.

See more of this Session: Industrial Applications of Computational Chemistry and Molecular Simulation III
See more of this Group/Topical: Computational Molecular Science and Engineering Forum

(393a) Characterizing the Pore Structure of Biochars: A New Approach Based On Multiscale Pore Structure Models and Reactivity Measurements

Hao Sun[1], Caroline A. Masiello[2] and Kyriacos Zygourakis[1], (1)Chemical and Biomolecular Engineering, Rice University, Houston, TX, (2)Earth Sciences, Rice University, Houston, TX

Biochar is charcoal generated for intentional soil amendment by pyrolyzing sustainable biomass feedstocks. Properly "engineered" charcoals can increase the water holding and cation exchange capacities of soils, improving the ability of plants to survive under drought conditions and reducing fertilizer runoff into watersheds. Fertilizer runoff has become a serious problem, because as much as 70% of fertilizer applied to crop fields is leached into the groundwater or lost to streams and rivers, eventually leading to large hypoxic 'dead' zones in the world's oceans (including the Gulf of Mexico). The environmental performance of biochars depends on their ability to adsorb, retain and release water and nutrients. These biochar properties are controlled by their porosity and surface chemistry, which can vary widely depending on the composition of the biomass feedstocks and on the pyrolysis conditions employed during biochar production.

Biochars have a complicated pore structure consisting of multiple interconnected networks of micropores, mesopores and macropores that span multiple length scales: from sub-nanometer micropores to macropores with sizes of the order of 10 microns. Such pore structures cannot be characterized by a single analytical technique. Comprehensive studies will use a combination of analytical techniques to bridge the vastly different length scales: adsorption of multiple gases (like nitrogen, carbon dioxide and water) for the micropores, mercury porosimetry for the mesopores and sectioning with optical microscopy and 3-D reconstruction techniques for the macropores.

We report here the development of a novel approach to probe the multiscale pore structure of biochars. This approach uses thermogravimetric measurements of the transient reactivity of biochars in air over a wide range of temperatures. At low temperatures, combustion proceeds in the regime of kinetic control and the entire surface area attributed to micropores is completely accessible to the reactant. As the temperature rises, the reaction regimes shifts to diffusion control and strong diffusional resistances start appearing first in the micropores and subsequently in the mesopores. Thus, larger and larger fractions of the micropore and mesopore structure will become inaccessible to oxygen as the temperature is raised. At sufficiently high temperatures, combustion will take place only on the micropore and mesopores "mouths," where they open up into the large macropore cavities identified in SEM or optical microscopy images.

The novelty of our approach lies in the use of sophisticated discrete models that can accurately describe the temporal evolution of the surface area of solid reactants with complicated pore structure. The new models are significant extensions of the continuous [1-2] and discrete models [3] that have appeared in the literature. Simulations start by generating solids with the desired pore structure on three-dimensional computational grids. Pores of various shapes and sizes are distributed on the grid in a random or orderly fashion to match the experimentally determined micro, meso- and macroporosities, as well as any available information about the shape of the pores. The generated porous solids are then eroded using rules that simulate non-catalytic reactions between a gas and a solid reactant. Access of the gas reactant into pores of progressively increasing sizes is restricted to model pore diffusional limitations and parallel simulations are carried out to handle the multiple scales of this problem. Multiple realizations of a solid with the same pore structural properties are generated and reacted to estimate the average pore surface and to establish confidence intervals. For a given pore structure, this process will generate a family of curves that give the evolution of pore surface area with conversion as the intraparticle diffusional resistances increase. A comparison of these surface evolution patterns to the char reactivity patterns measured at different temperatures can provide important information about the complex pore structure of carbonaceous materials [1-4].

Simulation results were compared with experimental data from the combustion of different biochars to validate this technique. Chars produced from different biomass feedstocks (apple wood or corn stover) exhibited different reactivity patterns in the kinetic control regime. The simulations showed that the reactivity patterns of corn stover chars are consistent with a random distribution of the micropores. On the other hand, apple wood chars exhibited reactivity patterns that indicate the presence of a subpopulation of orderly distributed micropores. These micropores must be the slit pores formed between the graphitic-like layers of aromatic carbon clusters that are turbostratically arranged in nanometer-size crystallites, a structure that has been confirmed with NMR and XRD measurements. The same corn stover and apple wood chars exhibited similar reactivity patterns when combusted with oxygen in the regime of strong diffusional limitations, when oxygen cannot penetrate deeply into the micro- and mesopores and the reaction takes place at the pore "mouths" where they open up into the large macropore cavities. These patterns were consistent with a macropore structure that consists of many large cavities separated by walls of similar thickness.

These results demonstrate how TGA data and pore structure models can be used to gain significant insights into the pore structure of biochars. The new approach has the potential to overcome some of the major difficulties encountered in characterizing the complex pore structure of biochars and other carbonaceous solids.

REFERENCES:

1. G. Ballal and K. Zygourakis, *Ind. Eng. Chem. Res.* **26**, 911-921 (1987).
2. S. Sotirchos, *Chem. Eng. Science*, **42**, 1262-1265 (1987).
3. K. Zygourakis and C.W. Sandmann, Jr., *AIChE Journal*, **34**, 2030-2040 (1988).
4. M. Sahimi, G.R. Gavalas and T.T. Tsotsis, *Chem. Eng. Science*, **45**, 1443-1502 (1990).

(393b) Modeling and Simulation of Globular Protein Crystal Growth

Michael Nayhouse[1], Panagiotis D. Christofides[2] and G. Orkoulas[2], (1)Chemical and Biomolecular Engineering, UCLA, Los Angeles, CA, (2)Department of Chemical and Biomolecular Engineering, University of California, Los Angeles, Los Angeles, CA

The production of highly-ordered, high quality protein crystals through batch crystallization processes is vital in devising proteins for therapeutic purposes. Despite extensive experimental and theoretical work on understanding protein structure and function, there is lack of a systematic framework that relies on fundamental understanding of the nucleation and growth mechanisms of protein crystals at the microscopic level and utilizes such information to model and operate protein batch crystallization processes at the macroscopic level. The present work focuses on investigating the growth of protein crystals via molecular dynamics and kinetic Monte Carlo simulations and using such knowledge to model and control batch protein crystallizers through population, mass and energy balances. The modeling and simulation work is tested against available experimental data.

The crystal growth of globular proteins is a non-equilibrium process and it is thus simulated using kinetic Monte Carlo methods. As is common practice in simulations of crystal growth, the solid-on-solid model, in which molecules are deposited on the growing crystal without voids or overhangs, is adopted. In the simulations, only molecular attachment and detachment events are considered and surface diffusion effects are ignored. The implementation of the kinetic Monte Carlo methodology requires knowledge of the binding energies, the impingement rate, and the crystallization driving force. Previous work assigned a range of values to these parameters until satisfactory agreement between the calculated and the experimental growth rates was achieved. In the present work, the parameters required for the growth of protein crystals are calculated from molecular dynamics simulations of coarse-grained models for protein solutions.

The calculated growth rates are used in macroscopic models describing the evolution of entire crystal size distributions to achieve optimal crystallizer design and operation for large scale production. The population balance modeling framework is adopted to describe globular protein crystallization in batch systems from supersaturated solutions. The mathematical description of batch crystallization also requires conservation equations for the solute (protein) and the energy. The previous mathematical model is used in conjunction with stochastic model predictive control methodologies that account for model uncertainty to achieve a desired crystal size distribution at the end of the batch protein crystallization process. In the model predictive control formalism, the cost function typically maximizes the expected value of the volume-averaged crystal size subject to input and state constraints. Extensive simulation results covering the various aspects discussed above will be presented.

See more of this Session: Multiscale Modeling: Methods and Applications

(393c) Modeling Excitation Energy Transfer In Photosynthetic Systems: Application to Peridinin-Chlorophyll-Protein Complex In Dinoflagellates

William P. Bricker, Energy, Environmental & Chemical Engineering, Washington University in St. Louis, St. Louis, MO and Cynthia S. Lo, Energy, Environmental and Chemical Engineering, Washington University in St. Louis, St. Louis, MO

Photosynthetic light-harvesting complexes (LHC) have evolved over many millions of years to become extremely efficient at funneling excitation energy into their respective reaction centers. If these funneling processes are better understood, we can apply this knowledge to building an efficient bio-solar photovoltaic cell. Peridinin-chlorophyll-protein (PCP) complex is a LHC containing twenty-four peridinin and six chlorophyll-a pigment molecules, as shown by its x-ray crystallography structure.[1] We are studying the excitation energy transfer rates throughout the PCP complex using time-dependent density functional theory (TD-DFT) to calculate the pigment excited states, and Förster theory to calculate the transfer rates between pigment excited states. Since Förster theory does not take into account delocalized excitation states, we will extend our study to other quantum dynamics theories that allow delocalization. The excitation transfer pathway in PCP has been shown experimentally to go from the S2 to S1 state in peridinin, and from the S1 state in peridinin to the Qy state in chlorophyll-a.[2] Researchers have also been successful in reconstituting different species of chlorophyll molecules into the PCP complex array to tune the transfer efficiencies.[3] Our goal is to validate the transfer efficiencies of these known complexes using the computational methods outlined above and to extend these methods to study unknown LHC systems.

[1] Hofmann, E., et. al. *Science* **272**, 1788-1791 (1996).

[2] Damjanovic, A., Ritz, T. and Schulten, K. *Biophysical Journal* **79**, 1695-1705 (2000).

[3] Polivka, T., et. al. *Photosynthesis Research* **86**, 217-227 (2005).

See more of this Session: Multiscale Modeling: Methods and Applications
See more of this Group/Topical: Computing and Systems Technology Division

(393d) A Generalized Runge-Kutta Framework for Explicit Tau-Leaping Algorithms

Leonard A. Harris and James Faeder, Computational and Systems Biology, University of Pittsburgh School of Medicine, Pittsburgh, PA

Gillespie's tau-leaping algorithm has received considerable attention as a promising method for performing accelerated-stochastic simulations of multiscale chemical and biological systems. The simplest tau-leaping algorithm has been shown to be analogous to the simple forward Euler method for numerically integrating ordinary differential equations. A "midpoint" tau-leaping algorithm has also been proposed which is analogous to an explicit second-order Runge-Kutta integrator. In this vein, we develop a generalized Runge-Kutta formulation of tau-leaping. We focus exclusively on explicit methods, describing how reaction firing, postleap checking and tau-selection are accomplished within a Runge-Kutta context and contrasting with the simpler forward-Euler case. The advantages of the approach are demonstrated via illustrative example systems simulated within the open-source modeling and simulation software package BioNetGen.

See more of this Session: Multiscale Modeling: Methods and Applications
See more of this Group/Topical: Computing and Systems Technology Division

(393e) Multiphysics Model of Diesel Injector Deposit Formation

Richard H. West, Amrit Jalan and William H. Green, Department of Chemical Engineering, Massachusetts Institute of Technology, Cambridge, MA

Formation of deposits inside the narrow nozzles of diesel engine fuel injectors disturbs fuel flow and adversely affects engine efficiency and emissions. It is thought that the deposit is formed by oxidation of a thin film of diesel on the nozzle wall. Fuels and additives used to mitigate this problem are currently evaluated using very expensive tests in running engines. Modeling insights into this problem could therefore greatly reduce the cost of fuel and engine development.

The system comprises several different phenomena which span multiple timescales. The free radical chemistry responsible for the deposit precursors occurs on a microsecond timescale while the fuel injections every tens of milliseconds cause washing and replenishment of the fuel film. The deposit itself accumulates over thousands of injection events which occur over hundreds of hours. The challenge lies in addressing the many physical and chemical phenomena in a way that is computationally efficient and gives meaningful results.

We start with a detailed kinetic model for the autoxidation of a diesel fuel surrogate mixture, consisting of hundreds of intermediate species and thousands of elementary reactions, built using Reaction Mechanism Generator (RMG) software [1]. Thermochemical, transport, and solubility parameters are estimated for each of the intermediate species based on their molecular structure. The kinetic model is embedded in a coupled model of evaporation, chemical reaction, phase-separation, and washing, and solved for multiple injection cycles using PyDAS, a new Python wrapper for the stiff DAE solver DASSL.

A model such as this necessarily contains many uncertain and variable parameters. It is useful to identify which of these parameters are most important. To do this we use a modified version of the Morris method [2] to find the global sensitivity of the overall model with respect to each parameter. This has been developed into a Python framework that makes full use of a parallel computing environment and is simple to apply to other models written in the Python programming language.

As well as providing insight into the specific problem of deposit formation in diesel injectors, this work demonstrates an approach suitable for other multi-scale models of chemical engineering systems.

[1] W. H. Green et al. RMG reaction mechanism generator open-source software package, 2011. http://rmg.sourceforge.net/

[2] M. D. Morris. Factorial sampling plans for preliminary computational experiments. Technometrics, 33(2):161–174, 1991.

See more of this Session: Multiscale Modeling: Methods and Applications
See more of this Group/Topical: Computing and Systems Technology Division

(393f) Numerical Simulation of Fuel NOx and Thermal NOx Emissions from an Industrial Burner Using Biomass-Derived Producer Gas

Sujith Sukumaran and **Song-Charng Kong**, Mechanical Engineering, Iowa State University, Ames, IA

Most of the industrial burners in the market are designed for natural gas or synthesis gas resulting from coal gasification. These burners are optimized to reduce the thermal NOx emissions but often do not address the fuel NOx problem. On the other hand, producer gas derived from biomass gasification contains traces of ammonia. Combustion of such producer gas results in high emissions of NOx due to the significant amount of fuel NOx in addition to thermal NOx. In this study we performed modeling of an industrial burner to analyze NOx emissions resulting from combustion of biomass-derived producer gas. Multidimensional CFD modeling of the burner was performed initially using a detailed mechanism, and the results were validated using the experimental results. However, such a detailed mechanism is highly computationally intensive to be used with 3D modeling. Thus, the detailed mechanism was reduced to a skeletal mechanism, without loss of accuracy in NOx prediction. Further CFD modeling of the actual burner geometry were conducted and results were validated using experimental data. The relative ratios of thermal NOx and fuel NOx are characterized under different operating conditions. The resulting skeletal mechanism and modeling strategies can be used in the future as a tool to design low-NOx burner using producer gas containing ammonia.

See more of this Session: Multiscale Modeling: Methods and Applications
See more of this Group/Topical: Computing and Systems Technology Division

KINETIC MODELING OF SOLID-GAS REACTIONS AT REACTOR SCALE: A GENERAL APPROACH

Loïc FAVERGEON [a], Jacques MORANDINI [b], Michèle PIJOLAT [a], Michel SOUSTELLE [a]
[a] Centre SPIN, Ecole Nationale Supérieure des Mines, Saint-Etienne, FRANCE
[b] ASTEK Rhône-Alpes, Echirolles, FRANCE

Abstract

Understanding the industrial reactors behavior is a difficult task in the case of solid state reactions such as solid-gas reactions. Indeed the solid phase is a granular medium through which circulate gaseous reactants and products. The properties of such a medium are modified in space and time due to reactions occurring at a microscopic scale. The thermodynamic conditions are driven not only by the operating conditions but also by the heat and mass transfers in the reactor.

We propose to numerically resolve the thermohydraulic equations combined with kinetic laws which describe the heterogeneous reactions. The major advantage of this approach is due to the large variety of kinetic models of grains transformation (~40) compared to the usual approach, especially in the case of surface nucleation and growth processes which need to quantitatively describe the grain conversion kinetics at a microscopic scale due to nucleation frequency and growth rate laws obtained in separate isothermal and isobaric experiments.

The heat and mass transfers terms entering in the balance equations at a macroscopic scale depend on the kinetics evaluated at the microscopic scale. These equations give the temperature and partial pressure in the reactor, which in turn influence the microscopic kinetic behavior.

Keywords: kinetics, simulation, heat and mass transfers, multi-scale

1. Introduction

Solid-gas reactions are of great interest in many industrial fields such as nuclear, chemistry, metallurgy, CO_2 capture, etc... Industrial reactors where these reactions take place are difficult to understand. Indeed the solid phase is a granular medium through which circulate gaseous reactants and products. The properties of such a medium are modified in space and time due to reactions occurring at a microscopic scale. The thermodynamic conditions are driven not only by the operating conditions but also by the heat and mass transfers in the reactor. Several models have been developed to account for the complexity of these transformations such as the grain model [1] and the pore model [2] and all the related incremented models. However most of these models are based on the law of additive reaction times of Sohn [3] for which the order respective to the gas in the kinetic rate equation must be equal to 1. However such condition is scarcely encountered in many gas-solid reactions so that erroneous results may be obtained using this simplified approach.

Moreover, such approaches necessitate to describe the reaction kinetics using the shrinking core model or similar laws, that is to say using rate equations for which $d\alpha/dt$ is the product of an Arrhenius term by a f(α) function [1,2]. Such a behavior is unfortunately not always encountered for several reasons, especially in the case of reactions involving surface nucleation and growth processes [4].

In order to overcome such problems, we have developed a multi-physic approach based on the finite elements method which combines the resolution of the thermohydraulic equations with the kinetic laws describing the heterogeneous reactions at the scale of dense particles.

We propose [4] a large variety of models at the microscopic scale based on a general equation which accounts for the real influence of the partial pressures P_i of the relevant gases:

$$\frac{d\alpha}{dt} = \phi(T, P_i, \ldots) S_m(t, \phi, \gamma, r_0) \qquad (1)$$

where $\phi(T,P_i,\ldots)$ is the areic reactivity of growth, function of the thermodynamic conditions only (temperature, partial pressure, etc…) and $S_m(t,\phi,\gamma,r_0)$ is a function of the initial and current morphology of the sample and therefore of time, of ϕ, of the areic frequency of nucleation γ, and of the grains dimensions (e.g.: r_0 is the radius for spherical grain).

Using equation (1), about forty kinetic models may be derived based on various assumptions on the nucleation and growth processes in order to simulate heterogeneous reactions. These models can be simulated using CIN3 software (N° IDDN FR001130014.000.SP.2009.000.30625). Nevertheless CIN3 is able to simulate reactions where temperature and partial pressure of gases are uniform inside the granular medium. In the case of heterogeneous reactors where mass and heat transfers induce spatial and temporal variations of both temperature and partial pressures, it is necessary to couple kinetics of a grains population with the resolution of mass and heat transfers. This is the aim of the multi-physic software CIN4.

Section 2 presents the various scales approach and the main assumptions made in CIN4. Then section 3 summarizes a simplified presentation of the different mathematical models used, for kinetic modeling at grains scale as well as for heat and mass transfers at reactor scale. And finally section 4 presents the simulation of a thermobalance experiment, including both kinetic curves and mass and heat variations inside the system.

2. General presentation

2.1 The scales for modeling

Several geometrical scales characterize all phenomena which occur in a reactor. A rigorous description of our modeling process requires a precise definition of these different scales. We have deliberately set these scales from the objects: mesh, grain, grains population, agglomerate and reactor. In this paper, only three scales will be discussed: grain, grain population and reactor. At the grain scale, the nuclei appear at the grain surface[1] and grow at the expense of the initial solid phase: the kinetics is governed by the shape of the grain, the type of growth, the localization of the rate limiting-step of growth, the direction of development of the new phase. Then a microscopic population is a set of grains in the same thermodynamic conditions, i.e. the gradients of temperature and gas partial pressure are extremely small inside the population. Statistical considerations allow to define average behavior of this population from grain's one. And finally at the reactor scale, the chemical reaction is coupled with mass and heat transfers.

[1] Avrami's laws cannot be used for solid-gas heterogeneous reactions involving powders since they are based on bulk nucleation in an infinite volume, which is not physically acceptable.

2.2 The assumptions of the model

Six assumptions are done in this model:

A1: gas production and consummation by the reaction do not disturb the total pressure distribution,
A2: the gases are diluted in a carrier gas,
A3: gases flow rates are low (less than 5 $l.h^{-1}$ for a circular section of 2 cm),
A4: the gases are considered as ideal,
A5: the growth process can be described by a succession of elementary steps, one of them controls the growth kinetics (rate-determining step approximation),
A6: the grains shape and size distribution are known.

3. Mathematical modeling

3.1 Kinetic laws at grain and microscopic population scale

As previously recalled, in order to model reactions which involve surface nucleation and growth of the nuclei, one can use equation (1). For isothermal and isobaric conditions, the S_m function can take various expressions calculated from assumptions on the growth type (isotropic or anisotropic), the development direction of the new phase (inwards or outwards), the localization of the rate determining step of growth and the grains shape (spheres, cylinders, plates). Soustelle [5] has detailed the laws establishment and the expressions of S_m for each model.

The software CIN4 is able to solve forty-two kinetic models. Some of these models are derived at grain scale: in this case nucleation is considered as instantaneous compared to growth, so all the grains take the same kinetic law. Other models are derived at microscopic population scale: both nucleation and growth are taken into account, which leads to two kinds of models. According to the growth type (isotropic or anisotropic), two methods are used to calculate the fractional conversion of the reaction:

- in the case of isotropic growth of the nuclei, we used a statistical approach based on the works of Johnson and Mehl [6] and Mampel [7], then of Delmon [8],
- in the case of anisotropic growth, each grain starts its transformation at various times and the whole microscopic fractional conversion can be obtained considering the nuclei appearing on each grain of the microscopic population.

Moreover, CIN4 enables the calculation for a real grain size distribution.

3.2 Heat and mass transport at reactor scale

The macroscopic scale is the center of heat and mass exchanges that are described by the thermal and hydrodynamic models, as well as by a mass transfers equation.

3.2.1 Thermal model

Heat transfer is described by the following equation valid throughout each part of the reactor Ω^T :

$$\left[(1-\varepsilon)\rho_s c_{ps} + \varepsilon\rho_g c_{pg}\right]\frac{\partial T}{\partial t} + \varepsilon\rho_g c_{pg}\vec{u} \bullet \vec{\nabla} T + \vec{\nabla} \bullet \left(-k\vec{\nabla} T\right) = Q^T \qquad (2)$$

where s and g indexes refer to solids and gases respectively.

This equation is accompanied by two kinds of boundary conditions:

- Neumann conditions which impose the value of the heat flow on $\partial\Omega_N^T$ boundary:

$$-k\vec{\nabla} T \bullet \vec{n} = \phi_{imposed} + h(T - T_a) + \varepsilon_r\sigma\left(T^4 - T_a^4\right) \qquad (3)$$

- Dirichlet conditions which impose the value of the temperature on $\partial\Omega_D^T$ boundary:

$$T = T_D \tag{4}$$

These equations are valid for each part of the reactor: solid (ε=0), gas (ε=1), porous media (0<ε<1).

3.2.2 Hydrodynamic model

As the granular medium is static and as we consider solid-gas reactions, mass transfers are only due to the flowing of the gas produced or consumed inside the porous medium. The modeling of this flow inside the porous medium is based on notions of hydrodynamic charge and flow rate considerations. Hydrodynamic charge is calculated according to:

$$\mathrm{H} = \mathrm{z} + \frac{\mathrm{P}}{\rho \mathrm{g}} + \frac{1}{2\mathrm{g}}\mathrm{u}^2 \tag{5}$$

The flow rate is not the rate of fluidic particles inside the pores but an equivalent continuous rate. This flow rate is linked to the mass gradient according to the Darcy's law, which is expressed by:

$$\vec{\mathrm{u}}_\mathrm{d} = -\frac{\mathrm{K}}{\eta}\vec{\nabla}(\rho \mathrm{g h}) \approx -\frac{\mathrm{K}}{\eta}\vec{\nabla}\mathrm{P} = -\mathrm{k}_\mathrm{p}\vec{\nabla}\mathrm{P} \tag{6}$$

The whole pressure field (or charge) is solution of a conservation equation. Neglecting source terms due to gas production and consummation (reaction is slow compared to the gas flow), hydrodynamic problem can be written as:

$$\vec{\nabla}\bullet\left(-\mathrm{k}_\mathrm{p}\vec{\nabla}\mathrm{P}\right) = 0 \tag{7}$$

the boundary conditions being $-\mathrm{k}_\mathrm{p}\vec{\nabla}\mathrm{P}\bullet\vec{\mathrm{n}} = -\mathrm{u}_\mathrm{n}$ on $\partial\Omega_N^T$ and $\mathrm{P} = \mathrm{P}_\mathrm{D}$ on $\partial\Omega_D^T$.

3.2.3 Mass transport

Gas phase is governed by the ideal gas law. Thus the total molar concentration inside the gas is equal to $\mathrm{C} = \frac{\mathrm{P_r}}{\mathrm{RT}}$. The molar convective flow is due to the transport by the gas displacement and is expressed by:

$$\vec{J}_i^C = C\vec{u} = -\frac{P_i}{RT}\vec{u} \tag{8}$$

Diffusion in porous medium is a complex phenomenon including diffusion flow, Knudsen diffusion, Soret effect, which is well explained by Reid et al. [9]. Here, only diffusion flow is taking into account:

$$\vec{J}_i^D = -D_{ij}^D\vec{\nabla}C_j = -\frac{D_{ij}^D}{RT}\vec{\nabla}P_j \tag{9}$$

and finally each partial pressure is solution of the following equation:

$$\frac{\varepsilon}{RT}\frac{\partial P_i}{\partial t} + \frac{\varepsilon}{RT}\vec{u}_d\bullet\vec{\nabla}P_i - \vec{\nabla}\bullet\left(\frac{D_{ij}^D}{RT}\vec{\nabla}P_j\right) = Q_i^C \tag{10}$$

with the boundary conditions $-\frac{D_{ij}^D}{T}\vec{\nabla}P_j\bullet\vec{n} = R\phi_i$ on $\partial\Omega_N^T$ and $P_i = P_{iD}$ on $\partial\Omega_D^T$.

3.3 Coupling between the microscopic population scale and the reactor scale

At microscopic scale, the reaction fractional conversion is followed for a representative population of grains. By finite difference it is possible to evaluate the reaction rate $\frac{d\alpha}{dt}$. This rate allows to evaluate heat and mass sources produced by the reaction. Thus the heat source density is equal to:

$$Q^T = \sum_{i=1}^{n_g} \nu_i M_A \frac{1-\varepsilon}{V_{mA}} \frac{d\alpha}{dt} \Delta H_i \qquad (11)$$

and the mass (partial pressure) source density for a gas i is equal to:

$$Q_i^C = \nu_i \frac{1-\varepsilon}{V_{mA}} \frac{d\alpha}{dt} \qquad (12)$$

So, using these sources terms, microscopic reactions have an impact on the spatial and temporal evolution of the thermodynamic processes at the reactor scale. Inversely since it modifies the areic frequency of nucleation and the areic reactivity of growth, thermodynamic influences fractional conversion of the microscopic reaction. It is worthwhile to notice that CIN4 allows to simulate the behavior of reactions for which the kinetic rate at time t (at the microscopic population scale) depends on the local thermodynamic conditions that were established between t=0 and t. CIN4 highly improves the previous approaches which were limited to kinetic model based at grain scale only, excluding surface nucleation and growth models. The equations used in CIN4 require numerous physical properties (porosity, viscosity, density, heat capacity, thermal conductivity, diffusivity). CIN4's user should be able to calculate these properties using empirical laws or mathematical models like for example those given by Bird et al. [11].

4. Numerical modeling

At grain and microscopic population scales, two types of modeling are possible according to the type of growth: anisotropic or isotropic. In the first case, the method allows to calculate the evolution of the fractional conversion for all grains in each granulometric class. When the growth is isotropic, we used the Mampel's method [7] for three grain geometries: spheres, cylinders and plates. At reactor scale, for each equation (2), (7) and (10) (i.e. heat, hydrodynamic and mass transport problems), two finite element formulations were built, depending on the system symmetry (cylindrical or Cartesian coordinates). A classical way is followed to obtain weak formulations for each state variable: temperature, charge and partial pressures.

5. Results

Figures 1 to 3 present a thermobalance simulation achieved with CIN4. This simulation involves a model at the reactor scale coupled with a model of the reaction at the microscopic grains population scale. There, the system studied consists of a vertical cylindrical chamber in which a flow of reactive gas (G1) from the bottom up is maintained. In the middle position, crucible, whose base is porous, contains a granular medium initially composed of grains of A which are transformed into B according to the reaction:

$$A(s) + \frac{1}{2} G_1(g) \rightarrow 3B(s) + 2G_2(g) \qquad (13)$$

The system being axi-symmetric, we have used weak forms written in cylindrical coordinates. Figure 1 shows the mesh used for solving heat and mass transfers' equations throughout the reactor by the finite elements method (the mesh was done by using Gmsh freeware [11]).

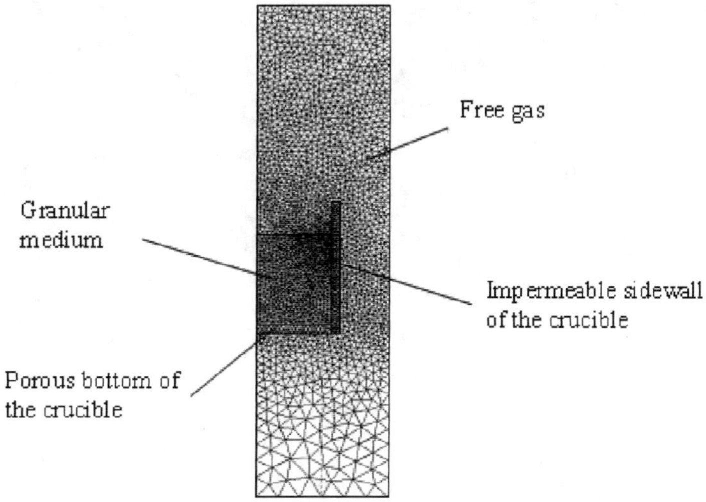

Figure 1: mesh of a thermobalance containing a crucible and a granular medium.

There are four areic regions in this mesh: the granular medium, the bottom of the crucible, the vertical wall of the crucible and the zone of free gas. We also the boundary conditions, which are:

- on the incoming boundary, the entering gaseous mixture is fixed: temperature, flow and composition (here 100% G1),
- on the outgoing boundary: free exit for the temperature, total pressure imposed at 1 atm,
- on the sidewall: the boundary is adiabatic and is impermeable for the mass flow,
- on the symmetry axis: the radial derivatives of all physical grandeurs are equal to 0.

In the following, the kinetic model is based on surface nucleation and anisotropic growth of the nuclei, the grains are spherical (r_0= 100µm), the rate determining-step of growth is located at the internal interface, with inward development.

The granular medium is the seat of heat and mass sources calculated by simulation of reaction phenomena at the microscopic level. Figure 2 shows the gas rate field obtained as well as some views of fields of temperature, partial pressure, fractional conversion and reaction rate for one instant. Figure 3 presents the whole fractional conversion versus time, and the whole rate of reaction versus fractional conversion.

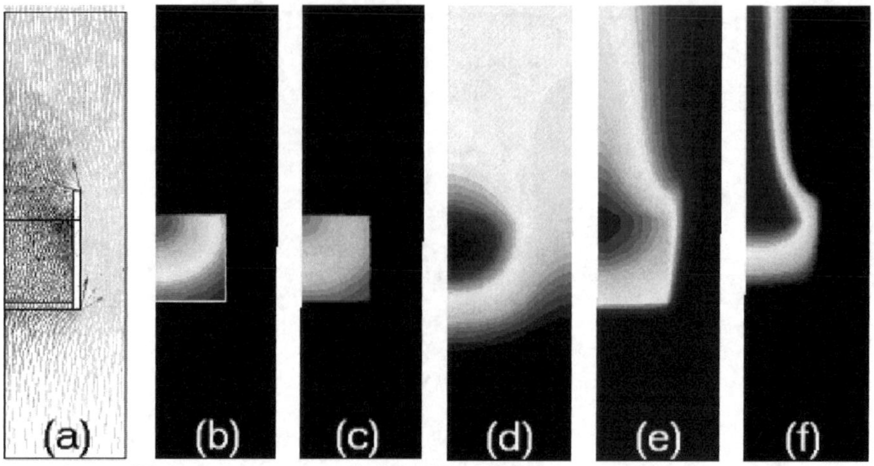

Figure 2: Representation of (a) the gas rate field, (b) the fractional conversion, (c) the rate of reaction, (d) the temperature, (e) G_1 partial pressure and (f) G_2 partial pressure in the thermobalance at a given time.

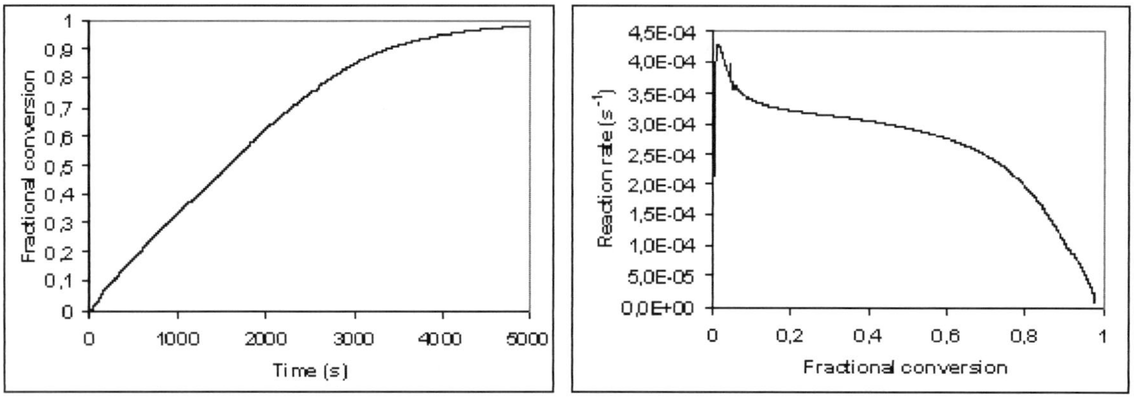

Figure 3: Whole fractional conversion vs. time, and rate of reaction vs. fractional conversion from the simulation.

6. Conclusions

We present a general approach for kinetic modeling of heterogeneous reactions by coupling kinetic models at grain and microscopic population scales with both heat and mass transfers by means of the finite element method. Kinetic models are based on a general method agreeing with physical reality, offering the possibility to simulate about forty models depending on the shape of the grains, the growth type (isotropic or anisotropic), the development direction of the new phase and the localization of the rate determining step of growth (internal interface, external interface or diffusion volume). Among these models, twenty-one include surface nucleation and growth processes. Heat and mass transfers are governed by differential equations and solved by a finite element method. This modeling process allows to simulate not only thermobalances but also industrial reactors.

Notations

α Fractional conversion of the powder β Fractional conversion of a grain C Concentration ($mol.m^{-3}$) c_p Heat capacity at constant pressure ($J.kg^{-1}.K^{-1}$) D Diffusivity ($m^2.s^{-1}$) D_{ij} Multicomponent diffusion coefficient ($m^2.s^{-1}$) S_m Space function ($m^2.mol^{-1}$) ε Porous medium porosity ε_0 Initial porous medium porosity ε_r Emissivity ϕ Areic reactivity of growth ($mol.m^{-2}.s^{-1}$) g acceleration of gravity ($m.s^{-2}$) γ Areic frequency of nucleation ($nuclei.m^{-2}.s^{-1}$) h Heat transfer coefficient ($W.m^{-2}.K^{-1}$) H Hydrodynamic charge (m) J^D Diffusion flow vector ($mol.m^{-2}.s^{-1}$) k Thermal conductivity ($W.m^{-1}.K^{-1}$) k_p diffusion coefficient ($m^2.Pa^{-1}.S^{-1}$) K Darcy's permeability (m^2)	M_i Molar mass for constituent i ($kg\ mol^{-1}$) η Dynamic viscosity of the fluid (Pa.s) P Total pressure (Pa) P_i Partial pressure of reactive gas (Pa) Q_T Heat sources density ($W.m^{-3}$) Q_C Concentration sources density ($mol.m^{-3}.s^{-1}$) R Ideal gas constant (= 8.314 $J.K^{-1}.mol^{-1}$) r_0 Initial radius of the grain (m) ρ Density ($kg.m^{-3}$) σ Stefan Boltzmann constant ($W.m^{-2}.K^{-4}$) T Temperature (K) T_a Ambient temperature (K) υ Rate vector ($m.s^{-1}$) υ_d Flowing rate vector ($m^2.s^{-1}$) u_n Incoming flow density ($m.s^{-1}$) V_{mA} Molar volume of A ($m^3.mol^{-1}$)

References

1. Szekely J. and Evans J.W. (1970) "A structural model for gas-solid reactions with a moving boundary", Chem. Eng. Sci., 25, pp 1091-1107.
2. Bhatia S. and Perlmutter D.D. (1980) "A random pore model for fluid-solid reactions: 1. Isothermal, kinetics control", AIChE J. 26, pp 379-386.
3. Sohn H.Y. (1978) "The law of additive reaction times in fluid-solid reactions", Metall. Trans., 9B, pp 89-96.
4. Pijolat M., Favergeon L. and Soustelle M. (2011) "From the drawbacks of "Arrhenius-f(α)" rate equation towards a more general formalism and new kinetic models for the kinetic analysis of solid-gas reactions", Thermochimica Acta, accepted on July 27th.
5. Soustelle M. (2010) "Handbook of Heterogeneous Chemistry", Edrs J. Wiley and ISTE Ltd.
6. Johnson W.A. and Mehl R.F. (1939) "Reaction kinetics in processes of nucleation and growth", Trans. Amer. Inst. Ming. Metal. Engrs. 135, pp 416-442.
7. Mampel, K.L. (1940) "Zeitumsatzformeln für heterogene Reaktionen an Phasengrenzen fester Körper", Z. Phys. Chem. A 187, pp 235-249.
8. Delmon B. (1969) "Introduction à la cinétique hétérogène", Eds. Technip, p. 397.
9. Reid R.C., Prausnitz J.M. and Poling B.E. (1987) "The properties of gases and liquids", MCGraw Hill, New-York.
10. Bird R.B., Stewart W.E. and Lightfoot E.N. (2002) "Transport phenomena", Wiley International Edition.
11. Geuzaine C. and Remacle J.F. (2009) "Gmsh : a three-dimensional finite element mesh generator with built-in pre- and post-processing facilities", Inter. J. for Numerical Methods in Engineering, 79, 11, pp 1309-1331.

(393h) Multiscale Methodology for Prototyping of Porous Catalysts

Milos A. Marek, Department of Chemical Engineering, Prague Institute of Chemical Technology, Prague, Czech Republic

The microstructure of the support determines a keyproperty of porous catalysts---effective diffusivity. Typically,supporting materials with bimodal pore size distribution are used thatinvolve both meso- and macro-pores. Spatial distribution of activemetal crystallites within the porous support then influences reactionrates and conversions. In order to optimize the catalyst supportmicrostructure and ultimately the whole catalyst, it is necessary torelate quantitatively the morphological features of the porousstructure both to its preparation conditions and to the final transportproperties and catalyst performance under reaction conditions.

In this paper we demonstratethe application of novel models based on the generalizedvolume-of-fluid method and 3D digital reconstruction of porousstructure. The procedure includes simulation of porous supportformation (virtual packing of primary particles of defined shapes andsizes), drying and crystallization of impregnated metal solution(growth of metal nano-particles), and solution of reaction andtransport within the final virtual catalyst structure to obtainvolume-averaged reaction rates that are then used in full-scale modelof catalytic monolith reactor. A parametric study is performed toinvestigate the effects of the sizes of primary particles (influencingthe meso- and macro-porosity and pore sizes) and active metalimpregnation conditions (influencing the distribution of activecatalytic surface area) on the macroscopic activity of a catalyticmonolith with Pt/g-Al2O3 washcoat used for automotive exhaust gasaftertreatment.

See more of this Session: Multiscale Modeling: Methods and Applications
See more of this Group/Topical: Computing and Systems Technology Division

(451a) First Principles Monte Carlo Simulations of Vapor-Liquid Equilibria: Investigation of Dispersion-Corrected Functionals

J. Ilja Siepmann, Depts. of Chemistry and of Chemical Engineering & Materials Science, University of Minnesota, Minneapolis, MN, Matthew J. McGrath, Department of Physics, University of Helsinki, Helsinki, Finland and I-F. Will Kuo, Chemical Sciences Division, Lawrence Livermore National Laboratory, Livermore, CA

Gibbs ensemble Monte Carlo simulations are used to compute the vapor-liquid phase equilibria for water, methanol, and methane using Kohn-Sham density functional theory. The effects of various dispersion corrections for common generalized gradient approximation functional are investigated. These dispersion corrections lead to much improved accuracy for the vapor-liquid coexistence curves of methanol and methane, but the situation is less promising for water. It appears that improved dispersion corrections will be required to increase the fidely of first principles simulations for molecular systems.

See more of this Session: First-Principles Simulation of Condensed Phases
See more of this Group/Topical: Computational Molecular Science and Engineering Forum

(451b) First Principles Monte Carlo Simulations of Elemental Fluid Phase Equilibria

Lev Gelb, University of Texas at Dallas, Dallas, TX

We use modern Monte Carlo sampling methods in concert with standard electronic structure codes (primarily NWChem) to perform liquid-vapor and liquid-liquid coexistence calculations from first principles. In order to improve computational efficiency an approximate "pre-sampling" potential is used to generate large moves with a high probability of acceptance. A hybrid approach in which two "levels" of pre-sampling is constructed, in which the first uses an empirical potential, and the second uses density functional theory but with a very low (and fixed) number of plane waves.

Issues of efficiency, correctness and load-balancing arising from the use of iterative density functional theory potentials are addressed
both directly and through analysis of model systems. System size effects are also tested using model systems. Results are presented
from Gibbs- ensemble calculations on group I and group II metals and mixtures. Extensions of these calculations to determination of elemental solubilities in molten metals are also presented.

See more of this Session: First-Principles Simulation of Condensed Phases
See more of this Group/Topical: Computational Molecular Science and Engineering Forum

(451c) Liquid to Vapor TRANSITION of WATER UNDER Hydrophobic CONFINEMENT

Sumit Sharma and Pablo Debenedetti, Chemical and Biological Engineering, Princeton University, Princeton, NJ

Liquid to Vapor transition of water UNDERHYDROPHOBIC confinement

Sumit Sharma, Pablo G. Debenedetti

Departmentof Chemical and Biological Engineering, Princeton University, Princeton, NJ08544

Introduction

Molecularorigins of phase separation of oil in water have been a topic of great interestas it is considered the basis of biological self assembly (Nicholls, Sharp et al. 1991; Sharp, Nicholls et al. 1991). Theoretical andsimulation studies predict that the behavior of water molecules nearhydrophobic solutes is solute-size-dependent (Lum, Chandler et al. 1999; Rajamani,Truskett et al. 2005). At ambient conditions, water molecules areable to retain their hydrogen (H) bonds around hydrophobic solutes ≤ 1 nm(Pangali, Rao et al. 1979). The solubility ofsmall hydrophobic solutes is shown to be proportional to the free energy ofcreating a cavity of the size of the solute in water (Hummer, Garde et al. 1998). For largersolutes, water molecules are unable to retain their H-bonds close to the soluteand hence form a soft, vapor-like interface (Wallqvist and Berne 1995). Simplethermodynamical arguments show that water confined between two large repulsivesurfaces separated by distance d should vaporize even when d isas large as 10^3 nm (Lum, Chandler et al. 1999). The fact that evaporationtransitions at such macroscopic length scales do not occur in finite times (Christenson and Claesson 2001) indicates thatkinetics of evaporation in hydrophobic confinement are of fundamentalimportance and merit computational scrutiny.

Inthis study, we estimate rates and activation energy for liquid to vaportransition of water confined between hydrophobic surfaces using Forward FluxSampling (FFS) (Allen, Frenkel et al. 2006). FFS allows a goodsampling of the liquid to vapor phase transition, which, being a rare event inthe simulation time scales, is hard to achieve in equilibrium simulations. Wefound that the activation energy, ΔG scales as d for thecase when the size of the hydrophobic solutes is ~ 1 nm. The transition stateensemble comprises of configurations in which almost the entire confined regionis devoid of water molecules.

Simulation system and methods

Weuse the SPC/E model of water (Berendsen, Grigera et al. 1987). The hydrophobicsurfaces were constructed with Lennard Jones (LJ) atoms arranged in a hexagonallattice. This arrangement of atoms mimics that of carbon atoms in graphene (Choudhury and Pettitt 2005). The LJ potentialparameters of carbon atoms for the Amber force field (Cornell, Cieplak et al. 1995) are: ε_{cc}= 0.086 kcal mol^{-1} and σ_{cc} = 3.4 . Thecarbon atoms of the surface interact with the oxygen atoms of water moleculeswith LJ potential. As per the Lorentz Berthelot rules,

$\varepsilon_{oc}= (\varepsilon_{cc} \cdot \varepsilon_{oo})^{0.5}$ (1)

$\sigma_{oc}= (\sigma_{cc} + \sigma_{oo})/2$ (2)

where, $\varepsilon_{oo}$is the SPC/E water oxygen oxygen LJ ε parameter and $\sigma_{oo}$is the SPC/E water LJ σ parameter. For SPC/E water, equations(1) and (2) give ε_{oc} = 0.1156 kcal/mol and σ_{oc}= 3.283 . This carbon oxygen potential leads to a very strong bindingof water molecules to the surfaces (Hummer,Rasaiah et al. 2001; Choudhury and Pettitt 2005).Hence, we simulated a weaker (more hydrophobic) wall oxygen potential with $\varepsilon_{ow}= \varepsilon_{oc} / 4$ and $\sigma_{ow} = \sigma_{oc}$.

Moleculardynamics (MD) simulations were conducted in the isothermal-isobaric (*NPT*)ensemble. 2329 water molecules were taken in a cubic simulation box, periodicin the x, y and z directions. Each wall / hydrophobic surface, made up of 60atoms, had the lateral dimensions of 9.05 x 0.05 . The walls were kept fixedand were were placed parallel at a distance *d* from each other.

Forward FluxSampling

Supposea system has two locally stable states designated *A* and *B* separatedby a free energy barrier. Let us identify a system property, say ρ,that changes in value from ρ_A at *A* to $\rho_B$at *B* (and let $\rho_A > \rho_B$).Now, we divide the phase space by interfaces, λ_i whichare loci of phase space points with the same value of ρ, say ρ_i.Let the interfaces be numbered such that $\rho_i > \rho_{i+1}$.Interface λ_A corresponds to $\rho_A$and λ_B to ρ_B.The rate of the transition from *A* to *B* will be,

$Rate= \varphi(\lambda_1|\lambda_A)\ \Pi_i P(\lambda_{i+1}|\lambda_i)$ (3)

Where$\varphi(\lambda_1|\lambda_A)$ is the flux oftrajectories which leave λ_A and reach λ_1.$P(\lambda_{i+1}|\lambda_i)$ is theprobability that a trajectory starting from λ_i wouldreach λ_{i+1} before reaching λ_A.

Inorder to calculate the rate of transition from liquid to vapor state for waterconfined between two hydrophobic surfaces, ρ is the density of thewater molecules in the confined region. From an equilibrium simulation (40 ns),we determine $\varphi(\lambda_1|\lambda_A)$ and collect Nconfigurations at λ_1, which correspond to the firstcrossing of λ_1 by a trajectory coming from λ_A.From each of the N configurations, C trajectories are shot (withslightly perturbing velocities) to determine $P(\lambda_2|\lambda_1)$. The configurations at the interface λ_2 arefurther propagated to the next interface. The process of propagating thetrajectories from one interface to the next is continued till the vapor phaseis reached.

Results and Discussion

Evaporationrates were calculated for d =9.0 to 14.0 . For d=9.0 , theconfined region sampled both liquid and vapor phase-space in the equilibriumsimulation. For $d \geq 9.8$, the confined region only sampled theliquid phase in the equilibrium simulation. The evaporation rate for d=9.0 was found to be 1.67 x 10^9 $nm^{-2}s^{-1}$, whereasfor d=14.0 , it was found to be 0.058 $nm^{-2}s^{-1}$. Therefore,an increase in d less than two times the diameter of a water moleculeleads to a decrease of the order ~10^{10} in evaporation rates. Thisimplies that the kinetics of evaporation of water in confined geometries isstrongly dependent on d.

Theactivation energy to the evaporation event for d=9.8 was determined bydoing evaporation rate calculations at T=298, 348 and 398 K. The activationenergy was found to be ~ 45 k_BT. Assuming a constant pre-factor,activation energies for other values of d were predicted from the calculatedevaporation rates. The predicted values of the activation energies werevalidated by calculating the activation energy at d=11.0 . Theactivation energy to evaporation was found to scale linearly with d. Thetransition state ensemble comprised of configurations in which the vapor bubbleis almost of the same volume as the confined region. This observation impliesthat the free energy barrier to evaporation should be proportional to the freeenergy required to create a vapor bubble of approximately the same volume asthe confined region, which explains the linear dependence of activation energyon d. A lattice gas model calculation to determine free energy barriersto evaporation between infinite plates showed a d^2 scaling ofthe activation energy (Luzar 2004). Therefore, it appears that thereshould be transition in the scaling factor from linear to quadratic as the sizeof the hydrophobic plates is increased.

References

Allen, R. J., D. Frenkel, et al. (2006)."Simulating rare events in equilibrium or nonequilibrium stochasticsystems." J. Chem. Phys. **124**(2): 024102.

Berendsen, H. J. C., J. R. Grigera, etal. (1987). "The Missing Term in Effective Pair Potentials." J.Phys. Chem. **91**(24): 6269-6271.

Choudhury, N. and B. M. Pettitt (2005)."On the mechanism of hydrophobic association of nanoscopic solutes." J.Am. Chem. Soc. **127**(10): 3556-3567.

Christenson, H. K. and P. M. Claesson(2001). "Direct measurements of the force between hydrophobic surfaces inwater." Adv. Colloid Interface Sci. **91**(3): 391-436.

Cornell, W. D., P. Cieplak, et al.(1995). "A 2nd Generation Force-Field for the Simulation of Proteins, Nucleic-Acids, and Organic-Molecules." J. Am. Chem. Soc. **117**(19):5179-5197.

Hummer, G., S. Garde, et al. (1998)."Hydrophobic effects on a molecular scale." J. Phys. Chem. B **102** (51):10469-10482.

Hummer, G., J. C. Rasaiah, et al.(2001). "Water conduction through the hydrophobic channel of a carbonnanotube." Nature **414**(6860): 188-190.

Lum, K., D. Chandler, et al. (1999)."Hydrophobicity at small and large length scales." J. Phys. Chem.B **103**(22): 4570-4577.

Luzar, A. (2004). "Activationbarrier scaling for the spontaneous evaporation of confined water." J.Phys. Chem. B **108**(51): 19859-19866.

Nicholls, A., K. A. Sharp, et al.(1991). "Protein Folding and Association - Insights from the Interfacialand Thermodynamic Properties of Hydrocarbons." Proteins: Struct.,Funct., Genet. **11**(4): 281-296.

Pangali, C., M. Rao, et al. (1979)."Hydrophobic Hydration around a Pair of Apolar Species in Water." J. Chem. Phys. **71**(7): 2982-2990.

Rajamani, S., T. M. Truskett, et al.(2005). "Hydrophobic hydration from small to large lengthscales: Understanding and manipulating the crossover." Proc. Natl. Acad. Sci.U. S. A. **102**(27): 9475-9480.

Sharp, K. A., A. Nicholls, et al.(1991). "Extracting Hydrophobic Free-Energies from Experimental-Data -Relationship to Protein Folding and Theoretical-Models." Biochemistry**30**(40): 9686-9697.

Wallqvist, A. and B. J. Berne (1995)."Computer simulation of hydrophobic hydration forces on stacked plates atshort range." J. Phys. Chem. **99**(9): 2893-2899.

See more of this Session: First-Principles Simulation of Condensed Phases
See more of this Group/Topical: Computational Molecular Science and Engineering Forum

(451d) Equilibrium Phase Dome and Critical Point of 1-Pentanol From Two-Phase Molecular Dynamics Simulations: Transferability of An Ab-Initio Pair-Potential Model

Sonal Patel, W. Vincent Wilding and Richard L. Rowley, Chemical Engineering, Brigham Young University, Provo, UT

All-atom, site-site pair-potential parameters fitted solely from *ab-initio* landscapes for small alcohol dimers are used in two-phase Molecular Dynamics simulations to determine coexisting liquid and vapor densities of 1-pentanol as a function of temperature. A Monte Carlo (MC) sampling method is used to extract the bulk vapor and liquid densities from the two-phase MD simulations and the critical point is determined from the resultant equilibrium densities in conjunction with the law of rectilinear diameters and the known critical exponent for the liquid-vapor density difference. Results obtained for the critical temperature, the critical density, and bulk vapor and liquid densities for the model are compared to experimental data. Errors due to transferability of the potential model are separated from errors due to the use of a true-pair potential by calculating energy landscapes for dimers of 1-pentanol using counterpoise (CP) corrected energies at the MP2/6-311 + G(2df,2pd) level and comparing these to the transferred potential models. The calculated landscapes were sampled at ten dimer separation distances for four different routes or relative orientations of the rigid monomers.

See more of this Session: First-Principles Simulation of Condensed Phases
See more of this Group/Topical: Computational Molecular Science and Engineering Forum

Polymer Decomposition Mechanisms for Polyethylene Determined from Reactive Molecular Dynamics

Kenneth D Smith, United Technologies Research Center, East Hartford, CT
Stanislav I Stoliarov, Department of Fire Engineering, University of Maryland, College Park, MD
Marc R Nyden, Fire Research Division, National Institute of Standards and Technology, Gaithersburg, MD
Phillip R. Westmoreland, Chemical and Biomolecular Engineering, North Carolina State University, Raleigh, NC

Introduction

Understanding molecular-level information is key to understanding the underlying chemistry and physics of condensed-phase pyrolysis. However, current experimental techniques can only evaluate macroscopic properties. Using a new Reactive Molecular Dynamics (RMD) approach that allows for smooth transitions between valence bond descriptions, the combination of bond scission and beta-scission reactions are considered for the decomposition of linear polyethylene.

Results and Discussion

To explore the decomposition of polyethylene, a series of scission and addition/beta-scission reactions were developed that smoothly transition between the valence bond reactant and product state. These transitions occur using our reactive force field RMDff [1], which is implemented in our RMD code, RxnMD [2]. The RMDff method monitors a characteristic reaction coordinate and once that reaction coordinate becomes active, smooth switching functions are used to transition the energetic and forces from the reactant state description to the product state description.

It has been previously shown [2] that as the polymer length increases, the kinetic rate per total number of backbone bond is increased. This work investigates the impact of beta-scission reactions on the overall decomposition kinetics. Beta-scission reactions were first analyzed by inserting bond scissions in the polymer, modeling the reactions as cases where the energy has had sufficient time to be redistributed from the bond scission event. The second method allowed the concurrent bond scission and beta-scission reactions in the polymer decomposition. The rate of beta-scission reactions were found to be independent of the polymer length. For cases where the energy had been redistributed, the beta-scission kinetics are very similar to those seen for small, gas-phase vinyl radicals. When energy redistribution was not enforced, it appears possible that at high temperatures the beta-scission rate may be influenced by torsion strain that has not had sufficient time to relax prior to the next reactive event.

A comparison between the microscopic kinetic rates obtained from the RMD simulations are also compared with the macroscopic mass loss curves through a basic kinetic Monte Carlo model. While the model may be missing overall details that result in some error, it appears that the overall influence of the microscopic kinetic rates may provide a basis for understanding the overall macroscopic kinetic decomposition.

Acknowledgements

K.D. Smith and P.R. Westmoreland were supported through Cluster F for Fire-Safe Polymers of the Center for UMass-Industry Research on Polymers (CUMIRP), a consortium sponsored by Boeing,

Boltaron Performance, Borealis, Federal Aviation Administration, Hickory Springs, Kleerdex Company, Multina, National Institute of Standards and Technology, SABIC, Solvay Advanced Polymers, and the US Army.

References

[1] K.D. Smith, S.I. Stoliarov, M.R. Nyden, P.R. Westmoreland, *Molecular Simulation* **33** 361-368 (2007).
[2] K.D. Smith, M. Bruns, S.I. Stoliarov, M.R. Nyden, O.A. Ezekoye, P.R. Westmoreland, *Polymer* **52** 3104-3111 (2011).
[3] http://rxnmd.sourceforge.net/

Contact information:

Kenneth D Smith (speaker)
United Technologies Research Center
411 Silver Lane, MS 129-90
East Hartford, CT 06108
Phone: 860-610-7162
Fax: 860-353-3379
Email: smithkd@utrc.utc.com
[Current address shown. This work is related to the PhD dissertation of KD Smith at the University of Massachusetts Amherst.]

Prof. Phillip R. Westmoreland (PI)
Executive Director, NCSU Institute for Computational Science and Engineering
Professor, Department of Chemical and Biomolecular Engineering
North Carolina State University
2036 Engineering Building I, Box 7905
911 Partners Way; Raleigh, NC 27695
Phone: (919) 515-7121; Fax: (919) 515-3465
Email: phil.westmoreland@ncsu.com

(451f) Improved Methods for Assigning Point Charges, Bond Orders, and Magnetic Moments In Complex Materials

Thomas A. Manz and David S. Sholl, School of Chemical and Biomolecular Engineering, Georgia Institute of Technology, Atlanta, GA

Net atomic charges (NACs), atomic spin moments (ASMs), and effective bond orders (EBOs) are important concepts for understanding many chemical systems. Although natural population analysis (NPA) provides reliable NACs, ASMs, and EBOs for nonperiodic materials, it is not yet available for periodic materials and does not specifically optimize NACs to reproduce the electrostatic potential V(**r**). Herein, we describe the development of density derived electrostatic and chemical (DDEC) population analysis for computing NACs, ASMs, and EBOs in both periodic and nonperiodic materials. This method analyzes a material's total and spin densities computed via density functional theory, coupled-cluster theory, or another *ab initio* method. To produce NACs suitable for constructing force-fields used in atomistic simulations and for understanding electron transfer during chemical reactions, the atomic charge distributions are simultaneously optimized to reproduce the chemical states of atoms and V(**r**) outside a material's electron distribution. Our method can be used to compute ASMs for both collinear and non-collinear magnetic materials. The EBO includes both traditional covalent bonding (e.g., H_2) and dispersion bonding (e.g., Kr_2). In this talk, we will describe the basic theory behind the DDEC population analysis method and use it to study (a) a spin-charge-ordering reconstruction anti-phase domain boundary on the magnetite 001 surface, (b) the non-collinear single molecule magnet $Fe_4C_{40}H_{52}N_4O_{12}$, (c) CO_2 adsorption in metal organic frameworks for carbon capture applications, and (d) other periodic and nonperiodic materials.

See more of this Session: First-Principles Simulation of Condensed Phases
See more of this Group/Topical: Computational Molecular Science and Engineering Forum

(451g) Two-Dimensional Crystalline Domains Embedded In Graphene Bilayers: Structure and Properties

Nick D'Eramo, Chemical Engineer, University of Massachusetts Amherst, Amherst, MA and **Dimitrios Maroudas**, Chemical Engineering, University of Massachusetts Amherst, Amherst, MA

Exposure of graphene bilayers to fluxes of atomic hydrogen leads to H chemisorption onto the outer surfaces of the graphene layers. Certain hydrogenation patterns favor the formation of covalent interlayer C-C bonds, which have been shown to stabilize the hydrogenated surfaces. The hydrogenation of graphene bilayers and the associated interlayer bonding may generate two-dimensional (2D) crystalline domains of sp^3-hybridized carbon atoms. These 2D crystals have structures that resemble those of various carbon allotropes under different plane orientation.

In this presentation, we report results of a systematic computational analysis of the atomic and electronic structures and properties of such 2D crystalline carbon nanomaterials. The analysis is based on structural relaxation and computation of electronic and mechanical properties of these materials. The atomic and electronic structure of the materials is determined according to first-principles density functional theory (DFT) calculations, while the mechanical properties are computed using classical molecular-dynamics (MD) simulations according to a reliable reactive bond order potential.

The crystalline structure of the interlayer-bonded configurations is determined by the plane stacking in the pristine bilayer and by the hydrogenation patterns on the surfaces. DFT calculations predict the configurations that lead to the most thermodynamically favorable phases. DFT also is used to compute the band structure of these thermodynamically stable phases; all of the 2D crystalline configurations examined constitute insulators with band gaps of approximately 3 eV, which is consistent with the electronic behavior of sp^3-bonded carbon allotropes. MD simulations of dynamic deformation, under conditions of constant tensile strain rate and temperature, allow for predictions of mechanical properties of these materials, such as their Young modulus, tensile strength, and fracture strain. These mechanical properties are dependent on the materials' crystalline structure and on the applied tensile stress.

See more of this Session: First-Principles Simulation of Condensed Phases
See more of this Group/Topical: Computational Molecular Science and Engineering Forum

(451h) Manganese Oxide Surface Chemistry and Adsorption of Cr(III)

Gloria A. E. Oxford and Anne M. Chaka, Physical Measurement Laboratory, National Institute of Standards and Technology, Gaithersburg, MD

Manganese oxides areprevalent in geological settings and play a significant role in the geochemicalcycling of heavy metals due to their high adsorption capacity. Of particular interest is their ability tooxidize Cr(III) to toxic Cr(IV). Whilemuch work has focused on the mechanism of oxidation, several questions remainabout the specific manganese-chromium interactions prior to and duringoxidation. An understanding of thestructure, chemical properties, and redox reactivity of manganese oxidessurfaces is an essential foundation for determining possible Cr(III) oxidationmechanisms.

Periodic densityfunctional theory calculations have been combined with *ab initio* thermodynamics to identify stable beta-MnO_2(110) and gamma-MnOOH (010) surfaceterminations and to probe their redox behavior in response to varying oxygenand water chemical potentials. Under ambientconditions, oxidation of the beta-MnO_2 (110) surface is notlikely, while oxidation of the gamma-MnOOH(010) surface is predicted to be favorable.Reduction of the clean surfaces leads to significant surfacereconstructions that result from the competition between optimal *d*-orbitaloccupation and manganese coordination geometry and lattice constraints of thebulk. Jahn-Tellereffects allow multiple oxidation states to exist at the surface. While these surfaces are not predicted to bestable at ambient conditions, they may be important in the oxidationmechanism. Hydration of the clean beta-MnO_2(110) and gamma-MnOOH (010) surfaces generallylowers the surface free energies significantly with hydrogen bonding networksformed at monolayer coverage. Adsorptionof Cr(III) has been investigated on the hydrated surfaces. The effect of manganese oxidation state onadsorption geometries and electron transfer will be discussed.

See more of this Session: First-Principles Simulation of Condensed Phases
See more of this Group/Topical: Computational Molecular Science and Engineering Forum

(456a) Temperature Distribution In Activated Carbon Bed During Adsorption of Nitrogen (model gas for hydrogen); Experiment & Mathematical Model

Agnieszka Truszkowska[1], **Christopher Loeb**[1], Richard Chahine[2], Bruce Hardy[3] and Goran Jovanovic[4], (1)Chemical, Biological and Environmental Engineering, Oregon State University, Corvallis, OR, (2)Hydrogen Research Institute, Trois-Rivieres, QC, Canada, (3)Computational Sciences, Savannah River National Laboratory, Aiken, SC, (4)Oregon State University, Corvallis, OR

Temperaturedistribution in activated carbon bed during adsorption of nitrogen (model gasfor hydrogen); experiment & mathematical model

Agnieszka Truszkowska[1],Christopher Loeb[1], Richard Chahine[2], Bruce Hardy[3]and

Goran N. Jovanovic[1]

[1]Oregon State University,Department of Chemical, Biological, and Environmental Engineering, Corvallis, OR 97330, USA[2]Hydrogen Research Institute at University du Quebec a Trois-Rivires, PO BOX 500, Trois- Rivires, QC, Canada, G9A 5H7

3•
Savannah RiverNational Laboratory•
ProcessModeling and Computational Chemistry Group •
Building 999-2W, Rm. 122 •
Aiken, SC 29808, USA •
•

Emails: truszkoa@onid.orst.edu, loebc@onid.orst.edu, goran.jovanovic@oregonstate.edu, bruce.hardy@srnl.doe.gov, richard.chahine@uqtr.ca

Providinghigh volumetric and mass storage density of hydrogen gas is a major challengein the progress towards hydrogen-fueled automobiles. Of the various storage media,liquid chemical systems, and metal hydrides, activated carbon adsorbents havesubstantial advantages in cost, weight, ease of handling and safety.

In this paperwe present experimental data (temperature and pressure) obtained during adsorptionof nitrogen (model gas for hydrogen) in an activated carbon bed. We alsopresent the results of a numerical simulation of the same process.

Experimentalapparatus used in this study consists of a cylindrical activated carbon bed (2in diameter x 1.5 in high) placed into a sealed high-pressure cylindricalvessel. The carbon bed, consisting of densely packed activated carbon materialand binding agent, is instrumented with six thermocouples placed at suitablelocations throughout the bed. In addition, a thermocouple, and three pressuretransducers are placed into the high-pressure vessel outside the bed. Dynamic change of temperature andpressure during the charging process are recorded, corresponding to compressionand adsorption phenomena. The charging process starts at 1.0 bar and ends at50.0 bar pressure. A parametric study of filling time is performed for time intervalsbetween 10 and 30 seconds, to observe its effect on the charging process.

The mathematicalmodel is developed for the carbon bed pressure vessel system, which enablesnumerical simulation of the charging process. Initially, numerical simulationsare used to analyze experimental data and to explore characteristic parametersof the carbon bed. The same numerical tool will be used in the future todesign carbon bed with optimal characteristics imposed by requirements forhydrogen storage media in motor vehicles.

See more of this Session: Hydrogen Storage Systems and Materials: Modeling, Simulation, and Optimization:
See more of this Group/Topical: International Congress on Energy 2011

(456b) Optimization of Adsorbent Tanks for Hydrogen Storage for Automotive Applications

Stephen L. Garrison[1], David A. Tamburello[1], Matthew R. Kesterson[1], Claudio Corgnale[2] and Bruce Hardy[1], (1)Computational Sciences, Savannah River National Laboratory, Aiken, SC, (2)Hydrogen Technology Center, Savannah River National Laboratory, Aiken, SC

A key technical hurdle to an energy economy focused on clean-burning hydrogen over non-renewable petroleum is on-board storage of hydrogen for automotive vehicular applications. Adsorbents provide one possible solution to the need for improved gravimetric and volumetric capacities. However, while the heat of adsorption released upon sorption of hydrogen is relatively low, so it the thermal conductivity. The subsequent rise in storage temperature results in decreases in maximum hydrogen storage capacity and loading rate at the short fill times mandated by the Department of Energy (DOE) targets. Novel designs for integrated heat exchangers are needed to maximize the heat transfer out of system to maximize storage rate and capacity.

While many literature studies focus on maximum capacity assuming an (impossible) infinite fill time, full, detailed modeling of the coupled heat and mass transfer is needed. The focus of this presentation will be our work developing detailed numerical models that couple heat and mass transfer via the general purpose finite element solver COMSOL Multiphysics® for various adsorbent media and bed designs and the automatic optimization of the designs by linking the COMSOL models with routines in MATLAB®.

See more of this Session: Hydrogen Storage Systems and Materials: Modeling, Simulation, and Optimization:
See more of this Group/Topical: International Congress on Energy 2011

(456c) Molecular Simulation of Hydrogen Adsorption In Porous Material Made up with Silsesquioxane Units and Metal Catalytic Sites

Nethika S. Suraweera[1], Michael E. Peretich[2], Joshua Abbott[2], Craig E. Barnes[2] and David J. Keffer[3], (1)Chemical & Biomolecular Engineering, University of Tennessee, Knoxville, TN, (2)Chemistry, University of Tennessee, Knoxville, TN, (3)Chemical and Biomolecular Engineering, University of Tennessee, Knoxville, Knoxville, TN

The adsorptive hydrogen storage of molecular hydrogen on microporous materials is widely studied as an effort for efficient utilization of hydrogen as a pollution free renewable energy source. Porous structures made up with spherosilicate building blocks and consist of metal catalytic sites exhibit promising results for the ability of adsorption of gases. Experiments are carried out for investigating the hydrogen adsorption ability of these materials. In this study we performed computer simulations to analyze hydrogen adsorption in structures which consist of interconnected silsesquioxane units. This porous material has cubic silicate building blocks (Si_8O_{20}), which are cross-linked by $SiCl_2O_2$ bridges and decorated with $OTiCl_3$ catalytic sites and $SiMe_3$ ends.

The structure for the simulations was developed first using a coarse grain model, followed by an energy optimization and replacing the coarse grains with atoms. The physical properties of the structure such as density, surface area, accessible volume and pore size distribution were compared with experimental vales and they are in good agreement.

Adsorption of hydrogen was modeled by Path Integral Grand Canonical Monte Carlo (PI-GCMC) simulations using standard force fields. From the simulations, hydrogen adsorption isotherms at 300K and 77 K were generated for low pressures up to 10 bars. The energy of adsorption of hydrogen was calculated at each case. Density distribution was developed and analyzed. Classical MD simulations were performed to calculate the self-diffusivity of hydrogen in the porous structure and activation energy was calculated.

Keywords: spherosilicate building blocks, hydrogen adsorption, PI-GCMC simulation, silsesquioxane units, coarse grain model, energy optimization, adsorption isotherm, density distribution, MD simulations, self-diffusivity, activation energy

See more of this Session: Hydrogen Storage Systems and Materials: Modeling, Simulation, and Optimization:
See more of this Group/Topical: International Congress on Energy 2011

Density Functional Theory Study of the Effect of 3*d* and 4*d* Transition Metals on $NaMgH_3$ Complex Metal Hydride for Hydrogen Storage Applications

Authors: Fernando Soto[1], Daniela S. Mainardi[2]

[1]Institute for Micro-manufacturing, Louisiana Tech University
[2]Institute for Micro-manufacturing, Chemical Engineering, Louisiana Tech University

In order to achieve the Hydrogen Economy, it is necessary to use hydrogen as an energy carrier. Hydrogen as an energy carrier is a possible solution to neutralize the climate changes occurred due to carbon emission. It is also highly attractive to ensure a smooth and efficient provision of energy without depending on other nations. Vehicles powered by fuel cells are one of the applications in which hydrogen can be used as an energy carrier. However, the absence of a commercially viable hydrogen storage technology is delaying the large-scale use of fuel cell powered vehicles. In order to promote research in this area, the U.S. Department of Energy (DoE) has set targets for the year 2010 and 2015 for a system of on board hydrogen storage. Previous results suggest that the addition of transition metals enhances the desorption kinetics of complex metal hydride and lowers the hydrogen dissociation temperature and activation energies.

In the present work, study of catalytic additives with improved performance over Ti allows us to generalize the work on the procedure for 3d/4d doping in NaMgH3 and use it in systems with higher gravimetric densities and more favorable thermodynamics. For this purpose, plane wave density functional theory (PW-DFT) calculations helps us examine bulk and surface models of NaMgH3, Mg2FeH6 and KMgH3 complex metal hydride (CMHs) with impurities from the 3d element block. In this respect, first- principles calculations determined the cohesive energies of pure and doped bulk/surface CMHs as well as the adsorption/substitution energies. Furthermore, DFT coupled Molecular Dynamics (DFT-MD) calculations at elevated temperatures elucidate the effect of Ti placed on different sites on surface models. The results indicate that 5 out of 7 additives at the bulk model are thermodynamically stable relative to the pure model; therefore, possible. The most favorable elements in this regard came out to be; Ti, V, and Co in that order. Afterwards, from dynamics calculations performed at surface slab models, we conclude that Ti additives replacing sodium lattice sites and Ti additives placed at a four-fold vacant position are the stable arrangement in this system with potential formation of TiMgxHx complexes. Meanwhile, alloyed systems with 4d transition metals may be extremely stable for on-board hydrogen storage applications. Also, dynamics calculations performed at the surface slab models suggests that Ti additives replacing sodium lattice sites and Ti additives placed at a four-fold hollow site are the preferred, stable arrangements in NaMgH3 with possible formation of TiMgxHx complexes. This conclusion may serve as a guideline for the future design of nano-structured complex hydride for on-board hydrogen storage utilization.

Keywords: density functional theory, hydrogen storage.

1. Introduction

Finding a feasible solution to the problem of hydrogen storage for vehicular applications is essential for the transportation sector because it accounts for two-thirds of energy consumption in the U.S. [1]. Also, it is necessary to ensure an energy supply and avoid energy dependence from regions such as the Middle East which are politically unstable [2]. It is imperative to reduce carbon emissions in order to improve the air quality and avoid harsh climatic changes [3]. In order to promote research in this area, the U.S. Department of Energy (DoE) has set targets for the year 2010 and 2015 for a system of on board hydrogen storage. The use of hydrogen within H2/O2 fuel cell systems requires an adequate and accessible hydrogen storage medium. The production of liquid or highly compressed hydrogen requires an tremendous amount of energy and may be dangerous in certain situations. Therefore, solid-state hydrogen storage is currently the safest and most effective way of storing hydrogen. As explained by Grochala et al. an ideal hydrogen storage material (HSM) will have a low-molar weight, be inexpensive, have rapid kinetics for absorbing and desorbing H2 in the 25-120 °C temperature region, and store large quantities of hydrogen reversibly [4]. As noted by Grochala et al., at the moment, no single material fulfills all of these requirements. High stability and poor kinetics are significant challenges, which have to be solved in order for any material to be considered for on-board automobile applications [5].Therefore, the identification of material that can meet this criterion is a grand challenge that has attracted the interest of the global scientific community.

Light chemical elements are essential to establish a reversible hydrogen storage system [6]. In order to achieve hydrogen gravimetric and volumetric densities required for mobile applications, host elements in the hydrogen storage system should be lighter than Al metal. Typical elements considered among the light elements are: Li, Be, B, C, N, O, F, Na, Mg, and Al. As explained by Wu et al., metal hydrides are either too stable for efficient hydrogenation cycling or too unstable that desorption occurs at or below room temperatures but adsorption requires high pressures of hydrogen [7]. Nonetheless, CMH have been identified as promising candidates for on board hydrogen storage applications. CMHs may enhance kinetics and reversibility and inhibit toxic-gas side products. Since the pioneer work of Bogdanovic and Schwickardi in 1997, the focus has shifted to CMHs such as alanates, borohydrides, amides/imides, and destabilized CMH [8, 9].

Since the work of Bogdanovic and Schwickardi in 1997, it has been established that certain element addition can help CMH meet the criteria for HSM and are capable of showing hydrogen reversibility at moderate conditions [8, 10, 11]. Thus, enhanced CMH are a promising class of material for hydrogen storage that needs further study because high stability and poor kinetics are important challenges which have to be solved for the CMH to be considered for on-board automobile applications [12]. Recently several types of CMHs have attracted the attention as promising enhanced CMHs candidates for HSM. These CMHs are: Ternary hydrides composed of transition metals and different alkaline earth metals [13], alkaline transition-metal borohydrides [14], and ABH_3perovskite structures [15]. Ti-doped CMHs have shown to improve the kinetics of complex metal hydride decomposition and the thermodynamic reactions relative to the pristine material. Thus, the role of Ti and other *3d*/*4d* transition elements in different types of CMHs need further studying in order to achieve the goals established by the DoE [9].

$NaMgH_3$

Recently, Mg-based hydrides have attracted interest as a hydrogen storage medium due to their light weight and low cost [10].Several Mg-based hydrides show the perovskite structure expressed as ABH_3, where A is an alkali element and B is magnesium. $NaMgH_3$ is the only magnesium based ternary hydride with Na and it has attracted the attention as a possible candidate for hydrogen storage material because it has high gravimetric hydrogen density (6 %) and high volumetric hydrogen density (88 kg/m^3) [11]. This magnesanate has not been widely investigated as a candidate for hydrogen storage. The crystal structure of this material is known and it derives from cubic $KMgH_3$ by an orthorhombic distortion and has a perovskite structure similar to $GdFeO_3$ [10]. It belongs to the *Pnma* space group with orthorhombic symmetry in the temperature range of 4 to 370 K [11,12].The primitive unit cell contains four formula units for a total of 20 atoms.The $NaMgH_3$ unit cell is shown in figure 1.

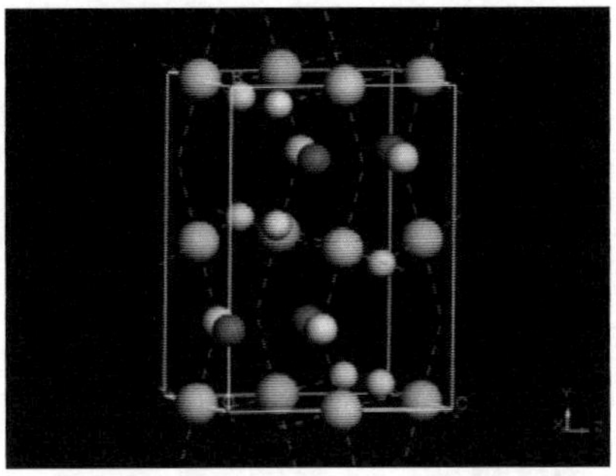

Figure 1. Unit Cell of Sodium Magnesium Trihydride.

Hydrogen disassociation in this material occurs at atmospheric pressure and approximately 670 K, which is well above the operable range for on board hydrogen storage applications. Thus, $NaMgH_3$ is more thermally stable than binary MgH_2, which has a hydrogen dissociation temperature ~550 K. The dehydrogenation reaction scheme takes place in two steps [12]:

$$NaMgH_3 \rightarrow NaH + Mg + H_2 \quad (1)$$

$$NaH \rightarrow Na + \frac{1}{2} H_2 \quad (2)$$

Numerous studies have been carried out to develop suitable Mg-based systems that show fast sorption kinetics at moderate temperatures. These studies achieved partial success in improving the kinetics by addition of dopants. Ma and co-workers demonstrated high affinity of transition metal cations toward hydrogen. Catalysts are usually mixed with the hydride by mechanical ball milling, which introduces defects and impurities. However, much research still needs to be done in order to identify appropriate catalysts, study the existence of stable intermediate phases and their crystal structure and understand the kinetics. Much insight can be gained from first principles simulations for the purpose of designing alloy catalysts.

For instance, Bouhadda et al. used ab initio calculations to report on the crystal, electronic and thermodynamic properties of $NaMgH_3$. Bouhadda et al. reported relaxed structures in agreement with experimental structures (Table 1).

Perovskite Structure	Experimental Lattice Constants (Å)	Calculated Lattice Constants (Å)
$NaMgH_3$	a=5.463	a=5.410
	b=7.703	b=7.628
	c=5.411	c=5.358

Table 1. Experimental and Calculated Cell Parameters

In addition, this author also reported that this structure is a wide band gap semiconductor with a calculated band gap of 3.4 eV. Bouhadda et al. explains that the interaction between Na and MgH_3 is ionic and that the interaction between Mg and H is both ionic and covalent.
To study the thermodynamic properties of the material, Bouhadda et al. calculated the standard enthalpy of formation H_f with the following reaction equation.

$$Na + Mg + 3/2\ H_2 \rightarrow NaMgH_3 \qquad (3)$$

The result of 151.83 KJ/mol is well in agreement with the results obtained from Ikeda et al. (145.05 KJ/mol, 147 KJ/mol) [15]. Thus, this formation enthalpy is considerably larger than that of metal hydride NaH (-114 KJ/mol) [15]. Klaveness et al. also calculated the enthalpy of formation for this material obtaining similar results as those presented by Bouhadda et al. and Ikeda et al. Thus, the first principles approach tends to reproduce experimental results very well, and this is true also for ionic hydrides such as $NaMgH_3$ [16,17].

Furthermore, Shane et al. reported that the high rates of H motion in $NaMgH_3$ are due to regions with poorly organized crystal structure [17]. The author concludes that if the disorder could be sustained, it could provide a route towards improved reaction kinetics. Hao et al. presented a theory-guided computational study explaining the existence of intrinsic defects in $NaMgH_3$. The author reported that schottky defects are the dominant defects in this material and that hydrogen diffusion is determined by mobility of schottky defects. Hao et al. concluded that the diffusion induced by schottky defects can be tweaked by adding suitable dopants such as Cu and Co that can adjust the material's Fermi level [18]. Komiya et al. studied the substitution energies in $NaMgH_3$ by replacing the Na ion with an element from the same 1A group element from the periodic table (Li, K, Rb, Cs) and replacing the Mg ion with an element from the 2A group elements (Be, Ca, Sr, Ba) [19]. Komiya et al. reported positive substitution energies on the Na site, indicating that doping the structure makes the material unstable, except when doping with the K ion. Also, the addition of Be, Sr, and Ba destabilizes the structure and doping with Ca increases the stability. These results show that addition of impurities has a positive correlation on the destabilization of the material.

Moreover, ABH_3 structures such as $NaMgH_3$ may enhance kinetics and reversibility and inhibit toxic gas side products compared to the parents' borohydride, amide, and alanate compounds. Also, higher H mobility in this material due to their perovskite structure cause improved hydrogenation rates compared to pure MgH_2. It is clear from the literature review presented that further investigation is needed to investigate the positive correlation between hydrogen mobility, addition of dopants and hydrogenation/dehydrogenation reversibility. Therefore, the aim of this paper is to study the stability and dynamics of pure and doped $NaMgH_3$ for on-board hydrogen storage applications.

2. Computational Methods

2.1 Bulk Calculations

In this study, we use the first principles pseudopotential total energy method through the generalized gradient approximation (GGA) calculations with the Perdew and Wang (PW91) functional as implemented in the CASTEP® module [5] of the Materials Studio® software by Accelrys, Inc. [6] to study bulk and surface models of $NaMgH_3$. Spin-polarized calculations on Fe-, Ni-, and Co-doped systems show no significant effect to be created energetically, with respect to the non-spin polarized calculations. Hence, further conducted geometry optimization calculations were spin-unpolarized, and the cut-off energy was 400 eV with a fine k-point mesh in all cases.

We created hydride models alloyed with 3*d* and 4*d* block transition metals to investigate their effects on the energetics of bulk/surface of $NaMgH_3$ systems. The alloying elements substitute one Na atom at a center lattice location of each structure, resulting in a concentration between 5.65 and 7.57 at.wt. % in the case of 3*d* transition metals. Afterwards, bulk hydrides and surface slabs relaxed entirely in terms of all atomic degrees of freedom to obtain the ground-state energy system. After complete relaxation of the systems was obtained, single-point energies were calculated for the optimized structures. Since the main aim of this work is to illustrate the effect of transition metals on the prototype $NaMgH_3$ system, calculation of cohesive energies (ΔE_{coh})of the doped systems were doneaccording to equation (4).

$$\Delta E_{coh} = E_{TOT}(Na(X)MgH_3) - [NE\,(Na) + NE(Mg) + NE(H) + E(X)] \qquad (4)$$

Where E_{TOT} is the total energy of the system, N is the number of atoms, E(Na) is the energy of a free neutral atom of Na, E(Mg) is the energy of a free neutral atom of Mg, E(H) is the energy of a single hydrogen atom and E (X) is the energy of a free neutral transition metal X.

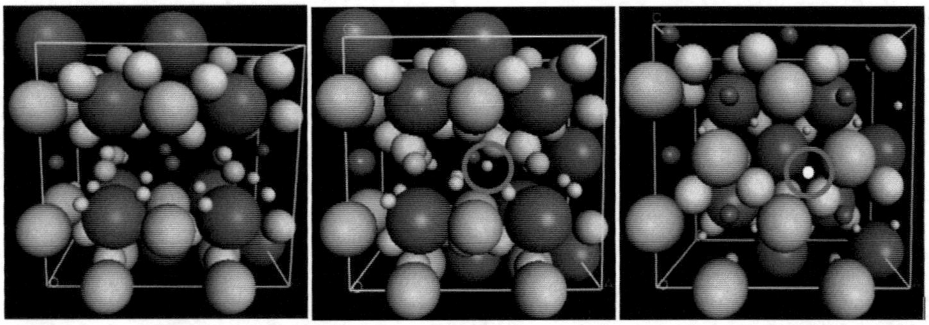

(a) (b) (c)

Figure 2. $NaMgH_3$ 2x1x2 Supercell consisting of five f.u. **(a)** pure model, **(b)**Ti_{sub}@Na in model, and **(c)**Vac_{sub}@Na in model.

2.2 Surface Calculations

Study of surface properties is vital to understand the catalytic behavior of impurities. An explanation for this is that the arrangement of atoms, distribution of electrons, and their response to external behavior is different at the surface when compared to the bulk.

Thus, besides the studies regarding bulk doped $NaMgH_3$ systems, we considered the adsorption/substitution behavior of Ti on the $NaMgH_3$ (001) surface. Accordingly, the calculations were performed for surface structures with a slab width of 2x1x2, and a gap of approximately 5 Å at the top of the $NaMgH_3$ surface. Moreover, k-point sampling is performed at the gamma point only. Doping on the surface structure proceeded in two distinct steps: first, one is the substitution adsorption, in which the dopant replaces a Na atom at the center lattice location of the uppermost layer of $NaMgH_3$ (001) surface. The second form of adsorption in the on-top of surface adsorption.

The substitutional/adsorption energy calculation for the (001) surface proceeds via the following equation:

(5)

Where N is the total number of atoms in the model and X is the transition metal added to the system.

3. Results and Discussion

The equilibrium cell parameters a, b and c of pure $NaMgH_3$ are 5.48, 7.68 and 5.40 Å, respectively. A brief comparison with experimental values showed that; these calculated values are of 0.20 % different than the initially set unit cell dimensions, which is a strong indication of the good agreement between the computational and experimental results. The calculated results for the lattice parameters are shown in Table 2.

Lattice Parameter (Å)	
experimental	**DFT (GGA/PW91)**
a = 5.463	5.47
b = 7.703	7.68
c = 5.411	5.39

Table 2. Experimental and Calculated Lattice Parameter

The calculated energy gap is 3.308 eV which is similar to the values found in literature [15]. Calculated band gap figure is shown in figure 3.

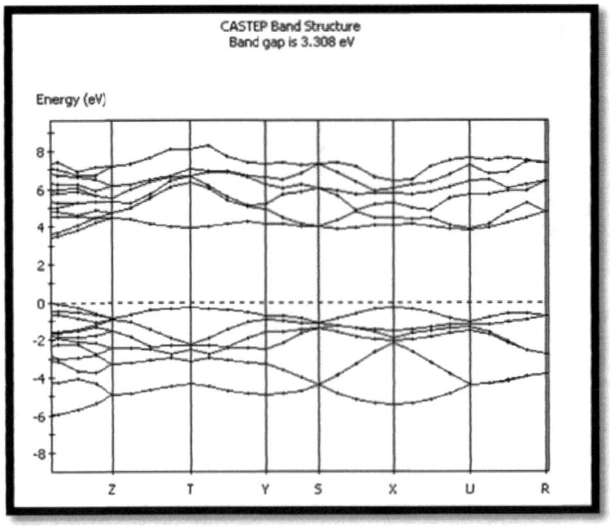

Figure 3. Calculated Band Gap as obtained from CASTEP®

Furthermore, electron density maps corroborate the results reported by Bouhadda et al., the interaction between Na and MgH_3 is ionic and that between Mg and H involves both ionic and covalent bonding schemes. In this respect, the theory level used is appropriate and the results encourage us to continue with the same theory level for subsequent ab initio calculations (Figure 4).

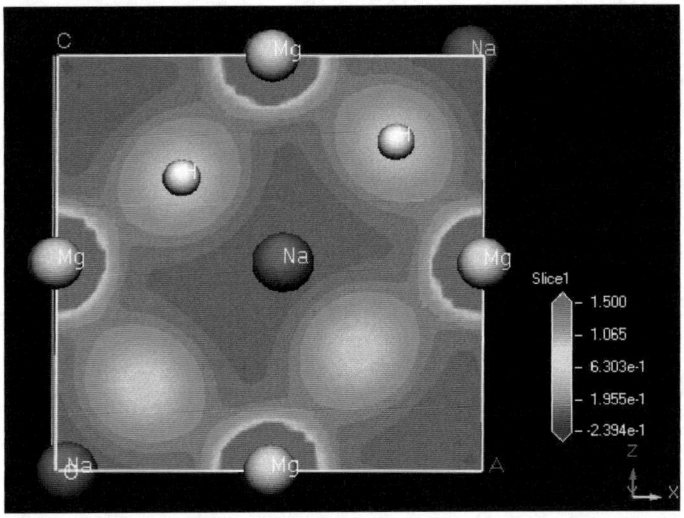

Figure 4. Total Charge Density of $NaMgH_3$ parallel to the x and z axis. Where the green balls are Mg atoms, purple balls are Na atoms and white balls are hydrogen atoms.

3.1.1 Bulk

Next, the primary goal in calculating the cohesive energy energies for each alloyed system was to investigate the effect of 3*d* and 4*d* block elements in the stability of the system. Modification in $NaMgH_3$ models took place via addition of selected impurity elements X (where X can be any element from 3*d* and 4*d* block: Sc, Ti, V, Nb, Y, Mo, etc) to the desired $NaMgH_3$ model. The cohesive energy of the pure model served as the reference value; thus, any energy above 0 eV may be considered as stable. It is interesting to note that several elements added as dopants in the $NaMgH_3$ structure are thermodynamically stable and; therefore, appealing for further consideration.

3*d* Transition Metals

In this respect, for the 3*d* block, Ti, V, Sc, Co, Ni and Cu in the decreasing order, are the most favorable elements with respect to the cohesive energy of the pure system for 3*d* transition metals (Figure 5). For instance, the cohesive energy due to the presence of the Ti dopant is 2.89 eV more favorable than the pure case where no dopants are available. Meanwhile, the removal of a Na ion from the pure model is 3.95 eV less preferred than the pure model. Also, V is 2.64 eV more energetically favorable, Sc has a value of +2.16 eV, and for Co and Cu those values are +1.62 eV and +0.62 eV respectively. On the other hand, Zn and Mn where found to be unfavorable in this system with cohesive energies of -2.58 eV and -0.68 eV, respectively. These results and their corresponding bond lengths with Mg and H are tabulated below.

Model	X in Bulk Site	System	*E_{coh} (eV)	E_{sub} (eV/atom)
Bulk	Na	$Na_{16}Mg_{16}H_{48}$	0.00	
	0	$Na_{15}Mg_{16}H_{48}$	-3.95	
	Sc	$(Sc)Na_{15}Mg_{16}H_{48}$	2.16	0.027
	Ti	$(Ti)Na_{15}Mg_{16}H_{48}$	2.89	0.036
	V	$(V)Na_{15}Mg_{16}H_{48}$	2.64	0.033
	Mn	$(Mn)Na_{15}Mg_{16}H_{48}$	-0.68	-0.009
	Co	$(Co)Na_{15}Mg_{16}H_{48}$	1.62	0.02
	Cu	$(Cu)Na_{15}Mg_{16}H_{48}$	0.62	0.008
	Zn	$(Zn)Na_{15}Mg_{16}H_{48}$	-2.58	-0.032

Table 3. Cohesive Energies of the different Bulks Models

X @Na site Bond Length (Å)	Na	Sc	Ti	V	Mn	Fe	Co	Ni	Cu	Zn
X-Mg	3.24	3.09	3.06	2.99	3.02	2.72	2.67	3.18	3.23	3.26
X-H	2.51	2.11	1.93	1.83	1.74	1.67	1.63	1.71	1.75	1.92
Mg-H	1.99	1.98	2.00	2.00	2.04	1.99	2.00	1.98	2.00	1.99

Table 4. Inter Bond Lengths of the different 3*d*Bulk Models

It should be pointed out that; that the cohesive energies of these 3*d* block elements in the bulk follow a strong correlation with the cohesive energies of their metallic elements counterpart. Also, the cohesive energies of the 3*d* block elements show a negative correlation with the ionization energies of the pure metallic elements and they also show a positive correlation with the melting point in Kelvin of the pure crystalline elements.

Thus, a high cohesive energy value of the metallic element helps to predict a high cohesive energy value in the bulk model, and transition elements with a high ionization potential energy appear to be responsible for the low thermal stability of the bulk model and the melting point of the metallic elements follow a strong positive correlation with the cohesive energy of the bulk suggesting that all these physical quantities have to be taken into account when destabilizing $NaMgH_3$ with partial substitution of the cation Na. Such prediction helps to identify additional elements that could improve on the effect of 3*d* block transition elements on CMHs.

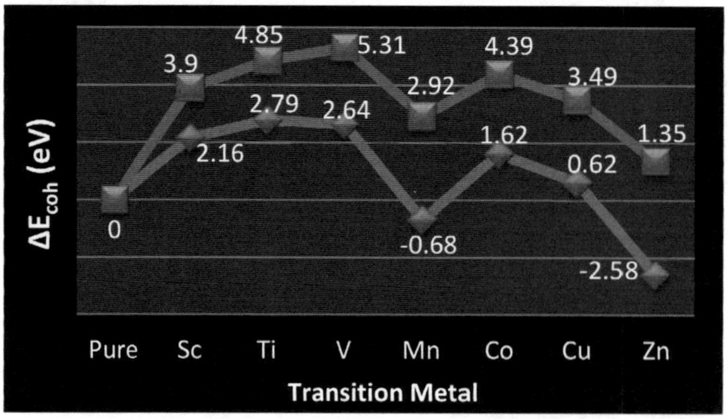

Figure 5. Cohesive Energy Results. Bulk substitution of transition metal in $NaMgH_3$(blue), pure crystalline transition metal (red).

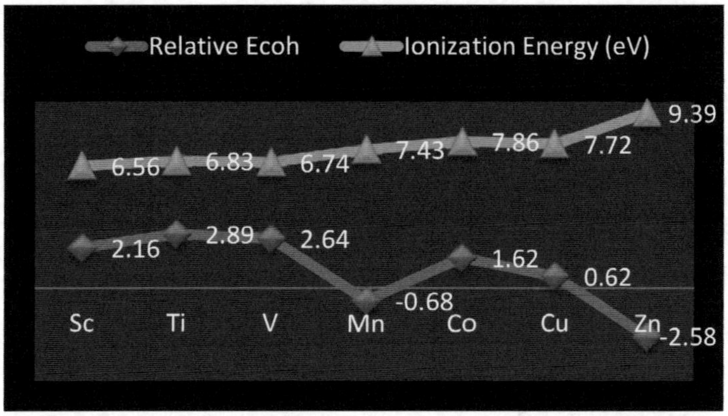

Figure 6. Cohesive Energy Results. Bulk substitution of transition metal in $NaMgH_3$ (blue), Ionization Energy (green)

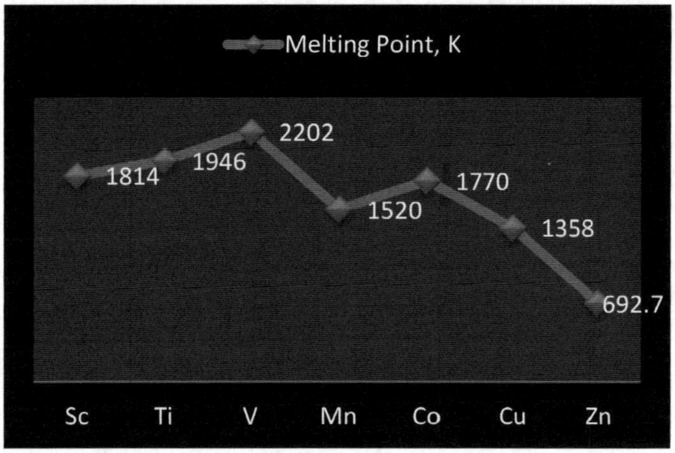

Figure 7.Melting point in Kelvin for transition metals in their pure crystalline state.

3d Total Charge Density

To further understand the effect of transition metals on bulk $NaMgH_3$ the total charge density of pure and Ti-doped models are shown below. The total charge density can be used to provide quantitative information at the atomic scale surrounding any atom or fragment of interest. The color corresponds to the total charge density, with the red regions corresponding to the highest density, which surrounds the atoms. The majority region of the model displays a very low charge density, colored blue. The plots suggest that the interaction between Na and Mg-H complex is mainly ionic in both models. However, the Ti-doped model shows an increased association of Ti with surrounding H and Mg atoms, as reported in table 4 above. Thus, a shortening of the X-H and X-Mg bond lengths implies an elongation and possible weakening of Mg-H bond lengths, making the desorption of H atoms more feasible.

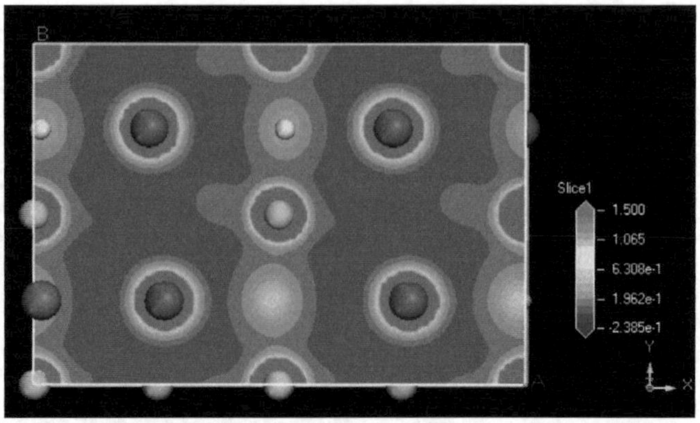

Figure 8. Total Charge Density of pure bulk $NaMgH_3$ parallel to the x and y axis. Where the green balls are Mg atoms, purple balls are Na atoms and white balls are hydrogen atoms.

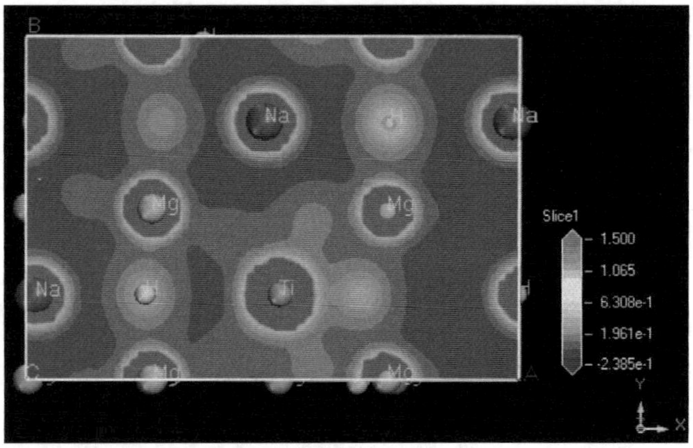

Figure 9. Total Charge Density of Ti-doped bulk $NaMgH_3$ parallel to the x and y axis. Where the green balls are Mg atoms, purple balls are Na atoms, white balls are hydrogen atoms and gray ball is Ti.

4*d* Transition Metals

In a similar fashion, cohesive energies for 4*d* block were calculated, Tc, Mo, Nb, Zr, and Y in the decreasing order, are the most favorable elements with respect to the cohesive energy of the pure . However, for the 4*d* block, all the models tested show a high positive energy with respect to the pure model. Thus, the alloyed systems are too stable for hydrogen storage applications. For instance, the cohesive energy due to the presence of the Tc dopant is 7.36 eV more favorable than the pure case where no dopants are available. This cohesive energy is 2.5 times greater than the cohesive energy of the system due to the presence of Ti. However, the cohesive energy of the system due to the presence of 4*d* block elements shows a strong positive correlation with the ionization energy of the crystalline element. Meanwhile, the correlation with the different melting points of the 4*d* element block and the cohesive energy of the crystalline elements is not as positive when compared to the 3*d* transition metal results.These results and their relationships are shown below.

Model	X in Bulk Site	System	*E_{coh} (eV)	E_{sub} (eV/atom)
Bulk	Na	$Na_{16}Mg_{16}H_{48}$	0.00	
	Y	$(Y)Na_{15}Mg_{16}H_{48}$	3.17	0.04
	Zr	$(Zr)Na_{15}Mg_{16}H_{48}$	5.15	0.064
	Nb	$(Nb)Na_{15}Mg_{16}H_{48}$	5.46	0.068
	Mo	$(Mo)Na_{15}Mg_{16}H_{48}$	6.21	0.077
	Tc	$(Tc)Na_{15}Mg_{16}H_{48}$	7.36	0.92

Table 5. Cohesive Energies of the different 4*d* Bulks Models

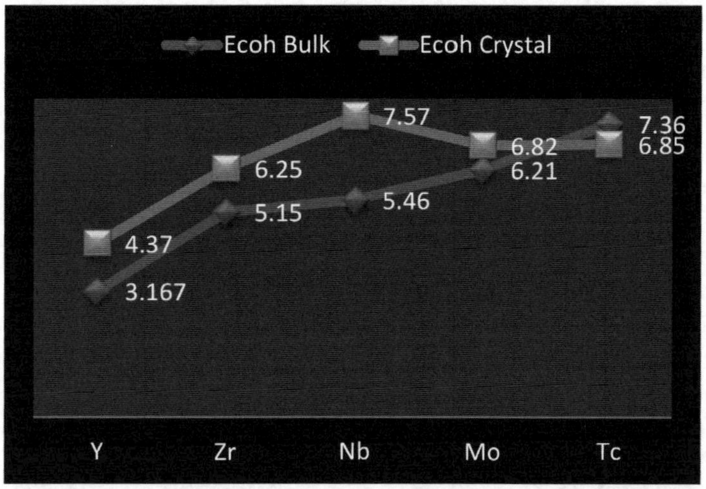

Figure 10. Cohesive Energy Results. Bulk substitution of 4*d* transition metal in $NaMgH_3$ (red), pure crystalline transition metal (blue).

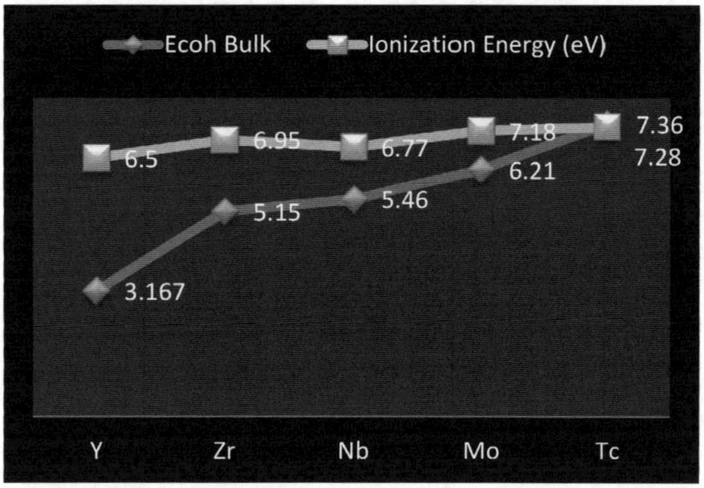

Figure 11. Cohesive Energy Results. Bulk substitution of 4*d* transition metal in NaMgH3 (blue), Ionization Energy (green)

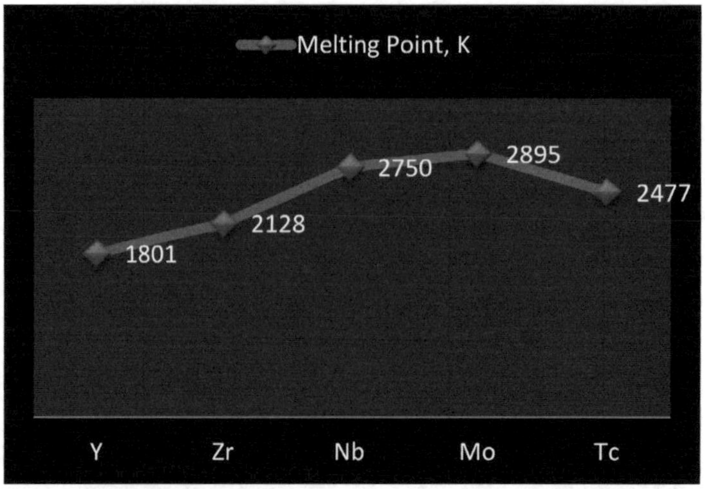

Figure 12. Melting point in Kelvins for 4*d* transition metals in their pure crystalline state.

4*d* Total Charge Density

The total density charge of the Nb-doped bulk model is taken as a sample from the 4*d* transition metal block. This plot also suggests that the addition of a dopant affects the local electron density of the model. This local modification of the 4*d* doped bulk model leads to an increase association of the dopants with the surrounding Mg and H atoms, thus; leading to a possible destabilization of the Mg-H complex as well.

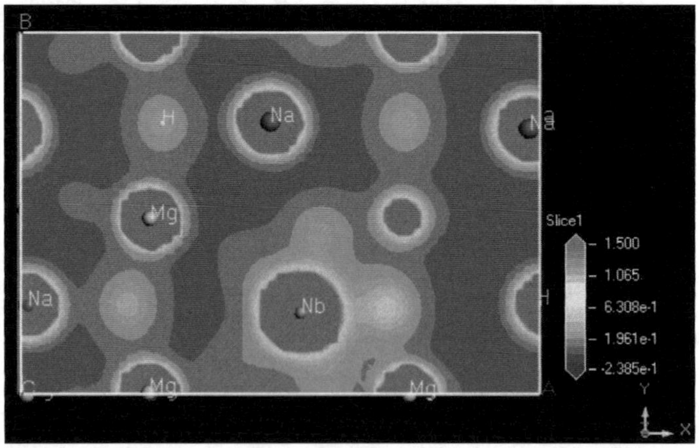

Figure 13. Total Charge Density of Nb-doped bulk $NaMgH_3$ parallel to the x and y axis. Where the green balls are Mg atoms, purple balls are Na atoms, white balls are hydrogen atoms and gray ball is Nb

3.2 Surfaces

Next, DFT calculations were conducted on five (001) surface models : Ti @ Na lattice site, Ti @ hollow site, Ti @ top of Na lattice site, Ti @ top of hollow site, and Ti @ Mg site. It is possible to observe the preferred Ti dopant site in terms of the substitutional/addition energy relative to the pure surface model. In this regard, possible formation of $TiMg_xH_x$ complexes was systematically investigated. The models used are showed below:

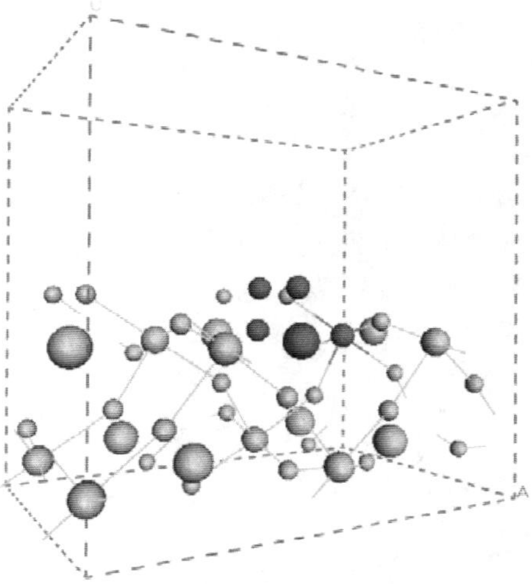

Figure 14 Model with Dopant (red) Sites

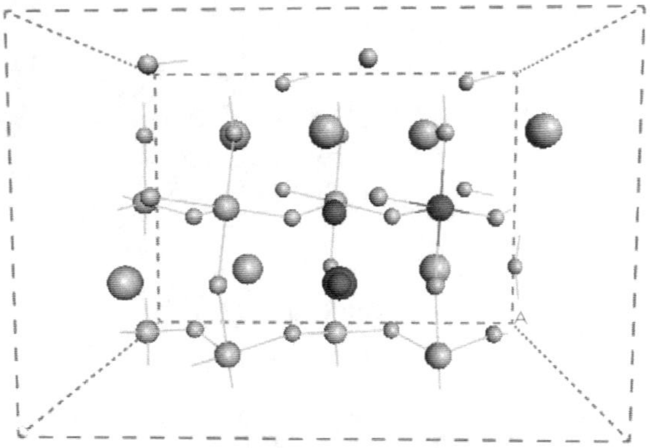

Figure 15 Model with Dopant (red) sites (Top View)

From our calculations, Ti @ Hollow site and Ti @ Na site are the preferred sites for the dopant in this system. Ti @ Hollow site has a cohesive energy of +5.47 eV and Ti @ Na site has a cohesive energy value of +4.60 eV (Table 5). Furthermore, Ti @ Mg shows a more favorable cohesive energy value than Ti @ Top sites with values of +4.23 eV and +3.22 eV, respectively. Simulation results also showed that it is more favorable to remove a hydrogen atom from a doped model than from the pure model, with values of +3.05 eV and –3.46 eV, respectively. Thus, for subsequent calculations only the most favorable models will be tested.

Site	**System**	$\mathbf{E_{coh}}$ **(eV)**	$\mathbf{\Delta E_{add}}$**(eV/atom)**
Ti@ Na site	$TiNa_7Mg_8H_{24}$	4.60	N/A
Ti @ Hollow Site	$TiNa_8Mg_8H_{24}$	5.47	0.13
Ti@ Top Na site	$TiNa_8Mg_8H_{24}$	3.22	0.08
Ti@ Top Hollow Site	$TiNa_8Mg_8H_{24}$	3.22	0.08
Ti@ Mg Site	$TiNa_8Mg_7H_{24}$	4.23	0.11
H removed from Ti@ Na site	$TiNa_7Mg_8H_{23}$	3.05	N/A
H removed from Pure model	$Na_8Mg_8H_{23}$	-3.46	N/A

Table 5. Surface Energy Results

Formation of Ti-Mg-H complexes at the surface as shown in Figure 17 and Figure 18 may improve the reaction mechanism, in order to observe this phenomena surface models are investigated using DFT-MD at different temperatures (423 K and 448 K).

Figure 17 Optimized Ti@ Na site surface model

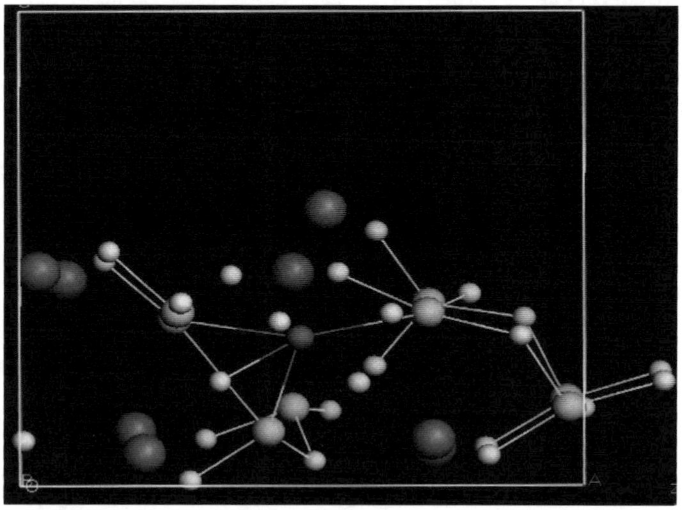

Figure 18 Optimized Ti@ Hollow site surface model

4. Dynamics

After completing the calculations regarding surface doped models and identifying the preferred sites of Ti dopants, the effect of high temperature on the Ti-Mg_x–H_x complexes is studied with ab initio molecular dynamics at 423K and 448 K in the canonical and micro-canonical regimes as mentioned earlier. First, via DFT coupled MD simulations, the effect of Ti impurity on surface models was studied to determine the stability of the complexes and to aid us in determining the role they play in hydrogen behavior at the surface. The calculations were spanned over an 8 ps time frame. The systems were considered to reach equilibrium once the drift in the constant of motion was calculated to be less than 8 meV/ atom. Also, dynamic equilibrium was analyzed through the use of radial distribution functions at the 1 ps interval. Electron density and electron density difference maps were generated to verify the distinguished effects of the Ti additive on the electronic structure of the system.

First, dynamic simulations were carried out for Ti@ Na site at 423 K for a total simulation time of 4 ps. At the end of the 4 ps the Ti-Mg_x–$H_x$was still present with a Ti-Mg_5 –H_7

formation. The calculated bond lengths as confirmed with radial distribution function plots were as follow: Ti-Mg 2.97 Å, Ti-H 1.87 Å. Then, the temperature is increased to 448 K to further study the temperature effect for a total simulation time of 8 ps. In a similar fashion, at the end of the 4 ps the $Ti\text{-}Mg_x\text{–}H_x$was still present with a $Ti\text{-}Mg_4\ \text{–}H_7$ formation. The calculated bond lengths as were as follow: Ti-Mg 3.10 Å, Ti-H 1.93Å.

Next, the electron density difference in the surface model due to the Ti substitution at the Na lattice site is investigated. The electron density difference illustrates how the charge density changes during a chemical reaction or on docking of a molecule to a surface.

The electron density difference with respect to atoms can be represented in the following manner:

$\Delta\rho = \rho Ti@\ Na\ site\text{-}\ \Sigma(\rho i)$ (6)

Where the subscript *i* runs over all atoms. This equation shows the changes in the electron distribution due to the formation of all the bonds. Furthermore, Figure 19 shows the formation of chemical bonds between Ti and the surrounding atoms by delocalization of the atomic electron density. In these plots, a loss of electron is indicated in blue, while regions rich in electrons are indicated in red. White colors indicate regions with small changes in the electron density. Based on this plots, it is clearly shown that Ti lost electrons; thus, confirming that electron affinity and ionization energy are important parameters in hydrogen storage research.

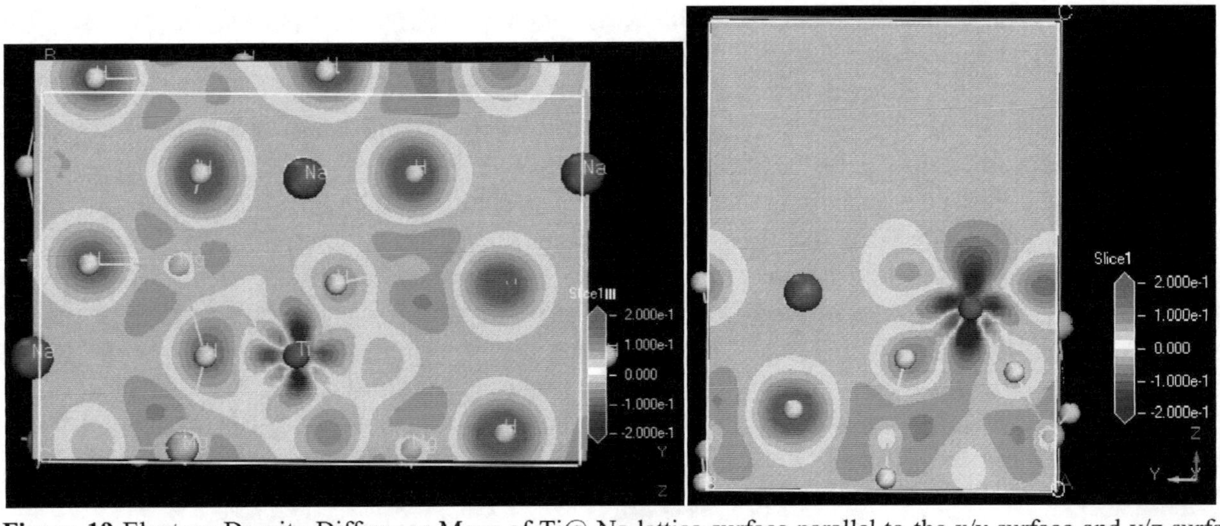

Figure 19 Electron Density Difference Maps of Ti@ Na lattice surface parallel to the x/y surface and y/z surface. Where red ball is Ti, green ball is Mg, purple ball is Na and white ball is H.

Then, following the same procedure DFT-MD simulations were carried out for Ti@ hollow site at 423 K with a canonical ensemble equilibrium run of 2 ps followed by a micro-canonical ensemble production run of 2 ps for a total simulation time of 4 ps. At the end of the 4 ps the $Ti\text{-}Mg_x\text{–}H_x$was still present with a $Ti\text{-}Mg_6\ \text{–}H_7$ formation. The calculated bond lengths were as follow: Ti-Mg 2.92 Å, Ti-H 1.91 Å. Then, the temperature is increased to 448 K to further study the temperature effect for a total simulation time of 8 ps. Similarly, at the end of the 4 ps the $Ti\text{-}Mg_x\text{–}H_x$ was still present with a $Ti\text{-}Mg_5\ \text{–}H_7$ formation. The calculated bond lengths were as follow: Ti-Mg 2.94 Å, Ti-H 1.87Å.

Comparing the dynamic results of both models, it seems that the complex formed at Ti @ Hollow site shows increased association as predicted by cohesive energy values. However, it is apparently necessary to increase the number of accessible neighbors to the dopant to accurately elucidate the complex formation and exact role it plays in absorption/desorption mechanisms in CMHs.

In light of the MD simulation results, bulk/surface geometry optimization results several observations are reached. To summarize, Ti-Mg_x–$H_x$are observed after geometry optimization of doped surface models, DFT-MD show the existence of these complexes even at high temperatures, cohesive energy calculations at the surface models show that Ti@Na lattice site and Ti@ hollow site are the preferred Ti locations in this system, and the mechanism in which Ti @ hollow site ejects Na ion from lattice site requires further study with an extended slab model.This phenomenon will be further studied to determine the role that this plays in creation of nucleation sites or possible Na or Mg rich sites that ultimately leads to saturation and desorption of H from the system.

4. Conclusion and Future Work

The cohesive energies of the 3*d* block elements in the bulk follow a strong correlation with the cohesive energies of their metallic elements counterpart. For the 3*d* block elements, Ti, V, Sc, Co, Ni and Cu in the decreasing order, are the most favorable elements with respect to the cohesive energy of the pure system. On the other hand, Zn and Mn doped models are unstable in comparison to the pure model. Also, 4*d* block elements are too stable in comparison to the pure model to be used as viable dopants for CMHs. Furthermore, 4*d* block elements show a positive correlation with crystalline elements ionization energy, suggesting that dopants replacing Na ions should have high ionization energies to further weaken the Mg-H covalent bond, and ultimately reducing the thermodynamic stability of the system.

From the cohesive energy calculation on surface models, we conclude that dopants at the hollow and Na lattice sites are the most favorable sites for dopant locations in the $NaMgH_3$ system. Thus, future studies will have more emphasis on those models. Ti @ Na lattice site shows a possible Ti-Mg_4H_6 complex formation upon completion of geometry optimization. In the same manner, Ti @ Hollow site shows a possible $TiMg_5H_7$ complex formation; thus, showing an increased association of the bond lengths when compared to the Ti@ Na lattice site model. The increased association of the dopants in the bulk and surface model leads to elongation of Mg-H bond lengths, thus; possible weakening that facilitates hydrogen desorption is feasible.

It is worth mentioning that the Ti placed @ the hollow site of $NaMgH_3$ ejects a Na ion from its lattice site. This phenomenon will be further investigated to determine the role that it plays in desorption of H from the system. Thus, two destabilization methods can be rationalize based on bulk and surface models: 1) additives with a mix of low ionization energy (high ionization energy for 4*d* transition metals), low melting point and low crystalline cohesive energy should be considered to locally destabilize the CMHs by increasing the association with Mg-H complexes which ultimately results in regions rich in H and Mg atoms that initiate nucleation/growth sites that facilitate H kinetics and 2) additives with high ionization energy or low electron-negativity affinity should be considered to lower the overall thermal stability of the system which may lead to a decrease in H desorption temperature.

Furthermore, as a second step in the research program, the role of 3*d* transition metals such as Ti and Mn in the improved dehydrogenation kinetics of $NaMgH_3$ needs to be understood. Reactants and products will be optimized using DFT codes within the GGA approximation to determine the reaction barrier height for the first reaction. Reaction pathways will be determined through transition state search using LST/QST methods.

Acknowledgements

Financial support provided by NSF through the "Louisiana EPSCoR Research Infrastructure Improvement" grant (EPS 1003897) is acknowledged.

References

1) Hirst, E; Moyers, J.C.; "Efficiency of Energy Use in the United States." Science. 1973. 179. 1299-1304.

2) Salameh, M.G; "Quest for Middle East oil: the US versus the Asia-Pacific region." Energy Policy. 2003.31.1085-1091.

3) Kone, A.C; Buke, T.; "Forecasting of CO2 emissions from fuel combustion using trend analysis." Renewable and Sustainable Energy Reviews. 2010. 14. 2906-2915.

4) Grochala, W; Edwards, P. P.; "Thermal Decomposition of the Non-Interstitial Hydrides for the Storage and Production of Hydrogen." Chem. Rev. 2004. 104. 1283-1315.

5) Alapati, S. V; Johnson, K. J.; Scholl, D. S. "Identification of Destabilized Metal Hydrides for Hydrogen Storage Using First Principles Calculations." J. Phys. Chem. B. 2006. 110. 8769-8776.

6) Choudhury, P.; Bhetanabotla, V.R.; Stefanakos, E.; "Manganese Borohydride as a Hydrogen –Storage Candidate: First-Principles Crystal Structure and Thermodynamic Properties." J. Phys. Chem. C. 2009. 113. 13416-13424.

7) Wu, H; Zhou, W.; Udovic, T.J.; Rush, J.J.;Yildirim, T.; " Crystal Chemistry of Perovskite-Type Hydride NaMgH3 : Implications for Hydrogen Storage." Chem. Mater. 2008. 20. 2335-2342.

8) Bogdanovic, B; Schwickardi, M.; "Ti-doped alkali metal aluminum hydrides as potential novel reversible hydrogen storage materials." J. Alloys. Compd. 1997.253. 1-9.

9) Ozolins, V; Majzoub, E.H.; Wolverton, C.; "First-Principles Prediction of Thermodynamically Reversible Hydrogen Storage Reactions in the Li-Mg-Ca-B-H System." J. Am. Chem. Soc. 2009. 131. 230-237.

10) Lovvik, O.M; Swang, O.; "Crystal structures and electronic structures of alkali aluminohexahydrides from density functional calculations."J. Alloys. Compd. 2005. 404. 757-761.

11) Aguayo, A; Singh, D.J.; "Electronic Structure of the complex hydride NaAlH4."Physical Review B. 2004. 69.103-108.

12) Alapati, S. V; Johnson, K. J.; Scholl, D. S. "Identification of Destabilized Metal Hydrides for Hydrogen Storage Using First Principles Calculations." J . Phys. Chem. B. 2006. 110. 8769-8776.

13) Lovvik, O.M; Opalka, S. M.; "Density Functional calculations of Ti-enhanced NaAlH4." Physical Review B. 2005. 71. 1-9.

14) Yvon, K; Bertheville, B.; "Magnesium based ternary metal hydrides containing alkali and alkaline-earth metals." J. Alloys Compd. 2006. 38. 1-8.

15) Bouhadda, Y.; Boudouma, Y.; Fennineche, N.; Bentabet, A.; " Ab initio calculations study of the electronic, optical and thermodynamic properties of $NaMgH_3$, for hydrogen storage." J. Phys. Chem. Solids. 2010. 71. 1264-1268.

16) Klaveness, A.; Swang, O.; Fjellvag, H.; "Formation enthalpies of $NaMgH_3$ and $KMgH_3$: A computational study." Europhys. Lett. 2006. 76. 285-290.

17) Shane, D.T.; Corey, R.L.; Bowman, R.C.; Zidan, R.; Stowe, A. C.; Hwang, S.; Kim, C.; Conradi, M. S.; " NMR Studies of the Hydrogen Storage Compound $NaMgH_3$." J. Phys. Chem. C. 2009.113. 18414-18419.

18) Hao, S.; Sholl, D.S.; "Role of Schottky Defects in Hydrogen and Metal Diffusion in NaH, MgH_2, and $NaMgH_3$." J. Phys. Chem. Lett. 2010. 1. 2968-2973.

19) Komiya, K.; Morisaku, N.; Rong, R.; Takahashi, Y.; Shinzato, Y.; Yukawa, H.; Morinaga, M.; " Electronic Structure Calculations of Perovskite-type Hydrides for Hydrogen Storage." J. Alloys. Compd.2008. 453. 157-160.

20) Accelrys, "Materials Studio®," 4.0 ed. San Diego: Accelrys Software, Inc., 2006.

21) M. D. Segall, P. J. D. Lindan, M. J. Probert, C. J. Pickard, P. J. Hasnip, S. J. Clark, M. C. Payne;"First-principles simulation: ideas, illustrations and the CASTEP code." Journal of Physics: Condensed Matter. 2002.14.2717-2743

(456e) A Computational Study of Steam Hydrolysis of NaBH4

Ping Li, University of Pittsburgh, Pittsburgh, PA, Karl Johnson, Chemical and Petroleum Engineering, University of Pittsburgh, Pittsburgh, PA and Michael A. Matthews, University of South Carolina, Columbia, SC

Complex hydrides are promising hydrogen storage materials and have received significant attention due to their high hydrogen storage capacity. $NaBH_4$ has high gravimetric (10.7 wt%) and large volumetric (115 kg of H_2 m^{-3}) hydrogen content. The hydrolysis reaction of $NaBH_4$ releases hydrogen with both fast kinetics and high extent of reaction under technical conditions by using steam deliquescence of $NaBH_4$. This catalyst-free reaction has many advantages over traditional catalytic aqueous phase hydrolysis. However, the reaction mechanism and kinetic barriers associated with steam hydrolysis have not yet been elucidated. We are developing a fundamental molecular level understanding of the steam hydrolysis reaction of $NaBH_4$ using a combination of experiments and first-principles density functional theory (DFT) calculations. We have applied finite temperature *ab initio* molecular dynamics (AIMD) to investigate the reaction mechanism. AIMD provides clues to elementary reaction steps. We have studied the effect of $NaBH_4$ concentration on the initial reaction rate, and found that high $NaBH_4$ concentrations lead to high initial reaction rates. The reason for this is appears to be due to a cooperative interaction between two BH_4^- groups. We have computed the free energy profiles of possible reaction pathways observed in AIMD using a cluster approach with an implicit solvent model and explicit water molecules. Explicit water is critical because proton transfer is a key step in the hydrolysis reaction.

See more of this Session: Hydrogen Storage Systems and Materials: Modeling, Simulation, and Optimization:
See more of this Group/Topical: International Congress on Energy 2011

(556a) History of the IFPSC and Description of the 7th Challenge

Jonathan D. Moore, Core R&D - Materials Science, The Dow Chemical Company, Midland, MI

The Industrial Fluid Properties Simulation Challenge is an open competition organized by the Computational Molecular Science and Engineering Forum (CoMSEF) of the American Institute of Chemical Engineers (AIChE), the American Chemical Society (ACS), Army Research Lab, Case Scientific, National Institute of Standards and Technology, DuPont, ExxonMobil, The Dow Chemical Company, 3M, and Wittenberg University. The goals of the competition are to drive improvements in the practice of molecular modeling, formalize methods for the evaluation and validation of simulation results with experimental data, and ensure relevance of simulation activities to industrial requirements.

This talk will provide an overview of the history of the IFPSC and a description of the current challenge problem.

See more of this Session: The Industrial Fluid Properties Simulation Challenge
See more of this Group/Topical: Computational Molecular Science and Engineering Forum

(556b) Presentations by the Challenge Participants

Jonathan D. Moore, Core R&D - Materials Science, The Dow Chemical Company, Midland, MI

Presentations by the challenge participants describing the methods they used to solve the problem.

See more of this Session: The Industrial Fluid Properties Simulation Challenge
See more of this Group/Topical: Computational Molecular Science and Engineering Forum

(556c) Announcement of the 7th IFPSC Champions

Jonathan D. Moore, Core R&D - Materials Science, The Dow Chemical Company, Midland, MI

The results of the challenge will be announced.

See more of this Session: The Industrial Fluid Properties Simulation Challenge
See more of this Group/Topical: Computational Molecular Science and Engineering Forum

(563a) Response of the UO2 (111) Surface to Varying Oxygen Partial Pressure

Gloria A. E. Oxford and Anne M. Chaka, Physical Measurement Laboratory, National Institute of Standards and Technology, Gaithersburg, MD

UO_2corrosion is a significant concern for the nuclear power industry as well asfor the weapons industry. Oxidation ofUO_2 can lead to significant expansion of the material, which inturn, can cause cracking and spallation.While limited studies of oxidation and reactivity of bulk $UO_2$and the UO_2 (111) surface have been performed using periodic densityfunctional theory (DFT), surface and subsurface oxidation as a function oftemperature and pressure has not been explored.We present periodic DFT calculations of the clean and oxidized UO_2(111) surfaces with and without subsurface oxidation. Generally, the generalized gradientapproximation (GGA) formulation used in this work gives highly exothermicoxidation energies, which is consistent with experimental data and in contrastto previous GGA+*U* studies,demonstrating that GGA describes the energetics of this system better. Using *abinitio* thermodynamics, the temperatures and pressures needed to oxidize thesurface and perform initial subsurface oxidation have been calculated. We also analyze the uranium magnetic andelectronic structure as well as local geometry in response to oxidation.

See more of this Session: Advances In Numerical Simulations Bridging Chemical and Nuclear Engineering Phenomena or Processes
See more of this Group/Topical: Nuclear Engineering Division

(563b) Continuation of Experts In Discrete Element Modeling (EDEM) Validation Study for Packed Bed Structural Properties

Vaibhav Khane, Chemical Engineering, Missouri University of Science & Technology, Rolla, MO, G E Mueller, Nuclear Engineering Dept., Missouri University of Science & Technology, Rolla, MO and Muthanna H. Al-Dahhan, Chemical Engineering Dept., Missouri University of Science & Technology, Rolla, MO

Experts in Discrete Element Modeling (EDEM)[1] is a discrete element method (DEM) based modeling software program, capable of providing a wealth of information about granular solid flows occurring in many industrial applications. The DEM is a numerical approach used to compute the motion of large number of particles[2]. A computational study involving the application of EDEM to a pebble bed nuclear reactor (PBR) application is being performed at Missouri S&T. Prior to full involvement with EDEM based numerical simulations, an effort has been made to validate the packing algorithm used in EDEM; particularly for cylindrical packed bed applications. This is essential, as the local flow and transport properties are closely coupled with the structural characteristics of a packed bed which in turn depends upon the packing of the particles inside the container. Also, the packing of particles inside the container is the first and main step in EDEM based analysis. Packing algorithms available with commercial codes such as EDEM are used as a 'Black-Box' and are without any detailed validation exercise. In most cases, average porosity results are compared to benchmark the numerical packing results with available experimental results, which is not sufficient. Detailed technical information about the packing algorithm used in EDEM software is not available in open literature and hence there is a need to perform validation studies of the packing algorithm used in EDEM. The packing algorithm used in EDEM demands accurate input of elasticity (material properties) and frictional (interaction properties) parameters which are not readily available in the literature for materials of interest. The radial porosity profile, along with average porosity values, will be used as a means for this validation study. This study also involves a parametric sensitivity study of material and interaction properties on structural properties of cylindrical packed beds. This sensitivity study will highlight important material and interaction properties which need to be provided as an input from a reliable simulation point of view. Also, a DEM based simulation study will optimize the values of material and interaction properties by mimicking an experimentally obtained radial porosity variation profile to a larger extent. Hence, obtained results will provide a new and better understanding about the packing algorithm used in EDEM as well as recommendations to simulate realistic packed bed structures using EDEM. The cylindrical packed bed has been chosen as a model for this validation study. There is much benchmark data available for packed bed structural properties from previous experimental measurements[3] and from high-fidelity numerical packing algorithms4 for a wide range of aspect ratios. Also, the slow and dense granular flow encountered in a PBR can be approximated by static packed beds[5]. The various structural properties of cylindrical packed beds which can be used for this validation study are mean porosity, the distribution of particle centers, and the axially averaged radial porosity profile which exhibits a typical shape of damped oscillation. The radial distribution of particle centers serves as an input to Computational Fluid Dynamics (CFD) analysis of packed beds in order to simulate them more accurately[6]. Radial and axial porosity profiles and associated solid phase distributions are required as an input to hydrodynamic and thermal models of packed beds[7]. It has already been demonstrated in our previous work that the friction between wall and particle plays an important role in the structural characterization of packed beds[8] and neglecting this friction results in simulations of tightly packed bed which are unreal.

DESCRIPTION OF THE ACTUAL WORK

A series of EDEM simulations is being carried out to perform a parametric sensitivity study to check the effect of static friction between particles. The Hertz-Mindlin (with no-slip) contact model[9] has been used to model particle-particle and container wall-particle contact forces. Higher values of static friction coefficient between particles damps out the porosity variation profile quickly while moving towards the center. Neglecting friction between particles under predicts porosity variation near the container wall, resulting in tightly packed beds which are evident from mean porosity values. The effect of other material and interaction properties on the structural properties of packed beds needs to be investigated further. Such study might help in the development of a mechanistic model for the radial porosity variation of packed beds which will require inputs of material properties besides the aspect ratio.

Acknowledgements

The authors acknowledge the financial support provided by Department of Energy (DOE) Nuclear Energy Research Initiative (NERI) project (**NERI-08-043**).

REFERENCES

[1] EDEM™ User Manual, DEM Solutions (USA) Inc., Edinburgh, Scotland (2010).

[2] P.A. CUNDALL, O.D.L.STRACK, A discrete numerical model for granular assemblies. Geotechnique, 29:47–65, 1979.

[3] G. MUELLER, Radial void fraction distributions in randomly packed fixed beds of uniformly sized spheres in cylindrical containers, Powder technology,72,pp.269-275, 1992.

[4] G. MUELLER, Numerically packing spheres in cylinders, Powder technology, 159, pp.105-110, 2005.

[5] C. du TOIT, The Numerical Determination of the Variation on the Porosity of the Pebble-Bed Core, Proceedings of the conference on High-Temperature Reactors, Petten, The Netherlands, European Nuclear Society, 2002.

[6] N.J. MARIANI, W.I. SALVAT, A. CAMPESI, , G.F. BARRETO, O.M. MARTINEZ, Evaluation of structural properties of cylindrical packed beds using numerical simulations and tomographic experiments, International Journal of Chemical Reactor Engineering 7, Art. No. A82.,2009.

[7]A.J. SEDERMAN, P.ALEXANDER and L.F. GLADDEN, Structure of packed beds probed by Magnetic Resonance Imaging, Powder Technology, 117, pp. 255-269, 2001.

[8] V. KHANE, G.E. MUELLER, and M. AL-DAHHAN, Experts in Discrete Element Modeling (EDEM) Validation for Packed Bed Structural Properties, AIChE-2010 Annual Meeting, Salt Lake City, Utah, November 7-12, 2010.

[9] Y. TSUJI, T. TANAKA and T. ISHIDA, Lagrangian numerical simulation of plug flow of cohesion-less particles in a horizontal pipe, Powder Technology, 71 239-250, 1992.

[#] A.G. POLSON, Distinct element method analyses of fuel spheres in the PBMR core using RFC3D, Proceedings of the 2004 International Congress on Advances in Nuclear Power Plants, ICAPP'04, June 13-17, 2004.

See more of this Session: Advances In Numerical Simulations Bridging Chemical and Nuclear Engineering Phenomena or Processes
See more of this Group/Topical: Nuclear Engineering Division

(563c) Structure of Rigid Hard-Ring Fluid

Mariam Nouri Dariani and Marc Robert, Chemical and Biomolecular Engineering, Rice University, Houston, TX

The structure of hard ring fluids is studied in two and three dimensions, using Monte Carlo computer simulations in the canonical ensemble. These fluids are composed of molecules consisting of hard-spheres linked together to form planar rigid rings. For rings of various sizes and for a wide range of densities, results are reported for the pair distribution function of the ring centers and for the pair distribution of the ring orientations. In a dense monolayer, where the rings are completely confined to a plane, a "shoulder", precursor of the freezing transition, is observed in the second peak of the pair distribution function of the ring centers, as previously seen in the simple hard-sphere fluid. In two dimensions, where the ring centers are confined to a plane but the rings themselves are free to rotate on all axes, nematic and tetratic ordering is observed at sufficiently high densities. Results are also presented for three dimensions.

See more of this Session: Advances In Numerical Simulations Bridging Chemical and Nuclear Engineering Phenomena or Processes
See more of this Group/Topical: Nuclear Engineering Division

(563d) Adsorption of Hard Spheres: Development of a Re-Summed Bridge Function Theory for the Structures and Work of Insertion At Moderate to High Densities ($\rho 0$ = 0. 5745 ~ 0.9135)

Lloyd L. Lee, Department of Chemical and Materials Engineering, The California State University Pomona, Pomona, CA and Giuseppe Pellicane, School of Physics, University of Kwazulu-Natal, 3209 Pietermaritzburg, South Africa

We propose a star-function bases density functional theory* for accurate determination of the density profiles in the hard sphere/hard wall adsorptive system. We transform the Euler-Lagrange equation into an equivalent bridge-function based equation. The bridge function is analyzed in terms of a functional Taylor expansion and the terms in the infinite expansion are re-summed into a closed expression** for calculation. This new closure is tested on the simple system of hard spheres. Results for bulk densities varying from ρ_0 = 0. 5745 ~ 0.9135 are compared with Monte Carlo data and very close agreement is obtained. The exact sum rule of wall density is incorporated into the closure. We also characterize the insertion work function *W(z)* (work required to insert a test particle into the fluid at position *z* away from the wall) as the density varies. Comparison is made with other theories on the hard sphere fluids.

*.. Lee L. L., J. Chem. Eng. Data, 2010, 55, 1897.

.. Lee L. L., Pellicane G., Chapman W.G., J. Supercritical Fluids 2010, **55, 524.

See more of this Session: Advances In Numerical Simulations Bridging Chemical and Nuclear Engineering Phenomena or Processes
See more of this Group/Topical: Nuclear Engineering Division

(563e) Force Field Parameterization and Monte Carlo Simulation of Alkoxy-Ethanol Surfactant

Sunghyun Jang, Junhyuk Lim and Hwayong Kim, Department of Chemical and Biological Engineering, Seoul National University, Seoul, South Korea

Alkoxy-ethanolis highly associating molecule which contains ether (-O-) and hydroxyl (-OH)group in the same molecule. It has both inter- and intra-molecular hydrogenbonding, thus the complex interaction makes it difficult to estimate thethermophysical properties of its fluid mixtures. When simulating the pure alkoxy-ethanolor its mixture system, the positive charge of hydroxyl hydrogen plays animportant role because it could form partial hydrogen bonding with both theproton accepters.

The TraPPE and NERDforce filed is the two most widely used force fields which provide the robustmolecular simulation framework for various thermophysical properties'estimation, especially phase equilibrium study. These two force fields use theunited atom based approach to represent the CH, CH2, CH3 groups which enables thetransferability of their force field parameters to a wide range of molecularspecies. With this simulation methodology, coexistence properties can bepredicted with relatively high levels of accuracy. Basically these two forcefields have very similar energy expression each other, but they don't considerthe hydrogen bonding term explicitly in its energy expression. Therefore theeffect of hydrogen bonding should be investigated for some specific system suchas alkoxy-ethanol and its mixture system.

On this unitedatom base, the force field parameters of alkoxy-ethanol have been optimized andthe molecular structure and coexisting properties of alkoxy-ethanol and itsmixture has been studied. The results are compared to simulation results usingNERD and TraPPE parameters.

See more of this Session: Advances In Numerical Simulations Bridging Chemical and Nuclear Engineering Phenomena or Processes
See more of this Group/Topical: Nuclear Engineering Division

(580a) Transforming Pharmaceutical Development and Manufacturing with Molecular Computations

Bernhardt L. Trout, Massachusetts Institute of Technology, Cambridge, MA

A great amount of resources have gone into molecular computational approaches for the rational discovery of new pharmaceutical compounds with marked success. However, little emphasis has been placed on molecular computational approaches for the rational design of formulations (development) and of pharmaceutical processes (manufacturing). This is despite that fact that total costs of these exceed that of discovery by a large fraction and industry-wide are over $150 billion per year.

We discuss two directions in which we have applied molecular computations to play a significant role in pharmaceutical development and manufacturing. One is in the area of stabilization of therapeutic antibodies. These molecules are inherently unstable with respect to aggregation, and we have developed new algorithms to predict these instabilities and identify sites responsible for them that can be mutated out to lead to enhanced stability. The second area is in small molecule pharmaceutical crystallization. Crystallization is a major part of both development and manufacturing, but it is currently more an art than a science. We present new algorithms and their application to understand the mechanisms of nucleation and crystallization and apply them to model pharmaceutical systems and therein develop better products and manufacturing processes.

See more of this Session: Computational Molecular Science and Engineering Plenary Session
See more of this Group/Topical: Computational Molecular Science and Engineering Forum

(580b) Elucidating the Shape Selectivity for Polycyclic Aromatic Hydrocarbons In Reversed-Phase Liquid Chromatography

J. Ilja Siepmann, Depts. of Chemistry and of Chemical Engineering & Materials Science, University of Minnesota, Minneapolis, MN, Jake L. Rafferty, Department of Chemistry, University of Minnesota, Minneapolis, MN and Mark R. Schure, Engineering Sciences Department, The Dow Chemical Company, Spring House, PA

Reversed-phase liquid chromatography (RPLC) is the foremost technique for the separation of analytes that have very similar chemical functionalities, but differ only in their molecular shape. This ability is crucial in the analysis of various mixtures with environmental and biological importance including polycyclic aromatic hydrocarbons (PAHs) and steroids. A large amount of effort has been devoted to studying this phenomenon experimentally, but a detailed molecular-level description remains lacking. To better understand the mechanism of shape selectivity in RPLC, particle-based simulations using efficient Monte Carlo sampling algorithms and accurate force fields were carried out. The retention of aromatic hydrocarbons ranging in size from benzene to 6-ring PAHs was examined for octadcylsilane RPLC systems with various surface coverages. The simulations yield precise and accurate separation factors for these complex analytes. Structural analysis of the simulation trajectories shows that the stationary phase acts as a very heterogeneous environment where analytes with different shapes prefer different spatial regions with specific local bonding environments of the ODS chains. However, these favorable retentive regions cannot be described as pre-existing cavities. Rather, the chain conformation in these local stationary-phase regions adapts to accommodate the analytes.

See more of this Session: Computational Molecular Science and Engineering Plenary Session
See more of this Group/Topical: Computational Molecular Science and Engineering Forum

(580c) Integrated Theory and Simulation Approach for Studying Polymer Functionalized Nanoparticles In Polymer Nanocomposites

Arthi Jayaraman[1], Nitish Nair[2] and Nathaniel Wentzel[1], (1)Chemical and Biological Engineering, University of Colorado at Boulder, Boulder, CO, (2)Chemical and Biological Engineering, University of Colorado, Boulder, CO

In the area of polymer nanocomposites there is significant interest in creating materials with controlled spatial arrangement of nanoparticles in a polymer matrix. By functionalizing or grafting polymers on to nanoparticle surfaces one can tailor the inter-particle interactions and control the assembly/dispersion of the particles in the polymer matrix. Systematically tuning the various molecular-level parameters, such as the graft and matrix polymers' composition, chemistry, molecular weight and grafting in order to understand their effect on the phase behavior of the functionalized nanoparticles involves spanning a large parameter space, a task best suited for theory and simulations. We have developed an integrated approach involving self-consistent Polymer Reference Interaction Site Model (PRISM) theory- Monte Carlo simulations to study polymer functionalized nanoparticles in an explicit polymer matrix. [1] The main advantage of this self-consistent PRISM-MC approach is that a) it is faster than pure MC simulations, especially in a system as computationally intensive as polymer grafted particles embedded in a dense polymeric solution, and b) it avoids some of the approximations of pure PRISM theory. In this talk I will present briefly the details of the method, and its use for studying copolymer functionalized nanoparticles [1] and homopolymer functionalized particles with polydispersity in graft length [2] embedded in polymer matrix.

1. N. Nair and A. Jayaraman ,'Self-Consistent PRISM Theory-Monte Carlo Simulation Studies of Potential of Mean Force between Functionalized Nanoparticles in a Homopolymer Matrix' *Macromolecules* 43 (19), pp 8251–8263 (2010)
2. N. Nair, N. Wentzel and A. Jayaraman ,'Effect of Bidispersity in Grafted Chain Length on Grafted Chain Conformations and Potential of Mean Force between Polymer Grafted Nanoparticles in a Homopolymer Matrix' *J. Chem Phys* in press (2011)

See more of this Session: Computational Molecular Science and Engineering Plenary Session
See more of this Group/Topical: Computational Molecular Science and Engineering Forum

(580d) Algorithms for Partition Function Calculation Applied to Self-Assembly of Patchy Particles

Eric Jankowski, Chemical Engineering, University of Michigan, Ann Arbor, MI and Sharon C. Glotzer, Department of Chemical Engineering, University of Michigan, Ann Arbor, MI

The direct calculation of partition functions can be useful in understanding the thermodynamics of self assembly. We present algorithms to efficiently calculate both the energies and entropies of clusters of building blocks. Our methods rely on shape-matching techniques [1] to distinguish clusters and numerical integration of vibrational and rotational entropic contributions of each cluster to the partition function. The assembly pathways we obtain by constructing a network of connected ensembles are then used to inform building block design.

[1] Keys AS and Iacovella CR and Glotzer SC, Annual Reviews of Condensed Matter Physics, 2 (2011)

See more of this Session: Computational Molecular Science and Engineering Plenary Session
See more of this Group/Topical: Computational Molecular Science and Engineering Forum

(580e) Methane Hydrate Nucleation: Choreography and Cadence From Molecular Simulations

Matthew R. Walsh[1], Carolyn A. Koh[1], E. Dendy Sloan[2], Amadeu K. Sum[1], David T. Wu[3] and Gregg T. Beckham[4], (1)Chemical Engineering Department, Colorado School Of Mines, Golden, CO, (2) Colorado School of Mines, (3)Chemical Engineering and Chemistry Departments, Colorado School Of Mines, Golden, CO, (4)National Bioenergy Center, National Renewable Energy Laboratory, Golden, CO

Methane hydrates are ice-like solids that form at low temperatures and high pressures, with methane molecules encaged in a hydrogen-bonded water lattice. They have been of long-standing interest to the oil and gas industry because of their tendency to plug pipelines, as a resource in the vast seafloor hydrate sediment deposits, and recently in the containment of the Gulf oil spill. The control of hydrate formation in these and other situations can be guided by moleular level knowledge of the hydrate nucleation process. The nucleation of hydrates is a rare event that occurs in nanoseconds at a random location on a nanometer length scale, and thus a molecular understanding of the process has been difficult to obtain experimentally or by simulation. By employing large-scale microsecond long molecular dynamics simulations, we captured information about the mechanism of spontaneous hydrate nucleation, revealing an intricate cooperative dance between methane and water molecules [1].

Cooperative organization is observed to lead to methane adsorption onto planar faces of water and the fluctuating formation and dissociation of early hydrate cages. The early cages are mostly face-sharing partial small cages, favoring structure II, however larger cages subsequently appear due to steric constraints and thermodynamic preference for the structure I phase. The resulting structure after nucleation and growth is a combination of the two dominant types of hydrate crystals (structure I and structure II), which are linked by uncommon 51263 cages facilitating structure coexistence without an energetically unfavorable interface. Knowledge of the phase boundary, necessary for determining the subcooling driving force, was calculated by thermodynamic integration for the molecular potentials used [2]. Analysis of trajectories under different driving forces and geometries allowed an estimate of the nucleation rate as a function of methane concentration.

References

1. Walsh M. R., Koh C. A., Sloan E. D., Sum A. K. and Wu D. T. Microsecond Simulations of Spontaneous Methane Hydrate Nucleation and Growth, *Science*, **326**, 1095-1098, 2009.
2. Jensen L., Thomsen K., von Solms N., Wierzchowski S., Walsh M. R., Koh C. A., Sloan E. D., Wu D. T. and Sum A. K. Calculation of Liquid Water-Hydrate-Methane Vapor Phase Equilibria from Molecular Simulations, *J. Phys. Chem. B*, **114**, 5775-5782, 2010.

(580f) Modeling Polymer Self-Assembly for Sub-Lithographic Patterning

Jed W. Pitera[1], Joy Cheng[1], Dan Sanders[1], Greg Doerk[1], Melia Tijo[1] and Su-Mi Hur[2], (1)Science & Technology, IBM Almaden Research Center, San Jose, CA, (2)Chemical Engineering, University of California Santa Barbara, Santa Barbara, CA

The semiconductor industry is rapidly approaching the limits of 193 nm optical immersion lithography with on chip features of 22 nm planned shortly. The high cost and complexity of extreme ultraviolet lithography is driving a search for alternative patterning technologies to create structures smaller than 14 nm. One promising alternative is a combination of top-down lithography with the directed self-assembly of block copolymers or blends. As part of an integrated experimental and computational research effort at IBM, we make use of parallel supercomputers, polymer field theory and Monte Carlo calculations to understand and guide the development of new DSA processes, polymers, and patterning designs.

See more of this Session: Computational Molecular Science and Engineering Plenary Session
See more of this Group/Topical: Computational Molecular Science and Engineering Forum

(661a) The Effects of Salt On Lower Critical Solution Temperature Transition of Thermo-Responsive Pnipam-Co-PEGMA Copolymer

Hongbo Du, Colorado State University, Ft Collins, CO, S. Ranil Wickramasinghe, Ralph E Martin Department of Chemical Engineering, University of Arkansas, Fayetteville, AR, Mathias Ulbricht, Lehrstuhl fr Technische Chemie II, Universitt Duisburg-Essen, Essen, Germany and **Xianghong Qian**, Department of Chemical Engineering, University of Arkansas, Fayetteville, AR

Classical molecular dynamics (MD) simulations were carried out to investigate the lower critical solution temperature (LCST) transition of PNIPAM–co–PEGMA, a thermo-responsive copolymer of poly (N-isopropylacrylamide) (PNIPAM) and poly (ethylene glycol) methacrylate (PEGMA). The simulations were carried out in 1 M LiCl, NaCl, KCl, CsCl and NaI aqueous solutions below and above their respective experimental LCSTs. The copolymer consists of 38 monomer units of NIPAM with 2 PEGMA chains attached to the PNIAPM backbone. The PNIPAM–co–PEGMA was observed to go through the hydrophilic-hydrophobic conformational change for simulations carried out above its LCST. Cations were found to bind strongly and directly with amide O, and more strongly with the O atoms on PEGMA chains, whereas anions only exhibit weak interaction with the polymer. A unique caged stable metal-organic complex involving Na^+ coordinated by 6 O atoms from the copolymer in NaCl solution was observed after a hydrophobic folded structure of the copolymer was formed. Interestingly, the binding affinity between the cation and the copolymer appears to depend not only on the size and charge of the cation, but also on the anion present. The cation–polymer interaction is found to be modulated by the cation–anion interaction.

See more of this Session: Molecular Modeling and Simulation of Complex Molecules I
See more of this Group/Topical: Engineering Sciences and Fundamentals

(661b) Molecular Modeling of the Regulation of Proteins and the Treatment of Alzheimer's: Amyloid Beta and Protein Kinase C

Jacob Weidman and **Brian Anderson**, Chemical Engineering, West Virginia University, Morgantown, WV

Alzheimer's Disease is one of the most prevalent forms of dementia that develops in elderly people with symptoms typically appearing after the age of 60. The most major effect of the disease is loss of memory, which can leave a person unable to remember specific details or even how to perform daily tasks. The effects on the brain are likely caused by amyloid plaques and neurofibrillary tangles which cause the loss of connection between neurons in the brain. The formation of plaques and tangles occurs in people without Alzheimer's as they age, but the amount of formation is much less than in patients with the disease and does not lead to problems with memory skills. Build-up of plaques and tangles continues with time to the point where communication between brain cells is so badly damaged that symptoms are noticeable in patients.

While a cure for Alzheimer's Disease has not been found, various form of treatment have been developed to slow the effects on the brain. Some of the drugs used include cholinesterase, which acts as a chemical messenger in the brain for memory function, and memantine, which can regulate chemical messengers. Many other drugs are used to downplay symptoms of the disease though they do not slow the damage to nerve cells.

Protein kinase C (PKC) is a superfamily of enzymes which uses transduction of cellular signals after prolonged activation for a wide variety of cellular functions ranging from control of processes at the cellular level, such as differentiation and proliferation, up to higher-level processes including memory. Activation of PKC has been linked to learning processes in animals, supporting the idea of its role as an essential aspect in development and maintenance of memory. One of the roles of PKC-ε, a particular isoform of PKC, is the down-regulation of amyloid beta (Aβ), a peptide formed by sequential cleavages of the amyloid precursor protein (APP) (Nelson, 2009). Aβ can form an oligomer which forms tangles on dendrites in the brain. This could be a cause of reduced signal transmission across synapses, leading to damage of neural function, especially loss of memory.

This work presents a molecular modeling study of the interaction of a pharmaceutical PKC-ε activator, bryostatin-1, and PKC-ε in order to identify the optimum binding site and to analyze its binding affinity to the target protein. The bryostatin-1 compound has been tested as a PKC-ε activator and shown to both decrease memory loss and increase learning ability in experiments on rats (Hongpaisan, 2007). Bryostatin has also been tested as an antidepressant and a way to slow the growth of tumors (Sun, 2006).

Hongpaisan, J., and D. L. Alkon. "A Structural Basis for Enhancement of Long-term Associative Memory in Single Dendritic Spines Regulated by PKC." *Proceedings of the National Academy of Sciences* 104.49 (2007): 19571-9576.

Nelson, Thomas J., Changhai Cui, Yuan Luo, and Daniel L. Alkon. "Reduction of Beta-Amyloid Levels by Novel PKC Epsilon Activators." *Journal of Biological Chemistry* 284.50 (2009): 34514-34521.

Sun, Miao-Kun, and Daniel L. Alkon. "Bryostatin-1: Pharmacology and Therapeutic Potential as a CNS Drug." *CNS Drug Reviews* 12.1 (2006): 1-8.

See more of this Session: Molecular Modeling and Simulation of Complex Molecules I
See more of this Group/Topical: Engineering Sciences and Fundamentals

(661c) Chemical and Mechanical Relaxations In Multicomponent Model Bitumens

Derek D. Li and **Michael L. Greenfield**, Dept. of Chemical Engineering, University of Rhode Island, Kingston, RI

The chemical, physical, and mechanical properties of multicomponent amorphous solids represent contributions from the various molecules in the system. A prime motivation is quantifying the mechanisms by which chemical changes invoke changes in mechanical properties, i.e. "chemo-mechanics". As an example, this work presents next-generation model systems for simulating asphalt/bitumen on the molecular level. Compared to prior work (Energy Fuels 2007, 21, 1712 and 2008, 22, 3363), the molecules included are of higher molecular weight and have been observed in geochemistry studies. Their structures, such as squalane and substituted hopanes, have biological origins. Model asphaltene molecules proposed by others have been altered to alleviate high energies that arose from poor architecture choices. Atomistic molecular dynamics simulations are employed to predict single molecule rotation and diffusion rates. Debye-Stokes-Einstein scaling is used to relate relaxation rates to temperature-dependent mechanical properties. Viscosity predictions show good agreement compared to representative measurements reported in the literature. Molecules across the employed size ranges show the same diffusion activation energy, indicating common collective barriers for translation.

See more of this Session: Molecular Modeling and Simulation of Complex Molecules I
See more of this Group/Topical: Engineering Sciences and Fundamentals

(661d) DNA Folding and Regulation: From Nucleosomes to Chromatin to Chromosomes

Gaurav Arya, Darren Yang, Balaji Iyer, Irina Dobrovolskaia and Martin Kenward, Department of Nanoengineering, University of California, San Diego, La Jolla, CA

In this talk, I will discuss modeling of chromatin at three different length scales, with the aim of enhancing our fundamental understanding of the organization, dynamics, and regulation of DNA within our cells. First, I will talk about high-resolution, all-atom molecular docking, molecular dynamics, and free energy studies of the H4 histone tail [1]. These studies provide novel insights into the structure of the tail bound to the acidic patch of neighboring nucleosomes and how the acetylation of lysine K16 on the histone tail could disrupt the binding interactions to trigger chromatin unfolding and subsequent transcriptional activation. Next, I will talk about an intermediate-resolution, mesoscale model [2] of chromatin that uncovers intriguing features like twist inversion and nucleosome flipping related to the propagation of torsional stresses across nucleosomes within the chromatin fiber [3]. Finally, I will talk about our recent efforts in developing low-resolution, lattice-animal models of chromosomes that represents chromosome loops at different length scales, leading to a self-similar fractal structure [4]. I will show how this model explains emerging results from genome-wide contact probability and spatial distance measurements from the HiC and FISH techniques, respectively.

[1] D. Yang and G. Arya, "Structure and binding of the H4 histone tail and the effects of lysine 16 acetylation," *Phys. Chem. Chem. Phys.* 13, 2911 (2011).

[2] S. Grigoryev, G. Arya, S. Correll, C. Woodcock, and T. Schlick, "Evidence for heteromorphic chromatin fibers from analysis of nucleosome interactions," *Proc. Natl. Acad. Sci.* 106, 13317 (2009).

[3] I. V. Dobrovolskaia, M. Kenward, and G. Arya, "Twist propagation in dinucleosome arrays," *Biophys. J.* 99, 3355 (2010).

[4] B. V. S. Iyer and G. Arya, "Lattice animal model of interphase chromosomes", submitted.

See more of this Session: Molecular Modeling and Simulation of Complex Molecules I
See more of this Group/Topical: Engineering Sciences and Fundamentals

(661e) Coarse-Grained Models of Structures and Dynamics of Nanoscale Ionic Materials

Bingbing Hong, Alexandros Chremos and Dr. Athanassios Z. Panagiotopoulos, Department of Chemical and Biological Engineering, Princeton University, Princeton, NJ

Nanoscale ionic materials (NIMs) provide design flexibility in terms of mechanical, optical and thermal properties relative to existing hybrid materials. Composed of ionically modified nanoparticles with large counter ions, NIMs can be engineered to be fluids at room temperature.

We use coarse-grained molecular dynamics over μ s time scales to simulate two NIMs models, tethered by linear counter ions (NIMs-L) and 3-arm star counter ions (NIMs-S), respectively, and one comparative uncharged model (NOHMs). We choose model parameters to match experimental systems. The pair correlation functions show interpenetration of attached-ion layers between cores and close core-counter ion contacts in NIMs-L. Due to steric hinderances and entropy loss, the corresponding feature peaks are greatly diminished or disappear in NIMs-S. Frequent intra- and inter-core hopping are observed, confirming the hypothesis from a recent experiment [ACS Nano 4, 3735-3742, 2010]. Inter-core hopping is also found to be a dominant factor for chain diffusions at time scales longer than ~10 ns. Finally the cores do not exhibit diffusive behavior until after ~ 0.3 μ s and the NIMs cores diffuse more slowly than the NOHMs due to the electrostatic interactions.

See more of this Session: Molecular Modeling and Simulation of Complex Molecules I
See more of this Group/Topical: Engineering Sciences and Fundamentals

(661f) Single-Stranded DNA-Carbon Nanotube Hybrid Interactions

Daniel Roxbury[1], Jeetain Mittal[2] and Anand Jagota[1], (1)Chemical Engineering, Lehigh University, Bethlehem, PA, (2)Department of Chemical Engineering, Lehigh University, Bethlehem, PA

It has been shown that, under certain conditions, single-stranded DNA (ssDNA) preferentially wraps in a helical fashion around carbon nanotubes (CNTs), forming a water-dispersible hybrid. Recently, its ability to sort mixtures of CNTs based on chirality has been demonstrated using special short ssDNA sequences. It is believed the emergence of DNA secondary structures, arising from DNA base-base hydrogen bonding, plays a crucial role in enabling this process.

Previously, through short molecular simulations, we have identified the possibility of hydrogen bond stabilized oligonucleotide secondary structures, deemed β-sheets and β-barrels, on graphene and carbon nanotubes, respectively. More recently, through the use of replica exchange molecular dynamics (REMD) simulation, we have conducted a study on the (6,5)-CNT, its respective DNA recognition strand, $(TAT)_4$, and other related sequences. Starting with one strand simulated on an infinitely long CNT, certain hybrid characteristics begin to emerge. In particular, a large majority of DNA-CNT hybrids (97% for $(TAT)_4$) adopt a right-handed helical configuration, with greater than 98% of all DNA bases adsorbing on the CNT sidewall. Furthermore, DNA strands are found to fully loop around the CNT and stitch to themselves via hydrogen bonds between bases. A thermodynamic model is presented to accurately predict types and probabilities of configurations to which single strands of DNA will conform on specific CNTs.

See more of this Session: Molecular Modeling and Simulation of Complex Molecules I
See more of this Group/Topical: Engineering Sciences and Fundamentals

(661g) Protein-Protein Binding and Structure Prediction In the Photosynthetic Antenna Complex of Green Sulfur Bacteria

Sndor . Kovcs and Cynthia S. Lo, Department of Energy, Environmental and Chemical Engineering, Washington University in St. Louis, St. Louis, MO

A molecular understanding of protein-protein interactions in biological systems is required for effective insight into their structure and function. For example, in photosynthetic organisms, protein-protein interactions dictate the rate and efficiency of energy transfer between chromophores. In this work, all-atom steered molecular dynamics (SMD) simulations are carried out to calculate the free energy of protein binding to sites previously identified by docking simulations. This enables an evaluation of the most likely configurations of the light harvesting antenna complex of green sulfur bacteria in the natural aqueous environment. The complex consists of several pigments, such as bacteriochlorophyll A (Bchl A), and proteins, including the Fenna Matthews Olson (FMO) protein and chlorosome protein A (CsmA). Our findings suggest, in conjunction with available experimental data, that the C-terminus of CsmA preferentially binds to FMO, facilitating the alignment of Bchl A in CsmA with Bchl A #8 in FMO. These simulations provide the necessary structures for modeling excited-state energy transfer in photosynthetic and biohybrid systems.

See more of this Session: Molecular Modeling and Simulation of Complex Molecules I
See more of this Group/Topical: Engineering Sciences and Fundamentals

(661h) Molecular Dynamics Simulation On the First Stage of Thrombus Formation

Lin Zhang, Jian Li and Yan Sun, Department of Biochemical Engineering, School of Chemical Engineering and Technology, Tianjin University, Tianjin, China

Molecular interaction between the protein von Willebrand factor (VWF) A1 domain and its platelet surface receptor glycoprotein Ibα (GPIbα) promotes the initial adhesion of platelets to the sites of vascular injury, which intiates thrombus formation. So, this molecular interaction is crucial in the thrombus formation. However, little is understood about this interaction at molecular level, and it is very difficult to investigate the details by experiments. In this work, all-atom models and molecular dynamics simulation are used to investigate the interactions between VWF A1 domain and GPIbα, and to examine the disassociation/binding processes. Two solution conditions, water and physiological saline are considered. In the simulations, faster binding of VWF A1 domain and GPIbα is observed in water than that in physiological saline, and larger electrostatic interaction than hydrophobic interaction is also observed in the binding process. Then, electrostatics analysis is performed using Adaptive Poisson-Boltzmann Solver (APBS) to examine the electrostatic surface potentials of VWF A1 domain and GPIbα. Oppositely charged regions are observed at the interface between VWF A1 domain and GPIbα. Moreover, the interaction free energy is analyzed in details using molecular mechanics-Poisson-Boltzmann surface area (MM-PBSA) approach. It is shown that the binding process is first driven by electrostatic interaction and then stabilized by hydrophobic interaction. The results indicate that electrostatic interaction is dominate for the complex formation of VWF A1 domain and GPIbα. This has provided the molecular insight into the initial adhesion of platelets.

See more of this Session: Molecular Modeling and Simulation of Complex Molecules I
See more of this Group/Topical: Engineering Sciences and Fundamentals

(661i) Loop Structure Prediction for Fixed-Stem Geometries In Proteins

Ashwin Subramani and Christodoulos A. Floudas, Chemical and Biological Engineering, Princeton University, Princeton, NJ

Loop structure prediction is an important intermediate step towards the tertiary structure prediction of proteins[1,2,3,4]. In the absence of an inherent local hydrogen bonding pattern, loop structure prediction can be considered a mini *ab* initio structure prediction problem. Here, we present an algorithm for the prediction of loop structures in fixed-stem systems. In fixed-stem geometries, the knowledge of the coordinates of the secondary structure elements bordering a loop is known, while the structure of the loop itself is unknown. The problem forms a critical intermediate step in the elucidation of protein structures using database driven methods like homology modeling and fold recognition.

Using a distribution of loop phi/psi angles generated from a large, non-redundant database, initial structures for the loop between two given secondary structure elements are generated. While initial bounds on the dihedral angles in the loop region are generic, distance and dihedral angle constraints derived from the knowledge of the structure of stem residues provide restrictions on the possible prediction of loop structures. Three fast side chain rotamer optimization steps are utilized to alleviate steric clashes in the randomly generated initial structures, thus providing better starting points for structure optimization. The structures are now subjected to energy minimization, using the all atom force field ECEPP/3. The problem is formulated as a constrained non-linear optimization problem; with the constraints representing distance and dihedral angle bounds on the loop and stem residues. At the end of the optimization step, the generated conformers are clustered using a traveling salesman problem based clustering algorithm, ICON [5]. ICON is an iterative algorithm which treats each conformer generated as a node in a TSP, and finds the best path to connect these nodes. An integer optimization (ILP) based model is implemented to rigorously find cluster boundaries, which lead to the elimination of sparse clusters at each iteration. The densest clusters are then collected, and improved dihedral angle bounds on the backbone angles of each of the loop residues are generated for the next iteration of the algorithm. At each iteration, new conformers are tested against previously generated conformers to eliminate duplication of initial structures. The algorithm is being tested on a large set of loops between the lengths of 5 and 20. The initial results and comparisons with experimentally elucidated structures show a high degree of accuracy in predicting tight bounds on the backbone angles of loop regions, thus displaying the method's efficacy in being used as the loop generation algorithm for database driven structure prediction algorithms.

Bibliography

[1] McAllister SR and Floudas CA (2010) An improved hybrid global optimization method for protein tertiary structure prediction, *Comput. Optim. Appl.* **45**, 377-413

[2] Floudas CA, Fung HK, McAllister SR, Monnigmann M and Rajgaria R (2006) Advances in Protein Structure Prediction and De Novo Protein Design: A Review, *Chem Engg. Sci.*, **61**, 966-988

[3] Floudas CA (2007) Computational methods in protein structure prediction, *Biotech. BioEng.*, **97**, 207-213

[4] Subramani A, Wei Y and Floudas CA (2011) ASTRO-FOLD 2.0: An enhanced framework for protein structure prediction, *submitted.*

[5] Subramani A, DiMaggio PA and Floudas CA (2009) Selecting high quality Protein structures from Diverse Conformational Ensembles, *Biophys J*, **97**, 1728-1736

See more of this Session: Molecular Modeling and Simulation of Complex Molecules I
See more of this Group/Topical: Engineering Sciences and Fundamentals

(758a) Electrochemical Redox Flow Devices for the Electrically Driven Scrubbing of Acid Gases

Michael C. Stern[1], Fritz Simeon[1], Kristin Vicari[1], Howard Herzog[2] and T. Alan Hatton[1], (1)Chemical Engineering, Massachusetts Institute of Technology, Cambridge, MA, (2)MIT Energy Initiative, Massachusetts Institute of Technology, Cambridge, MA

The concept, thermodynamicanalysis, and preliminary experimental results are presented for anelectrochemical redox flow system capable of scrubbing acid gases from eitherdilute streams (such as air) to more concentrated industrial flue gases. Theability to electrically scrub acid gas offers significant advantages overexisting thermal and pressure swing techniques. Thermal scrubbing is anineffective strategy when not coupled with a Rankine cycle due to the narrowtemperature range and resulting poor Carnot efficiency. Pressure swingsystems, in contrast, are able to operate through electrical energy, but sufferfrom the inability to separate gas sorbates from dilute streams due to the weakbinding enthalpies inherent to physical sorption processes.

An electrochemical scrubbingprocess offers the possibility of a truly "drop-in" solution for powerplantsthat requires no reconfiguration of their internal turbine systems.Additionally, electrochemical scrubbers can be installed anywhere with accessto electricity in order to supply new markets for CO_2 such asphotosynthetic biofuels or electrofuels that will have large CO_2demands but may not want to be located near an existing powerplant.

The system consists of anelectrochemical flow cell (which closely resembles a redox flow battery) intandem with a traditional gas absorber unit. A theoretical framework forevaluating redox-responsive sorbents for redox flow separations based on theirkinetic and thermodynamic properties is presented. Experimental results of thesorbents have been analyzed with this framework to demonstrate the feasibilityof the electrochemical redox flow scrubbing process.

See more of this Session: Thermodynamics of Energy Systems
See more of this Group/Topical: Engineering Sciences and Fundamentals

(758b) Heat Transfer Coefficient Enhancement In Oil-Based Nanofluids

Kan Liu and **Gerold A. Willing**, Chemical Engineering, University of Louisville, Louisville, KY

Nanofluids have been the subject of intense research efforts since their discovery over a decade ago mainly because of the enhancements to the thermal conductivity over that of the base fluid. Unfortunately, many of the enhancements seen in thermal conductivity have not been seen in the overall heat transfer coefficient, which takes into account the convective flow of the fluid. In this work, we investigate the enhancement of the overall heat transfer coefficient of an oil based nanofluid containing copper nanoparticles of varying size and concentration. The results show that temperature and Reynold's number have the greatest influence on the enhancement of the heat transfer coefficient, with the largest enhancements seen under conditions that appear to limit brownian motion.

See more of this Session: Thermodynamics of Energy Systems
See more of this Group/Topical: Engineering Sciences and Fundamentals

Accurate estimation of thermochemical parameters for branched and strained organics: the enthalpy of formation of 2,2,4,4,6,8,8-heptamethylnonane

Gregory R. Magoon and William H. Green

Department of Chemical Engineering, Massachusetts Institute of Technology

77 Massachusetts Avenue, Cambridge, MA 02139

Abstract ID#: 237313

Session: Thermodynamics of Energy Systems

Abstract

Accurate thermochemistry is critical for many applications, particularly predictive kinetic modeling. The heat of formation of highly-branched hydrocarbons, such as the diesel primary reference fuel 2,2,4,4,6,8,8-heptamethylnonane (HMN, or "isocetane"), is significantly affected by steric crowding. Multiple approaches have been devised to account for these effects with group-additive methods. Several such approaches were investigated, with focus on two of the most widely-used approaches: Benson's approach (using group parameters from 1992) and the "methyl repulsion" method of Domalski and Hearing (1988). Our study of several highly branched alkanes suggests that Benson's approach to steric corrections is at least comparable, and possibly superior, in accuracy to the "methyl repulsion" approach.

The investigation, which also includes supporting evidence from MM4 and CBS-QB3 calculations, suggests that gas-phase ΔH_f(298 K) values for HMN used in the literature and reported by some group-additivity approaches are as much as 3-8 kcal/mol too low. The use of accurate thermodynamic parameters for HMN will be important in ongoing efforts to construct detailed chemical kinetic models for its combustion.

This work is part of a larger effort to improve parameter estimation during automated construction of detailed chemical kinetic models, particularly for combustion and pyrolysis. Although Benson's method is excellent for branched aliphatics, Benson's method for accounting for ring strain, particularly for fused ring systems, is less satisfactory. Our methods for estimating the thermochemistry of strained ring species using explicit three-dimensional molecular structures in the context of automated reaction mechanism generation will also be summarized.

Introduction

Accurate estimation of ideal-gas thermochemical parameters is a key consideration in many engineering applications. Thermodynamic parameters (enthalpy of formation, entropy, and heat capacity as a function of temperature) are particularly important in kinetic modeling. Over the years, many approaches have been developed to estimate thermodynamic quantities in a manner that is accurate, general, and fast or "inexpensive" (using as few human or computer resources as possible). The often competing nature of these aims leads to a wide range of viable methodologies that are each, in some sense, near-optimal for a given application.

In the case of kinetic modeling for gas kinetic applications (covering pyrolytic and combustion systems), accurate thermodynamic parameters (with errors on the order of kcal/mol) covering a wide temperature range (typically at least 300 K – 1500 K) must be estimated for hundreds of chemical species. Frequently, these chemical species can be branched or cyclic (both aromatic and non-aromatic) and can also include radical sites (unpaired electrons). The process of automatically building kinetic models using a software system such as RMG (Reaction Mechanism Generator)[1] is even more demanding, as tens or hundreds of thousands of chemical species are enumerated in the process of determining which are suitable for inclusion in the final kinetic model, and thermodynamic parameters must be estimated for these species as well. These needs call for fast, general thermodynamic estimation procedures with reasonable accuracy. Traditionally, group additivity approaches, such as the method designed by Benson,[2] have been used. These approaches require group values obtained through regression analysis of "known" target property values (e.g. standard heat of formation at 298.15 K) for a representative set of molecules. These pre-computed group values may then be extended to estimate properties for molecules outside the training set. Despite the long history of these methods, they continue to be an active area of research.[3] These approaches are not without limitations, as will be described later, but often perform satisfactorily for a wide range of molecular species.

Steric effects arising from non-next-nearest neighbor effects require special considerations in these approaches. These effects are particularly pronounced in highly branched alkanes. The Benson group additivity approach[2] and the "methyl repulsion" group additivity approach of Domalski and Hearing[4] are two of the most popular approaches for obtaining quick connectivity-based estimates using group additivity while taking these non-nearest-neighbor effects into account. The Benson and "methyl repulsion" methods are similar in principle, but differ in their approach to treating non-nearest neighbor steric effects, as well as in the raw group values.

The Benson group additivity approach has a long history, being first proposed and parametrized, including destabilizing non-nearest neighbor "gauche interactions", in 1958;[5] it was then extended in 1969.[6] The values from their 1976 update are amongst the most widely used, and included the introduction of new non-nearest-neighbor effects in the form of destabilizing "1,5-interactions."[7] Cohen and Benson later updated Benson's gauche interaction counting scheme and updated the group values in 1992.[8]

In the Benson scheme, gauche interactions are 1,4-interactions in branched species; the gauche interaction counting schemes of the 1976 and 1992 Benson approaches are summarized in Table 1. Examples of gauche and 1,5-interactions, are illustrated in Figure 1 and Figure 2. The 1976 and 1992 alkane group values are reproduced in Table 2.

Table 1. Comparison of Benson gauche counting schemes. Adapted from Cohen and Benson.[8]

Benson (1976) gauche counting scheme

C_a atom group	C_b atom group	# gauche interactions
P	P, S, T, or Q	0
S	S (trans/gauche)	0/1
S	T (gauche/syn)	1/2
S	Q	2
T	T (trans/cis)	2/3*
T	Q	4
Q	Q	6

*TTT will have 5 gauche interactions

Revised Benson (1992) gauche counting scheme

C_a atom group	C_b atom group	# gauche interactions
P	P, S, T, or Q	0
S	S	0
S	T	1
S	Q	2
T	T	3
T	Q	5
Q	Q	8

P=primary; S=secondary; T=tertiary; Q=quaternary

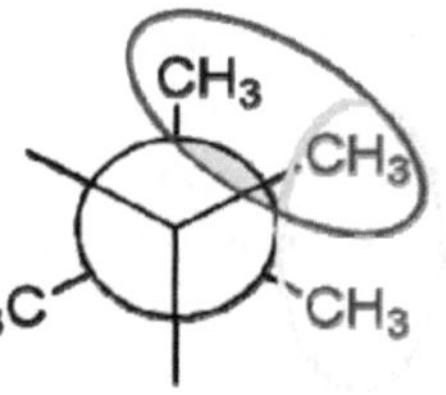

Figure 1. Example illustrating two gauche or 1,4-interactions

Figure 2. Example branched alkanes with two and one 1,5-interactions, respectively.

Table 2. Benson group additivity group values for alkanes from 1976 and 1992. Adapted from Cohen and Benson.[2]

Group	**$\Delta\Delta H_{f,298}$ (kcal/mol)** **(1976 values)**	**$\Delta\Delta H_{f,298}$ (kcal/mol)** **(1992 values)**
$C\text{-}(C)(H)_3$ = P	-10.2	-10.00
$C\text{-}(C)_2(H)_2$ = S	-4.93	-5.00
$C\text{-}(C)_3(H)$ = T	-1.90	-2.40
$C\text{-}(C)_4$ = Q	0.50	-0.10
gauche (1,4) correction = G	0.80	0.80
1,5-correction = F	1.5	1.60

In 1988, Domalski and Hearing attempted to address limitations in Benson's 1976 approach, particularly with regard to steric effects in branched hydrocarbons.[4] They proposed the "methyl repulsion" method. The name is somewhat of a misnomer as, in this approach, methyl ligands actually provide a stabilizing effect relative to larger ligands. Rather than applying *post hoc* steric corrections as in the Benson approach, the "methyl repulsion" method appears to incorporate the steric effects in the raw group values, which take the degree of the carbon center (i.e. primary, secondary, tertiary, or quarternary) into account; stabilizing corrections are then added when the substituent is small (i.e. methyl). The Domalski and Hearing approach, with its "methyl repulsion" method of accounting for steric effects in branched hydrocarbons is widely used; it is, for example, the recommended approach for estimating heat of formation in the latest *Perry's Chemical Engineer's Handbook, 8th ed.*[9] Alkane group values for Domalski and Hearing's "methyl repulsion" method are summarized in Table 3.

A comparison of the raw group values between Benson and "methyl repulsion" methods shows that the most significant difference is for quaternary carbons, which likely arises mainly from the different approach to steric corrections.

The appropriate interpretation of the "tertiary", "quaternary", "tert/quat", and "quat/quat" is not immediately obvious for certain molecules. In particular, the questions arise: 1) Should a methyl on a tertiary carbon always use the "tertiary" value or use the "tert/quat" value if a quaternary carbon is nearby? and 2) Do the "tert/quat" and "quat/quat" designations refer to quaternary carbons directly adjacent, or are carbons one spot removed considered as well? The table format and text of Domalski and Hearing[4] seem to imply that methyls on a tertiary carbon should always use the "tertiary" value and that "tert/quat" and "quat/quat" refer to tertiary or quaternary carbons directly bonded to a quaternary carbon. However, careful examination of the estimated values and a worked example for 2,2,4,4-tetramethylpentane suggests an alternative counting scheme, wherein methyls on a ternary carbon can use the "tert/quat" value if a quaternary carbon is nearby and "tert/quat" and "quat/quat" can refer to carbons either immediately adjacent or one spot removed. In the text that follows, the counting scheme suggested by the tables/text will be termed A, while a counting scheme in the spirit of the treatment implied by the alkane estimates and example in Domalski and Hearing[4] will be termed B.

Table 3. Gas-phase heat of formation group values in Domalski and Hearing's "methyl repulsion" method. Adapted from Domalski and Hearing.[10]

Group	**$\Delta\Delta H_{f,298}$ (kcal/mol)**
$C-(H)_3(C)$	-10.10
$C-(H)_2(C)_2$	-4.931
$C-(H)(C)_3$	-0.280
$-CH_3$ corr (tertiary)	-0.540
$C-(C)_4$	+4.589
$-CH_3$ corr (quaternary)	-1.090
$-CH_3$ corr (tert/quat)	-0.430
$-CH_3$ corr (quat/quat)	-0.153

The heat of formation for 2,2,4,4,6,8,8-heptamethylnonane presents an interesting case study for these alternative approaches to treating steric effects.

Heat of formation for 2,2,4,4,6,8,8-heptamethylnonane

2,2,4,4,6,8,8-heptamethylnonane (otherwise known as isocetane, and henceforth referred to as HMN) is a highly-branched alkane, most widely known for its role as a primary reference fuel component and benchmark for the cetane scale for diesel fuels. There have been recent efforts to construct detailed chemical kinetic models for HMN combustion.[11] The structure of this highly branched hydrocarbon is shown in Figure 3.

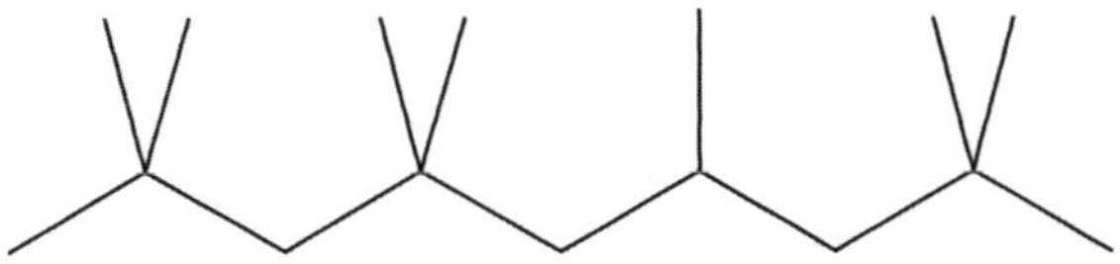

Figure 3. Structure of 2,2,4,4,6,8,8-heptamethylnonane (HMN)

Several approaches were used to estimate the standard heat of formation of HMN at 298.15 K, as shown in Table 4, below. In addition to using the previously-discussed Benson and "methyl-repulsion" methods, we found an estimate published by DIPPR[12], reportedly based on Benson group additivity (though the value suggests that older group values were used or some modifications were applied). Additionally, we have included an estimate[13] from the program Thergas (which uses a group additivity approach, likely based on the "methyl repulsion" method). We also estimated the heat of formation by applying the "2C" correlation from Somayajulu and Zwolinski, a connectivity-based approach for acyclic alkanes, which takes non-local connectivity effects into account.[14] Additionally, the MM4 force field of Allinger *et al.*[15] was applied to the problem. The MM4 approach has been found to reliably reproduce alkane heats of formation.[15] A conformation search with the MM4 program with default parameters (2000 "pushes") identified 76 distinct low-energy conformations. The minimum energy MM4 conformation was used as a reference onto which corrections were applied. A B3LYP/CBSB7-optimized version of this conformation is shown in Figure 4. MM2 rotor scans in Chem3D (part of the ChemBioOffice suite) suggested that one of the (non-methyl) rotors had a barrier that was below 5 kcal/mol, prompting a "TOR" correction of +0.5715 kcal/mol. Following the recommended approach of Allinger *et al.*,[16] contributions from the higher energy conformations were taken into account; through this, we estimated a "POP" contribution of +0.4098 kcal/mol. (Note that for some of the conformations, MM4 did not include contributions from low frequency vibrational modes; we expect that the error introduced by this is very probably less than 0.25 kcal/mol, and likely much less, so we did not make an effort to correct for it.) For the approach labeled CBS-QB3-BAC, we first applied the CBS-QB3 method[17] in Gaussian03[18] to the minimum energy conformation from the MM4 conformation search; for this relatively large $C_{16}H_{34}$ molecule, the calculation is quite computationally demanding. We then applied an atomization scheme with bond-additivity corrections, as determined by Sabbe *et al.*[19] to arrive at the final value.

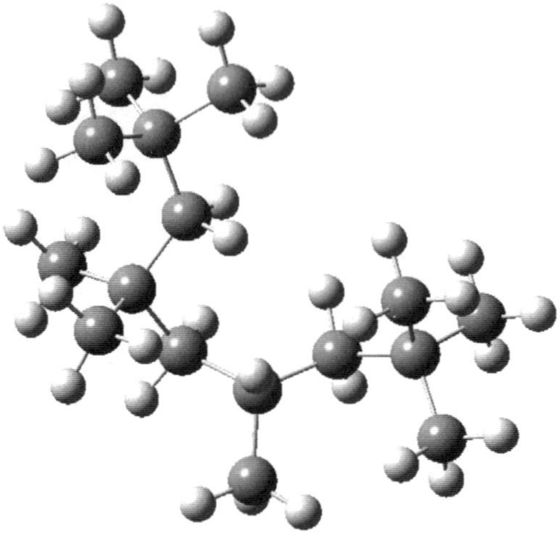

Figure 4. Minimum energy configuration for HMN from MM4 search, subsequently optimized with B3LYP/CBSB7

Table 4. Estimates for standard heat of formation of HMN

Method/Source	**$\Delta H_{f,298}$ (kcal/mol)**
Benson group additivity (1992 rev.; *no steric corrections*)	-107.7
"methyl repulsion" method (A/B)	-101.5/-94.7
DIPPR estimate[12]	-98.8
Thergas estimate[13]	-96.3
Benson group additivity (1992 rev.)	-93.3
Somayajulu and Zwolinski correlation[14]	-93.7
MM4	-93.5
CBS-QB3-BAC	-93.2

The fact that the last four approaches give results that agree within about 0.5 kcal/mol provides a strong indication that they all offer a reliable estimate of the true value. The difference in Benson values in the table also illustrates the significant magnitude (over 10 kcal/mol) that the destabilizing steric effects have in this molecule. The results suggest that the Benson method is significantly more accurate than the "methyl repulsion" (A) method in this case.

Although the agreement of the last four approaches appears to provide strong validation for these values and the Benson approach, we were not aware of an experimental value for the heat of formation for HMN for comparison, and we decided to seek further validation from two structurally similar molecules with experimental data for heat of formation: 2,2,4,6,6-pentamethylheptane and 2,2,4,4,6-pentamethylheptane. These are shown in Figure 5, below:

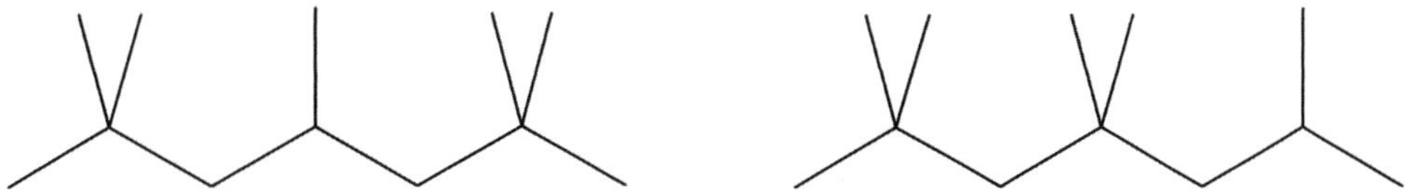

Figure 5. Structures of 2,2,4,6,6-pentamethylheptane and 2,2,4,4,6-pentamethylheptane

Results of application of various approaches, along with experimental data from Desai[20] and TRC recommended values (which are apparently based on both experiment and a Somayajulu-based estimate)[21], are shown in Table 5, below.

Table 5. Estimates for standard heat of formation of structural analogs for HMN

Method/Source	**$\Delta H_{f,298}$ (kcal/mol)**	
	2,2,4,6,6-pentamethylheptane	2,2,4,4,6-pentamethylheptane
"methyl repulsion" method (A/B)	-78.7/-74.7	-78.2/-73.3
Benson group additivity (1992 rev.)	-74.6	-72.2
Somayajulu and Zwolinski correlation[14]	-74.9	-72.15
experiment[20]	-75.45	-72.1
TRC[21]	-75.4	-71.6

In addition to matching closely with the experimental values, the Benson estimates capture both the direction and most of the magnitude of the difference in heat of formation between 2,2,4,6,6-pentamethylheptane and 2,2,4,4,6-pentamethylheptane.

Although the "methyl repulsion" (A) method also captures the relative instability of 2,2,4,4,6-pentamethylheptane, it significantly underestimates the magnitude of this difference. The "methyl repulsion" (B) method appears to be somewhat better, though still not as good as the Benson approach, at capturing this difference.

It is noted that the differences between Benson and "methyl repulsion" (A) approaches for these three molecules are of the right direction and magnitude to suggest that large portions of the apparent errors in the "methyl repulsion" approach may be due to the lack of destabilizing 1,5-interactions. It is therefore hypothesized that a refined version of the "methyl repulsion" approach, reparametrized with an additional "group" for 1,5-interactions, may produce significantly improved results, assuming the "A" counting scheme is used.

Without such reparametrization, however, it would appear that counting scheme B is more appropriate than counting scheme A. A summary of the methyl group counts for these alternative counting schemes is included in Supplementary Table 1.

Benson vs. "methyl repulsion" approach for other branched alkanes

Although a comprehensive comparison between the Benson and "methyl repulsion" methods is beyond the scope of this work, a comparison of a few cases where the "methyl repulsion" method did better than the original 1976 Benson approach is warranted, as such cases provided a key original motivation for the use of the "methyl repulsion" approach.

Table 6 shows estimates for the 16 branched alkanes considered in Appendix C of Domalski and Hearing[4], where they argued that the "methyl repulsion" approach outperformed Benson's 1976 parametrization in accuracy.

Table 6. Comparison of heat of formation estimates for 16 branched alkanes. (Values reported as A/B when the "methyl repulsion" counting approaches A and B differ; Supplementary Table 1 includes the differing group counts for these cases)

	Standard heat of formation at 298.15 K (kcal/mol)				**Error (experiment – estimate) (kcal/mol)**		
Molecule	experiment[4]	Benson (1976)	Benson (1992)	"methyl repulsion"	Benson (1976)	Benson (1992)	"methyl repulsion"
2-methylpropane	-32.1	-32.5	-32.4	-32.2	0.4	0.3	0.1
2,2-dimethylpropane	-40.2	-40.3	-40.1	-40.2	0.1	-0.1	0.0
2-methylbutane	-36.7	-36.6	-36.6	-36.6	-0.1	-0.1	-0.1
3-methylpentane	-41.1	-40.8	-40.8	-41.0	-0.4	-0.3	-0.1
2,2-dimethylbutane	-43.5	-43.6	-43.5	-44.0	0.1	0.0	0.5
2,3-dimethylbutane	-42.6	-43.0	-42.4	-43.1	0.4	-0.2	0.5
3-ethylpentane	-45.3	-44.9	-45.0	-45.4	-0.4	-0.3	0.1
2,2,3-trimethylbutane	-48.9	-48.4	-48.5	-48.6/-48.3	-0.5	-0.4	-0.3/-0.5
3,3-diethylpentane	-55.5	-53.6	-53.7	-55.5	-1.9	-1.8	0.0
2,2,3,3-tetramethylbutane	-53.9	-55.4	-53.8	-52.3	1.5	-0.1	-1.6
2,2,3,3-tetramethylpentane	-56.6	-58.7	-57.2	-57.1	2.2	0.6	0.5
2,2,3,4-tetramethylpentane	-56.6	-58.2	-56.9	-59.5/-59.2	1.6	0.3	2.9/2.5
2,3,3,4-tetramethylpentane	-56.4	-59.7	-56.9	-59.6/-59.2	3.3	0.5	3.2/2.7
2,2,4,4-tetramethylpentane	-57.8	-58.9	-58.8	-62.9/-57.3	1.2	1.0	5.1/-0.5
3,3-dimethylpentane	-48.1	-47.0	-46.9	-47.9	-1.1	-1.2	-0.2
3-ethyl-3-methylpentane	-51.4	-50.3	-50.3	-51.7	-1.1	-1.1	0.3
Root-mean-squared error					**1.33**	**0.72**	**1.74/1.06**

The table shows that four of the five largest errors from the 1976 Benson approach (2,2,3,3-tetramethylbutane, 2,2,3,3-tetramethylpentane, 2,2,3,4-tetramethylpentane, and 2,3,3,4-tetramethylpentane) are significantly mitigated when the updated group values and gauche counting scheme of the 1992 Benson approach are used. In fact, the root-mean-squared deviation between estimate and experimental values for these 16 molecules is lower with the 1992 Benson approach than with the "methyl repulsion" approaches. Even so, the "methyl repulsion" approach still outperforms the 1992 Benson approach in several cases, most notably 3,3-diethylpentane, 3,3-dimethylpentane, and 3-ethyl-3-methylpentane. The fact that the "methyl repulsion" approach outperforms the Benson approach in these cases with ethyl branching suggests that the general principle of treating methyl differently than larger ligands (e.g. ethyl) has a solid empirical basis. As mentioned earlier, further improvement on the "methyl repulsion" approach may be possible by reparametrizing with the inclusion of 1,5-repulsion effects.

Although further study is needed, the results for HMN and other branched molecules discussed here provide evidence that the Benson method (with the updated 1992 group values and gauche counting scheme) is at least comparable in accuracy to "methyl repulsion" (B) for estimating the heat of formation of branched alkanes.

Additionally, based on these findings, we suggest that when using the "methyl repulsion" method, care should be taken to implement appropriate methyl corrections, using an approach like counting scheme B. Alternatively, with counting scheme A, a refinement to the "methyl repulsion" method that considers 1,5-corrections may be appropriate.

An approach to circumvent limitations of group additivity approaches: on-the-fly explicit-structure-based methods

As the examples above demonstrate, group additivity can produce quite reliable estimates in many cases; however, in some cases, the approach is ill-suited to the task at hand. Two noteworthy limitations of group additivity procedure are: 1) it relies on a connectivity based molecule representation and 2) it requires availability of accurate group values to cover molecules of interest.

Although the fact that group additivity relies on a connectivity-based representation of a molecule may also be interpreted as a strength of the method, it also limits the application to structures without significant steric effects (outside the previously discussed branching considerations); in particular, group additivity will produce quite inaccurate results for cyclic species, unless *ad hoc* ring corrections are applied. Appropriate corrections, are not, in general, available, and accurate application of group additivity approaches is limited to simple (e.g. cyclopentane) or well-studied (e.g. adamantane) ring structures.

The fact that group additivity requires accurate pre-computed group values is not typically a significant issue for the well-studied classes of molecules involving only carbon, hydrogen, and/or oxygen. However, as soon as heteroatoms (e.g. sulfur, silicon, nitrogen, and halogens) are added to the set of target molecules, the limited availability of appropriate group values can become a significant issue.

An approach that, to a large extent, bypasses these two limitations is to generate explicit three-dimensional molecular structures and perform on-the-fly force field or quantum mechanics calculations on them to obtain the desired thermodynamic quantities.

Broadbelt and co-workers first introduced this approach in 1994.[22] However, use of their approach has been limited by factors including robustness and speed.[23] We have developed a next-generation system within RMG that attempts to circumvent some of these limitations.[24] In addition to making extensive use of open-source tools, the approach takes advantage of recent developments from several fields, including progress in three-dimensional geometry embedding, force fields, and chemical structure representation, along with enhanced robustness of quantum chemistry codes, and improvements in computer speed and disk storage capabilities. A diagrammatic summary of the system is shown in Figure 6.

Although details of the approach are beyond the scope of this document, a couple of interesting aspects are briefly mentioned here. Firstly, a key challenge in the process is obtaining a reliable initial guess three-dimensional geometry. A number of approaches were evaluated, but we ultimately decided to use a distance-geometry-based approach, as implemented in the open-source cheminformatics toolkit, *RDKit*.[25] In order to increase the robustness of the process, we allow RMG to try a number of different keywords/options, optionally using an alternative starting geometry, if calculations fail due to convergence issues or imaginary frequencies. We have also implemented experimental on-the-fly hindered rotor scan capabilities to account for conformational flexibility when the MM4 force field is used. In order to speed mechanism development and reduce the CPU burden, the results are stored on disk so they don't need to be recalculated when the same molecule is re-encountered in the same run or subsequent mechanism generation runs. In order to uniquely identify the molecule associated with each set of such stored results, we make use of a modified version of the recently-introduced InChI and InChIKey.[26]

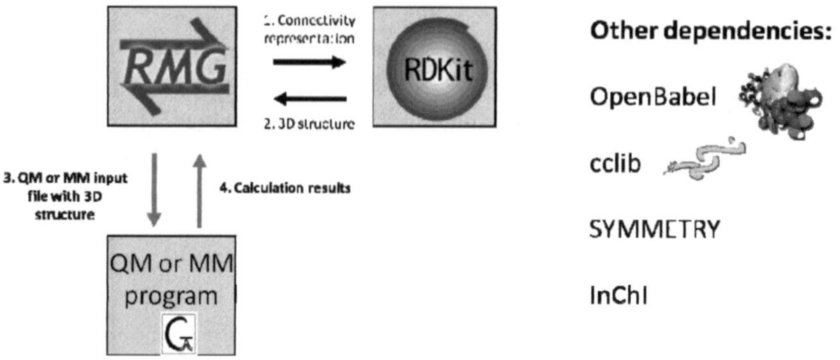

Figure 6. Summary of on-the-fly thermodynamic estimation process using explicit three-dimensional structures.

It is noted that the use of explicit three-dimensional structures in conjunction with force field or quantum mechanics calculations is not limited to thermodynamic property estimation; it can also be used to estimate other important properties, such as transport properties and vibrational frequencies.

Acknowledgements

We would like to acknowledge Dr. William J. Pitz of Lawrence Livermore National Laboratory who initially brought disparate estimates for HMN heat of formation to our attention and provided us with the value from Thergas. We also acknowledge helpful discussions regarding on-the-fly thermodynamic parameter estimation with Professor Linda Broadbelt of Northwestern University. We would like to acknowledge funding for the development of RMG by the Department of Energy, Office of Basic Energy Sciences through grant DE-FG02-98ER14914 and by the National Science Foundation under Grants No. 0312359 and 0535604.

References

1. *Allen, J. W.; Ashcraft, R. W.; Beran, G. J.; Goldsmith, C. F.; Harper, M. R.; Jalan, A.; Magoon, G. R.; Matheu, D. M.; Petway, S.; Sumathy, R.; Sharma, S.; Van Geem, K. M.; Song, J.; Wen, J.; West, R. H.; Wong, A.; Wong, H.-W.; Yelvington, P. E.; Yu, J.; Green, W. H. RMG (Reaction Mechanism Generator), RMG (Reaction Mechanism Generator): 2009.*

2. *Cohen, N.; Benson, S. W., Estimation of Heats of Formation of Organic Compounds by Additivity Methods. Chemical Reviews* ***1993,*** *93, 2419-2438.*

3. *(a) Sabbe, M. K.; De Vleeschouwer, F.; Reyniers, M.-F.; Waroquier, M.; Marin, G. B., First Principles Based Group Additive Values for the Gas Phase Standard Entropy and Heat Capacity of Hydrocarbons and Hydrocarbon Radicals. J Phys Chem A* ***2008,*** *112 (47), 12235-12251; (b) Khan, S. S.; Yu, X.; Wade, J. R.; Malmgren, R. D.; Broadbelt, L. J., Thermochemistry of Radicals and Molecules Relevant to Atmospheric Chemistry: Determination of Group Additivity Values using G3//B3LYP Theory. J Phys Chem A* ***2009,*** *113, 5176-5194.*

4. *Domalski, E. S.; Hearing, E. D., Estimation of the Thermodynamic Properties of Hydrocarbons at 298.15 K. J Phys Chem Ref Data* ***1988,*** *17 (4), 1637-1678.*

5. *Benson, S. W.; Buss, J. H., Additivity Rules for the Estimation of Molecular Properties. Thermodynamic Properties. J Chem Phys* ***1958,*** *29.*

6. *Benson, S. W.; Cruickshank, F. R.; Golden, D. M.; Haugen, G. R.; O'Neal, H. E.; Rodgers, A. S.; Shaw, R.; Walsh, R., Additivity rules for the estimation of thermochemical properties. Chemical Reviews* ***1969,*** *69 (3), 279-324.*

7. Benson, S. W., In Thermochemical Kinetics, 2nd ed., Wiley: New York, NY, 1976.
8. Cohen, N.; Benson, S. W., The thermochemistry of alkanes and cycloalkanes. In The Chemistry of Alkanes and Cycloalkanes, Patai, S.; Rappoport, Z., Eds. John Wiley & Sons Ltd: 1992.
9. Green, D. W.; Perry, R. H., Perry's Chemical Engineers' Handbook (8th Edition). McGraw-Hill: 2008.
10. Domalski, E. S.; Hearing, E. D., Estimation of the Thermodynamic Properties of C-H-N-O-S-Halogen Compounds at 298.15 K. J Phys Chem Ref Data **1993,** 22 (4), 805-1159.
11. (a) Oehlschlaeger, M. A.; Steinberg, J.; Westbrook, C. K.; Pitz, W. J., The autoignition of iso-cetane at high to moderate temperatures and elevated pressures: Shock tube experiments and kinetic modeling. Combust Flame **2009,** 156 (11), 2165-2172; (b) Westbrook, C. K.; Pitz, W. J.; Mehl, M.; Curran, H. J., Detailed chemical kinetic reaction mechanisms for primary reference fuels for diesel cetane number and spark-ignition octane number. P Combust Inst **2011,** 33 (1), 185-192; (c) Ranzi, E.; Frassoldati, A.; Faravelli, T.; Cuoci, A., Lumped Kinetic Modeling of the Oxidation of Isocetane (2,2,4,4,6,8,8-Heptamethylnonane) in a Jet-Stirred Reactor (JSR). Energ Fuel **2009,** 23 (10), 5287-5289; (d) Pitz, W. J.; Mueller, C. J., Recent progress in the development of diesel surrogate fuels. Progress in Energy and Combustion Science **2011,** 37 (3), 330-350.
12. Design Institute for Physical Properties, S. b. A., DIPPR Project 801 - Full Version. Design Institute for Physical Property Research/AIChE.
13. Pitz, W. J.; Herbinet, O., personal communication. 2008.
14. Somayajulu, G. R.; Zwolinski, B. J., Generalized Treatment of Alkanes. Transactions of the Faraday Society **1966,** 62, 2327-2340.
15. Allinger, N. L.; Chen, K.; Lii, J.-H., An Improved Force Field (MM4) for Saturated Hydrocarbons. J Comput Chem **1996,** 17 (5&6), 642-668.
16. Allinger, N. L.; Schmitz, L. R.; Motoc, I.; Bender, C.; Labanowski, J. K., Heats of formation of organic molecules. 2. The basis for calculations using either ab initio or molecular mechanics methods. Alcohols and ethers. J Am Chem Soc **1992,** 114 (8), 2880-2883.
17. Montgomery, J. A.; Frisch, M. J.; Ochterski, J. W.; Petersson, G. A., A complete basis set model chemistry. VI. Use of density functional geometries and frequencies. J Chem Phys **1999,** 110 (6), 2822-2827.
18. Frisch, M. J.; Trucks, G. W.; Schlegel, H. B.; Scuseria, G. E.; Robb, M. A.; Cheeseman, J. R.; Scalmani, G.; Barone, V.; Mennucci, B.; Petersson, G. A.; Nakatsuji, H.; Caricato, M.; Li, X.; Hratchian, H. P.; Izmaylov, A. F.; Bloino, J.; Zheng, G.; Sonnenberg, J. L.; Hada, M.; Ehara, M.; Toyota, K.; Fukuda, R.; Hasegawa, J.; Ishida, M.; Nakajima, T.; Honda, Y.; Kitao, O.; Nakai, H.; Vreven, T.; Montgomery, J., J. A.; Peralta, J. E.; Ogliaro, F.; Bearpark, M.; Heyd, J. J.; Brothers, E.; Kudin, K. N.; Staroverov, V. N.; Kobayashi, R.; Normand, J.; Raghavachari, K.; Rendell, A.; Burant, J. C.; Iyengar, S. S.; Tomasi, J.; Cossi, M.; Rega, N.; Millam, N. J.; Klene, M.; Knox, J. E.; Cross, J. B.; Bakken, V.; Adamo, C.; Jaramillo, J.; Gomperts, R.; Stratmann, R. E.; Yazyev, O.; Austin, A. J.; Cammi, R.; Pomelli, C.; Ochterski, J. W.; Martin, R. L.; Morokuma, K.; Zakrzewski, V. G.; Voth, G. A.; Salvador, P.; Dannenberg, J. J.; Dapprich, S.; Daniels, A. D.; Farkas, Ö.; Foresman, J. B.; Ortiz, J. V.; Cioslowski, J.; Fox, D. J. Gaussian 03, Revision D.01, Gaussian, Inc.: Wallingford CT, 2003.
19. Sabbe, M. K.; Saeys, M.; Reyniers, M.-F.; Marin, G. B.; Van Speybroeck, V.; Waroquier, M., Group Additive Values for the Gas Phase Standard Enthalpy of Formation of Hydrocarbons and Hydrocarbon Radicals. J Phys Chem A **2005,** 109 (33), 7466-7480.
20. Desai, P. D. Enthalpies of Combustion of Eight Branched Isomeric Alkanes from C-9 to C-16. Texas A&M University, 1968.
21. TRC thermodynamic tables. Hydrocarbons. Thermodynamics Research Center, Texas Engineering Experiment Station, Texas A&M University System: College Station, TX, 1985, with various loose-leaf updates.
22. Broadbelt, L. J.; Stark, S. M.; Klein, M. T., Computer generated reaction networks: on-the-fly calculation of species properties using computational quantum chemistry. Chemical Engineering Science **1994,** 49 (24, Part 2), 4991-5010.
23. Broadbelt, L. J., personal communication.
24. Magoon, G. R.; Green, W. H., Development of a next-generation software system for on-the-fly quantum and force field calculations in automated reaction mechanism generation. **(in preparation)**.
25. Landrum, G. RDKit, http://rdkit.org.
26. InChI, http://www.iupac.org/inchi/.

Supplementary Table I. Methyl correction group counts used in "methyl repulsion" counting schemes A and B, for cases where the two schemes differ

	Scheme A				Scheme B			
Molecule	-CH3 (tert)	-CH3 (quat)	-CH3 (quat/tert)	-CH3 (quat/quat)	-CH3 (tert)	-CH3 (quat)	-CH3 (quat/tert)	-CH3 (quat/quat)
2,2,4,4,6,8,8-heptamethylnonane	1	8	0	0	0	0	4	5
2,2,4,6,6-pentamethylheptane	1	6	0	0	0	0	7	0
2,2,4,4,6-pentamethylheptane	2	5	0	0	0	0	2	5
2,2,3-trimethylbutane	2	0	3	0	0	0	5	0
2,2,3,4-tetramethylpentane	3	0	3	0	0	0	6	0
2,3,3,4-tetramethylpentane	4	0	2	0	0	0	6	0
2,2,4,4-tetramethylpentane	0	6	0	0	0	0	0	6

(758d) Molecular Dynamic Studies of the Stability of CO_2 and CH_4 Hydrates In the Presence of Undersaturated Fluids

Srinath Velaga[1], Venkata Vedam[1] and **Brian Anderson**[2], (1)Chemical Engineering, National Energy Technology Laboratory/West Virginia University, Morgantown, WV, (2)Chemical Engineering, West Virginia University, Morgantown, WV

Methane hydrates are globally distributed in sediments along the continental margins and potentially contain more energy than all fossil fuel reserves. However, methane is also a potential greenhouse gas which could play a major role in global climate change. Accurately characterizing the stability of methane and CO_2 hydrates in water can help us understand their effects on earth's environment and also fesibilty of long term CO_2 sequestration in the sediments under the ocean floor. Hydrate stability can be better predicted by understanding the phenomena related to hydrate dissolution in water. Under hydrate stability conditions, the concentration difference between the hydrate and water phases of hydrate-forming gases should be an important factor affecting hydrate stability as oceanic hydrates are exposed to under-saturated sea water. In this work, the dissolution of methane and CO_2 hydrates have been studied and compared to one another in the presence of water using molecular dynamics simulations. The average lattice constant for structure I and II hydrate was calculated for two different potentials of methane OPLS and GROMACS and compare to the experimental value for validation. Methane hydrate dissociation in the presence of water was also studied using these two potentials in addition to the Anderson et al. model. The Harris and Yung model for CO_2 as well as our own *ab initio*-derived CO_2 potential were used in the simulations of CO_2 hydrate in the study. In order to provide comparison of the CO_2 potentials, the reference chemical potentials, densities of pure CO_2 liquid and vapor, and solubility of CO_2 in water were calculated using each of the potential models.

See more of this Session: Thermodynamics of Energy Systems
See more of this Group/Topical: Engineering Sciences and Fundamentals

(758e) Theory and Experiment for Thermodynamics of the CO2/H2O System

Allan H. Harvey[1], Mark O. McLinden[1] and Richard J. Wheatley[2], (1)Thermophysical Properties Division, National Institute of Standards and Technology, Boulder, CO, (2)School of Chemistry, University of Nottingham, Nottingham, United Kingdom

Mixtures containing CO_2 and H_2O are important in many applications. In the electric power industry, they are two of the primary components of combustion gases. In advanced power cycles that facilitate CO_2 sequestration, these components may appear in synthesis gas, and they will certainly appear in CO_2-rich streams being compressed and transported for sequestration. Of key importance in many of these applications is the vapor-phase thermodynamic properties, which determine, for example, the dew point where liquid water will condense out of a CO_2 stream being compressed for sequestration.

We have undertaken a comprehensive program to study the gas-phase thermodynamics of this mixture, particularly at high temperatures where experimental data are scarce and scattered. We have used *ab initio* quantum mechanics to develop an intermolecular potential-energy surface of low uncertainty for the CO_2/H_2O pair; this surface has been used to calculate the cross second virial coefficient B_{12} as a function of temperature. In addition, we have constructed a high-temperature, single-sinker magnetic suspension densimeter and used it to make density measurements on CO_2/H_2O mixtures at temperatures from 500 K to 620 K and pressures up to 17 MPa. Values of B_{12} derived from the mixture density measurements agree within mutual uncertainties with those derived from theory. Similar agreement between theory and experiment has been obtained for the N_2/H_2O system, which is also of interest in energy applications.

See more of this Session: Thermodynamics of Energy Systems
See more of this Group/Topical: Engineering Sciences and Fundamentals

THERMOPHYSICAL PROPERTY ESTIMATIONS OF MOLTEN SALTS

Amy Sun, Sandia National Laboratories, Albuquerque, NM
Scott M.Davison, Sandia National Laboratories, Albuquerque, NM
Saivenkataraman Jayaraman, Sandia National Laboratories, Albuquerque, NM
Joe G. Cordaro, Sandia National Laboratories, Livermore, CA
Alan M. Kruizenga, Sandia National Laboratories, Livermore, CA

Abstract

Projection of molten salt performance in thermal storage systems, whether based on sensible heat or latent heat, is highly dependent on the predictions of thermophysical properties. In the absence of experimental data, heat transfer properties rely on theoretical estimations. This work focuses on thermodynamic predictions of mixture properties for molten salts supportive of ongoing advanced heat transfer fluid research at the Sandia National Laboratories. Thus far, the candidate mixtures studied experimentally and theoretically at Sandia are made up of either ternary or quaternary nitrate and mixed nitrate/nitrite salts of various compositions. Experimentally, mixture properties such as melting points and heat of fusion are obtained by thermal gravimetric analysis (TGA) and differential scanning calorimetry (DSC). Classical thermodynamics are applied to resolve phase transitions of molten salt mixtures as well as mixture properties. The Wilson equation, developed originally for organic mixtures, is used to study phase boundaries of molten salts in this work. Molecular thermodynamics (MD), where atomistic simulation forms the basis for constructing the equation of state, are conducted where our fundamental understanding and experimental knowledge are lacking.

Introduction

Molten salts have been identified as a suitable class of working fluids for concentrated solar power systems. While achieving higher efficiencies for heat storage relative to organic heat transfer fluids, they are limited by high melting points and thermal stability. Numerous past experiments have focused around measurements of heat transfer properties, such as heat capacity and thermal conductivity. However, seeking potential candidates by experimentation is a challenging and expensive endeavor. When the number of components in a molten salt mixture increases, one must rely on predictive methods to expedite searches for thermophysical properties that exhibit desired heat transfer characteristics.

Sandia National Laboratories have conducted experiments and simulations to explore properties of molten salt mixtures, mostly centered around phase transition studies. The candidates that have been studied experimentally are made up of either ternary or quaternary nitrate and mixed nitrate/nitrite salts of various compositions [1,2]. Just to map out melting points of these mixtures across combinatorial composition-temperature-pressure space would be a time-consuming and expensive endeavor by experimentation alone.

This study describes the progress achieved thus far, and the continuing pursuit of harmonizing past and current efforts towards better design of heat transfer materials.

Melting Point Studies

Combined experimental and modeling studies explore solid-liquid transition nitrate and nitrate/nitrite mixtures. In particular, the study aims to deduce the eutectic composition where the transition from solid to liquid is at the lowest temperature point. The numerical studies of the phase equilibria of nitrate salts were based on classical thermodynamics to develop a rapid modeling capability to reduce the breadth of experimental mixtures studies required. The theory is derived from Prausnitz et al. and relates the free energy change of the solid-liquid phase transition to the ratio of fugacities and the activity coefficient as shown below [3]:

$$ln\frac{1}{\gamma_i x_i}=\frac{\Delta_{\mathrm{fus}}\mathrm{h}_{\mathrm{i}}}{RT_m}\left(\frac{T_m}{T}-1\right).$$

Where γ_i is the activity coefficient, x_i is the solubility, $\Delta_{fus}h_i$ is the heat of fusion, R is the universal gas constant, T_m is the melting temperature, and T is the mixture temperature. The Wilson equation is used to calculate the activity coefficients for each component in the mixture [4]. The Wilson equation to calculate the activity coefficient of component k of m components is

$$ln\gamma_k=-ln\left(\sum_{j=1}^{m}x_j\Lambda_{\mathrm{kj}}\right)+1-\sum_{i=1}^{m}\left(\frac{x_i\Lambda_{ik}}{\sum_{j=1}^{m}x_j\Lambda_{ij}}\right).$$

In the Wilson equation Λ_{ij} and Λ_{ji} are defined as the interaction coefficients of *i-j* and *j-i* pairs. The interaction coefficients are used as fitting parameters in order to fit the numerical predictions of binary mixtures to experimental results of the phase equilibria. Using the binary interaction coefficients, the Wilson equation is extended to predict the phase equilibria of mixtures of three and four components. One goal of the study was to find mixtures of nitrate salts with the lowest melting point. When combining three salts, the lowest melting point is a combination of 37% (by mole) $LiNO_3$, 36% KNO_3, and 27% $CsNO_3$ which melted at 115°C. With four salts a mixture of 28% (by mole) $LiNO_3$, 14% $NaNO_3$, 33% KNO_3, and 25% $CsNO_3$ is predicted to melt at 100°C. Figure 1 illustrates an example of quaternary phase diagram for mixture of lithium nitrate, potassium nitrate, sodium nitrate, and cesium nitrate [5].

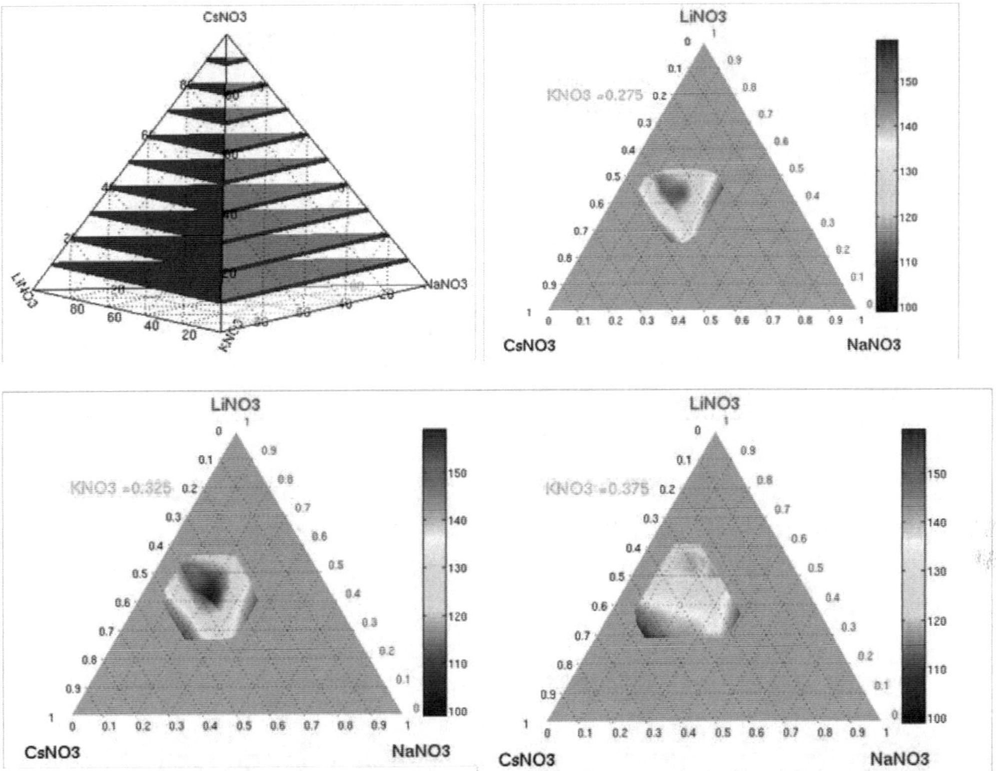

Figure 1 Quaternary phase diagram of lithium nitrate ($LiNO_3$), sodium nitrate ($NaNO_3$), potassium nitrate (KNO_3), and cesium nitrate ($CsNO_3$).

Experimentally, the effect of adding nitrite to the mixture is illustrated in Figure 2, where the melting point is plotted against the concentration of nitrate anion. Increasing lithium cation concentration lowers the melting temperature until it passes 34%-35%. Beyond that concentration, the melting point becomes insensitive to lithium cation. The melting point also decreases with increasing concentration of nitrate ion. The correlation to nitrate ion appears to be linear.

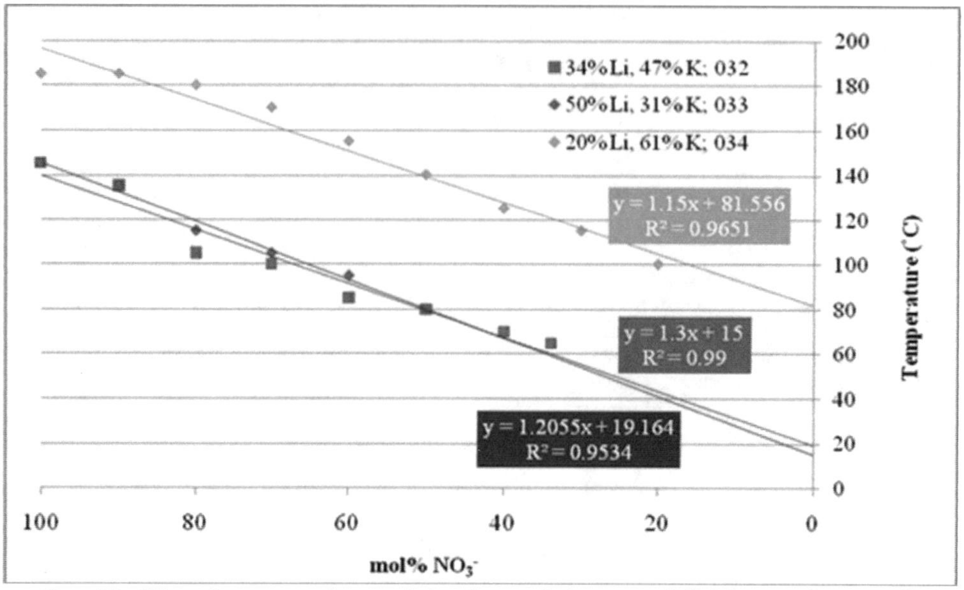

Figure 2 – Melting temperature data for mixtures of lithium nitrate ($LiNO_3$), potassium nitrate (KNO_3), sodium nitrate ($NaNO_3$), sodium nitrite ($NaNO_2$), and potassium nitrite (KNO_2), at various concentrations of lithium cation and nitrate anion.

Heat of Fusion

Similar to the melting point studies, experimental and theoretical effort are conducted to analyze pure component thermophysical properties. Constant temperature molecular dynamics simulations were conducted on the solid and liquid phases and the enthalpies of the solid and liquid phases were computed. The enthalpy difference between solid and liquid phases for pure component calcium nitrate $Ca(NO_3)_2$ is plotted as a function of temperature in Figure 3. The experimentally measured enthalpy of fusion is higher than the theoretical value predicted from Figure 3 at the experimental melting point of pure $Ca(NO_3)_2$. The theoretical value underestimates the enthalpy of fusion due to insufficiencies in the interatomic potential used [6]. Notwithstanding inaccuracies, these simulations offer valuable tools to estimate pure-component properties that are either difficult to measure or are non-existent.

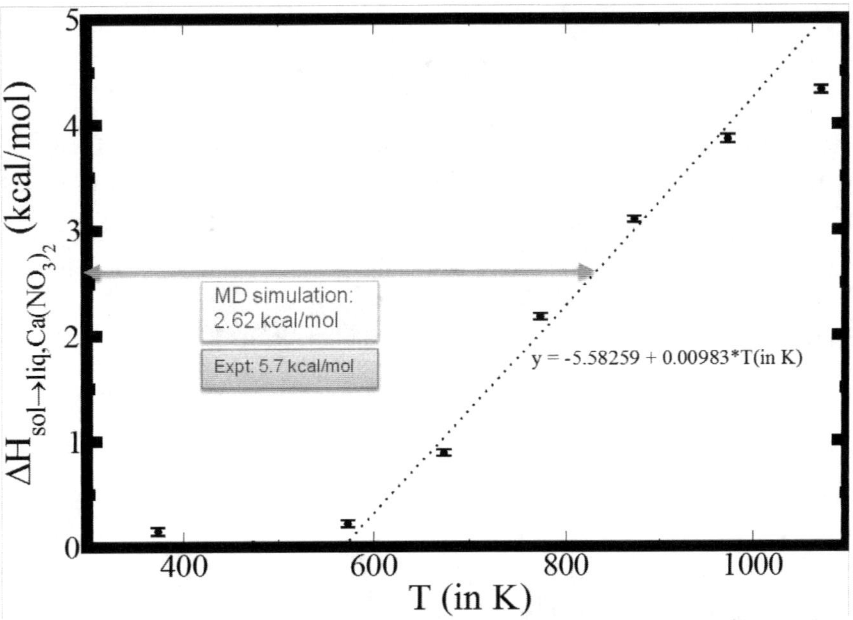

Figure 3 –Enthalpy between pure solid and pure liquid calcium nitrate as a function of temperature.

The heat of fusion is also measured experimentally for different binary mixtures. Figure 4 shows experimental data for Na/Ca nitrate mixtures. As the mol% of calcium increases, the apparent heat of melting decreases. Because pure calcium nitrate cannot be melted without undergoing decomposition, measuring the apparent heat of fusion for mixtures may enable us to extrapolate to 100% calcium nitrate. Binary salt mixtures based on lithium and potassium have also been investigated. These data, when looked at together, allow us to understand mixing behavior of salts and potentially predict thermodynamic properties based on classical models.

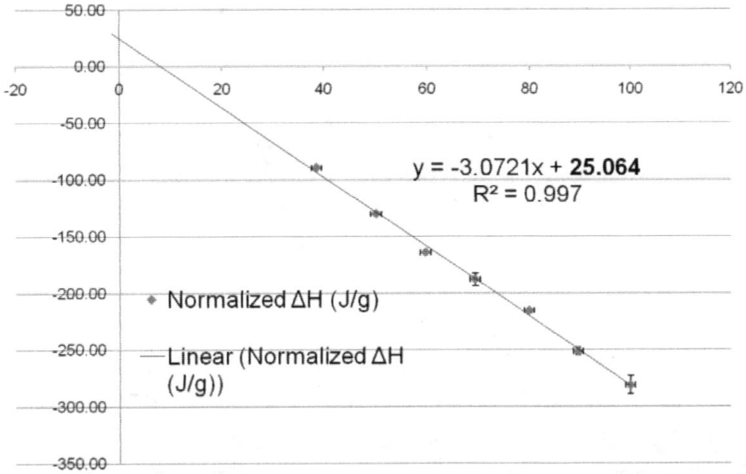

Figure 4 – Heat of fusion from solid to liquid for binary lithium nitrate/sodium nitrite mixture. The x-axis represents percentage of lithium nitrate.

Harmonization Study

There are numerous concurrent effort investigating candidate fluids that offer desirable heat transfer performance. Historical experiments that focused on discovery of thermal storage media collected over the years are also revisited to understand the variability of thermophysical properties. Figure 5 illustrates just a few examples of the relationship between melting points and thermal energy density. Table 1 summarizes the historical reports used to generate the data.

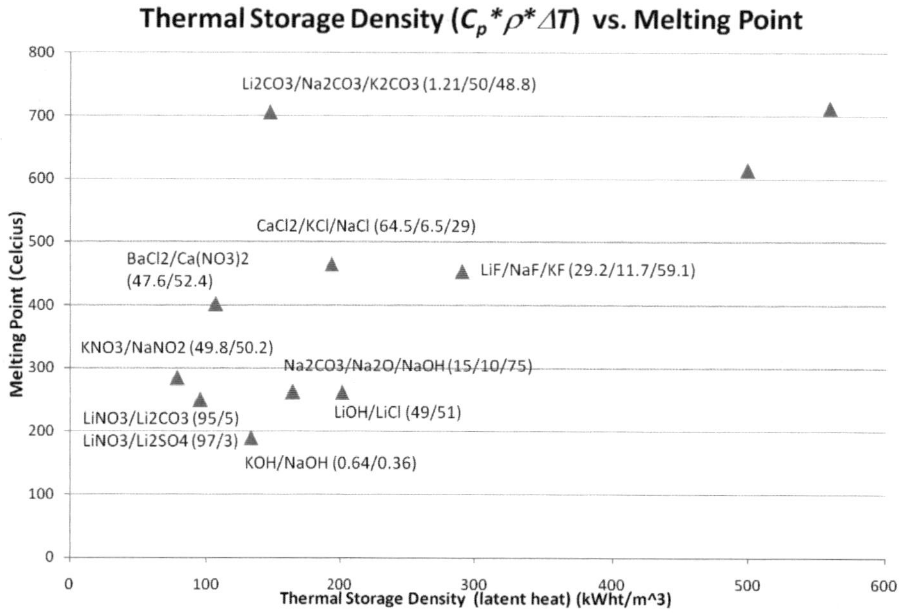

Figure 5 – Plot of melting points of various inorganic salt mixtures versus thermal storage density, defined as the product of liquid heat capacity, liquid density, and working temperature range. References are given in Table 1.

Concurrently, Halotecnics is using a proprietary high-throughput method to systematically screen combinatorial mixtures of molten salt [7]. Concurrent to this DOE-funded effort, similar experimental studies are underway at the University of Alabama and University of Wisconsin [8].

Summary

This work summarizes the effort towards understanding of molten salts heat transfer characteristics conducted at Sandia. Theoretical and experimental efforts are conducted internally to support the National Solar Thermal Testing Facility (NSTTF). Molecular dynamics and classical thermodynamics are both used to elucidate pure component as well as high-order mixture phase transition properties. Experiments are also conducted to explore novel mixtures that yield best heat transfer characteristics.

Acknowledgement

The authors would like to acknowledge Aidan Thompson and Nathan Siegel for their assistance on this work. Sandia is a multiprogram laboratory operated by Sandia Corporation, a Lockheed Martin Company, for the United States Department of Energy National Nuclear Security Administration under contract DE-AC04-94AL85000.

Table 1 – References for Figure 5.

Author(s)	Title	Date	Publisher	ID
Ferrara, Angelo, George Yenetchi, Robert Haslett, Robert Kosson	Thermal Energy Storage Heat Transfer	Oct-77	NASA	NASA CR 135244
Ferrara, Angelo	Thermal Energy Storage Heat Exchanger	Sep-77	ORNL	CONF-770955
Haslett, R., J. Alario, and G. Yenetchi	Active Heat Exchanger System Development for Latent Heat Thermal Energy Storage System	Dec-78	SNL/CA	CONF-781231
LeFrois, R.T., G.R. Knowles, A.K. Mathur, and J. Budimir	Active Heat Exchange System Development for Latent Heat Thermal Energy Storage	Feb-79	Honeywell/NASA/DOE	DOE/NASA/0038-79/1
Alario, Joseph, Robert Kosson, and Robert Haslett	Active Heat Exchange System Development for Latent Heat Thermal Energy Storage	Jan-80	Grumman/NASA/DOE	DOE/NASA/0039-79/1
Petri, Randy.J., Terry D. Claar, and Estela T. Ong	High-Temperature Molten Salt Thermal Energy Storage Systems for Solar Applications	Jul-83	IGT/NASA/DOE	DOE/NASA/0156-83/1
	Survey and Selection of Inorganic Salts for Application to Thermal Energy Storage	Jun-75	USRDA	ERDA-59
Heine, D., F. Heess, and M. Groll	Physical and Chemical Properties of Phase Change Materials for Application in Solar Tower and Solar Farm Plants	Aug-90		SOEMDH 24(1-4) 1-782(1991)

References

1. Bradshaw, R.W. and D. A. Brosseau, "Improved Molten Salt Formulations for Heat Transfer Fluids in Parabolic Trough Solar Power Systems," Proceedings to SolarPACES International conference, 2008.
2. Bradshaw, R.W. and N.P. Siegel, "Molten Nitrate Salt Development for Thermal Energy Storage in Parabolic Trough Solar Power Systems," Proceedings of ES2008, ES2008-54174, Jacksonville, FL, 2008.
3. Prausnitz, J.M., R.N. Lichtenthaler, and E. Gomes de Azevedo, "Molecular Thermodynamics of Fluid-Phase Equilibria," 3rd Edition, Prentice Hall, NJ, 1999.
4. Wilson, G.M., "Vapor-Liquid Equilibrium. XI. A New Expression for the Excess Free Energy of Mixing," J. Am. Chem. Soc., 86(2), 127, 1964.
5. Davison, S, Sun, A., "Thermodynamic analysis of solid-liquid phase equilibria of nitrate salts", *In Review*, I&EC Research, 2011.
6. Signorini, G. F., Barrat, J., and Klein, M.L., "Structural relaxation and dynamical correlations in a molten state near the liquid-glass transition: A molecular dynamics study", J. Chem. Phys, 92, 1294-1303, 1990.
7. Raade, J.W., "Deep Eutectic Salt Formulations Suitable as Advanced Heat Transfer Fluid," DOE-CSP review, Halotecnics, May, 2010.
8. Reddy, R.G., "Novel Molten Salts Thermal Energy Storage for Concentrating Solar Power Generation, " Phase I DOE review, 2010.

(758g) A New Auto-Ignition Temperature Model for Pure Hydrocarbons

M. Bagheri[1], **B. J. Neely**[2], K. M. Yerramsetty[2] and K. A. M. Gasem[2], (1)School of Chemical Engineering, University of Tehran, Tehran, Iran, (2)School of Chemical Engineering, Oklahoma State University, Stillwater, OK

Computer-aided molecular design (CAMD) can provide significant improvements in process safety by developing new materials with targeted properties. Auto-ignition temperature (AIT) is one of these properties, which relates to hazard assessment. AIT is the lowest temperature at which the compound will produce hot-flame ignition in air at atmospheric pressure without an external energy source. Accurate values for AIT are critical for the safe design and economical operation of many chemical processes. For example, AIT values has been used widely to characterize the hazard potential of chemicals in operations involving manufacture, handling, transport and storage of combustible materials,

Reported AIT values can vary by hundreds of degrees, and the experimental measurement of AIT is expensive, and laborious. In fact, for toxic, volatile, explosive and/or radioactive compounds, the measurement is difficult and may even be impossible. Thus, the development of a reliable predictive model is desirable. In this work, we have developed a three-parameter model for the accurate prediction of the auto-ignition temperature of 813 DIPPR hydrocarbons covering 72 diverse chemical classes. The model provided auto-ignition temperature predictions with a root-mean square error of 43 K, which is within the experimental uncertainty of the hydrocarbon database considered in this study.

See more of this Session: Thermodynamics of Energy Systems
See more of this Group/Topical: Engineering Sciences and Fundamentals

AUTHOR INDEX

AUTHOR INDEX

AUTHOR INDEX

AUTHOR INDEX

AUTHOR INDEX

AUTHOR INDEX